28th Annual IUPAP Conference on Computational Physics (CCP 2016)

Journal of Physics: Conference Series Volume 905

Pretoria, South Africa
10 - 14 July 2016

Editors:

Marius S. Potgieter

ISBN: 978-1-5108-5036-1
ISSN: 1742-6588

Printed from e-media with permission by:

Curran Associates, Inc.
57 Morehouse Lane
Red Hook, NY 12571

Some format issues inherent in the e-media version may also appear in this print version.

Printed by Curran Associates, Inc. (2017)

For permission requests, please contact the Institute of Physics at the address below.

Institute of Physics
Dirac House, Temple Back
Bristol BS1 6BE UK

Phone: 44 1 17 929 7481
Fax: 44 1 17 920 0979

techtracking@iop.org

Additional copies of this publication are available from:

Curran Associates, Inc.
57 Morehouse Lane
Red Hook, NY 12571 USA
Phone: 845-758-0400
Fax: 845-758-2633
Email: curran@proceedings.com
Web: www.proceedings.com

Table of contents

Volume 905

28th annual IUPAP Conference on Computational Physics (CCP2016)
(previous edition: CCP2015)

10–14 July 2016, Pretoria, South Africa

Accepted papers received: 6 October 2017
Published online: 22 October 2017

Preface

Papers

IOP Conf. Series: Journal of Physics: Conf. Series **905** (2017) 011001 doi:10.1088/1742-6596/905/1/011001

Editorial

The 28th annual IUPAP Conference on Computational Physics was held at the St. Georges Hotel in Pretoria, South Africa 10-14 July 2016. CCP2016 was officially hosted by the South African Institute of Physics (SAIP) on behalf of the International Union for Pure and Applied Physics (IUPAP) Commission for Computational Physics, C20. For more than 25 years since its establishment, this important conference had not come to Africa. In 2013 the South African Institute of Physics (SAIP) successfully bid for this conference to be brought to Africa for the first time.

The purpose of the conference series is to bring together computational scientists, leading academics, researchers and research scholars to exchange and share their experiences and research results about all aspects of Computational Physics. The conference also provides the premier interdisciplinary and multidisciplinary forum for researchers, practitioners, and educators to present and discuss the most recent innovations, trends, and concerns, practical challenges, and solutions adopted in the field of Computational Physics. CCP differs from many other similar conferences because presenters are encouraged to highlight their relevant computational details and innovative new computational ideas, and in so doing go beyond simply using computing methods as a tool to produce scientific results.

There were 125 participants from 27 countries at CCP2016. We had 26 invited talks including 7 plenary talks. The plenary speakers were Prof Russ Taylor, University of Cape Town and the University of the Western Cape, South Africa (*The Square Kilometre Array: Big Telescope, Big Science, Big Data*), Prof. Predrag Cvitanović, Georgia Institute of Technology, USA (*Recurrent flows: The clockwork behind turbulence*), Prof. Phuti Ngoepe, University of Limpopo, South Africa (*Simulated synthesis, characterisation and performance of nano-architectured energy storage materials*), Prof. Shiping Feng, Beijing Normal University, China (*Kinetic-energy driven superconductivity in cuprate superconductors*), and Prof Shobhana Narasimhan, Jawaharlal Nehru Centre for Advanced Scientific Research, India (*Understanding and designing novel nanomaterials from first principles*). A highlight was the awarding of the CCP2015 Young Scientist Prize to Dr. Wei-Min Wang, Chinese Academy of Sciences, Beijing, China (*Development of particle-in-cell code KLAPS and its applications in laser plasma physics*) and the CCP2016 Young Scientist Prize to Prof. Jianwei Sun, The University of Texas, El Paso, United States of America (*Efficient and accurate first-principles predictions of structures and energies for Materials Science, Chemistry and Biology*).

The main areas of Computational Physics were broadly covered during the conference, including Statistical Physics, Complex System and non-linear dynamics, Soft Matters, Polymers and Biological Physics, Computational Biology, Quantum Many Body and Strongly Correlated Systems, Material and Nano Sciences, Quantum Computing, Astrophysics and Space Physics, Plasma, Gravitation and Cosmology, Atomic, Molecular and Optical Physics, High Energy, Nuclear and Particle physics, Fluid Dynamics, Oceanography and Climate modelling, Geophysics and Geomechanics, Computational Physics Education, Software and Hardware Development. With South Africa co-hosting the Square Kilometre Array (SKA) which, when completed in 2022, will be the world's largest scientific experiment, there was a special interest in computational astrophysics and big data astronomy.

The CCP2016 conference organising committee is deeply indebted to Mr. Brian Masara, Executive Officer of the South African Institute of Physics for his exemplary leadership in organising the conference.

IOP Conf. Series: Journal of Physics: Conf. Series **905** (2017) 011001 doi:10.1088/1742-6596/905/1/011001

We also express our gratitude to the main editor Prof Marius Potgieter for his role in managing the manuscripts review process for the proceedings. The main editor wishes to thank the following reviewers for their assistance in evaluating the manuscripts; Drs J Adler, E Bakunina, P Bashkarev, E Benecha, M Boetcher, V Buyadzhi, V Cooper, R Dave, T Dietel, A Dudinov, K Dzhumagulova, L Einkemmer, I Gledhill, A Glushkov, M Dragowski, P Kilian, Y Kruglyak, Xi Lou, L Ludmila, E Maluta, E Mapasha, P Monceau, G Orjubin, A Rashid, G Rampho, T Salagaram, H Sithole, K Smit, F Spanier, R Strauss, M Takahashi, and T Vojta.

We are grateful to the IUPAP Commission for Computational Physics as principal sponsor of the conference. The conference would not have been possible without the financial support from the South African Institute of Physics, the National Research Foundation of South Africa, the Centre for High Performance Computing in Cape Town, the DST/NRF Centre for Strong Materials, the American Physical Society Division for Computational Physics. The metropolitan City of Tshwane was especially generous in hosting the conference attendees at a fabulous banquet where African music, food, and culture were much enjoyed.

Local Organising Committee

1. Prof. Nithaya Chetty (South Africa) – Chair
2. Prof Marius Potgieter (NWU) – Proceedings Editor
3. Prof Romeel Dave (UWC)
4. Prof Spencer Wheaton (UCT)
5. Dr. Happy Sithole (CSIR - CHPC)
6. Prof. Igle Gledhill (CSIR)
7. Mr. Brian Masara (SAIP)
8. Ms. Ndanganeni Mahani (SAIP)

International Advisory Board:

1. Prof Hai-Qing Lin Beijing Computational Science Research Center, China
2. Prof Alex Hansen NTNU, Norway
3. Prof David P. Landau The University of Georgia, USA
4. Prof Joan Adler Physics Technion, Israel
5. Prof Richard Liska Czech Technical University in Prague
6. Prof Priya Mahadevan SNBNCBS, India
7. Prof Hans Fangohr University of Southampton, UK
8. Prof Maria-Roser Valent Johann Wolfgang Goethe-University, Germany
9. Prof Masatoshi Imada University of Tokyo, Japan
10. Prof Lev Shchur Landau Institute for Theoretical Physics, Russia
11. Prof Suklyun Hong Sejong University, Korea
12. Prof Georg Kresse University of Vienna, Austria
13. Prof Marc Etienne Brachet CNRS, France
14. Prof Ronald Dickman Universidade Federal de Minas Gerais, Brazil
15. Prof Constantia Alexandrou University of Cyprus, Cyprus
16. Prof James Gubernatis Los Alamos National Laboratory, USA
17. Prof Hideake Takabe Osaka University, Japan
18. Prof Wei-Min Wang Institute of Physics, China
19. Prof Mathieu Salanne Université Pierre et Marie Curie, France

20. Prof Sitangshu Bikas Santra, IIT Guwahati, India
21. Prof Purusattam Ray IMSc Chennai, India
22. Prof Anders Sandvik Department of Physics, Boston University
23. Prof David Campbell Department of Physics, Boston University
24. Prof David Coker Department of Chemistry, Boston University
25. Prof Hideaki Takabe Osaka University, Japan
26. Prof Kimihiko Hirao RIKEN, (AICS), Kobe, Japan
27. Prof Robert Rosner University of Chicago, USA
28. Prof Ravil Nazirov Institute for Space Research, Russia
29. Prof Mikhail Romanovsky Prokhorov General Physics Institute, Russia

Conference Logistics Management

1. Brian Masara (SAIP)
2. Maria Swart (NRF)
3. Ndanga Mahani (SAIP)
4. Lizzy Sathekge (SAIP)

Conference Chairperson

Prof. Nithaya Chetty
Department of Physics
University of Pretoria
Private Bag X20
Hatfield 0028
South Africa

Email: Nithaya.Chetty@up.ac.za

Proceedings Editor

Prof. Marius S. Potgieter
Centre for Space Research
North-West University
2520 Potchefstroom
South Africa

Tel. +27 18 299 2406
Email: Marius.Potgieter@nwu.ac.za

IOP Conf. Series: Journal of Physics: Conf. Series **905** (2017) 011002 doi:10.1088/1742-6596/905/1/011002

Peer review statement

All papers published in this volume of *Journal of Physics: Conference Series* have been peer reviewed through processes administered by the proceedings Editors. Reviews were conducted by expert referees to the professional and scientific standards expected of a proceedings journal published by IOP Publishing.

IOP Conf. Series: Journal of Physics: Conf. Series **905** (2017) 012001 doi:10.1088/1742-6596/905/1/012001

Simulating the synchrotron emission of AGN with grid based relativistic hydrodynamics

I P van der Westhuizen, B van Soelen and P J Meintjes

Department of Physics, University of the Free State, Bloemfontein, 9301, SA

E-mail: vanderwesthuizenip@ufs.ac.za

Abstract. Grid based numerical hydrodynamic codes have been used to simulate the evolution of a variety of astrophysical environments, ranging from stellar winds to accretion disks in binary systems. These simulations can be used as valuable tools to correlate the theoretical models and observational data. In this study a grid based relativistic hydrodynamic simulation is created, using the freely available numerical code PLUTO ver 4.2, to study possible causes of variability in AGN. The code uses shock-capturing Godunov-type methods to solve the fluid dynamic conservation equations on a structured mesh grid. To simulate the production of a jet similar to those observed in AGN a three dimensional static grid was created containing $256 \times 256 \times 512$ computational cells. A uniform background medium was assigned to this grid, while a nozzle was defined on the $z = 0$ boundary to inject a pressure matched relativistic jet, with $\Gamma = 10$ ($v = 0.995c$), into the medium. The simulation showed the formation of a central collimated relativistic beam along with a surrounding cocoon region. In addition, a post-processing emission modelling code is being developed, in order to determine the emission produced by the relativistic jet. The emission model utilizes the data produced by the hydrodynamic code to calculate intensity maps due to synchrotron self absorption. The emission code takes into account the effects produced by the relativistic motion of the emitting regions (excluding light travel time) which can lead to effects such as the Doppler boosting observed in blazars. Synchrotron based emission models can be applied to other astrophysical objects, such as X-ray binary systems, which means that the application of the emission code can be adapted to model a variety of astrophysical sources.

1. Introduction

Relativistic astrophysical jet-like outflows have been associated with many different sources such as Active Galactic Nuclei (AGN), microquasars and gamma-ray bursts (GRB). In the case of AGN the jets can stretch over Mpc scales, carrying large amounts of energy away from the central engine and producing radiation over a wide range of the electro-magnetic spectrum. The spectral energy distribution of AGN are characterized by a double hump structure consisting of a low energy and a high energy component. The emission in the low energy component is dominated by synchrotron emission from non-thermal electrons, while the high energy component has been modelled by a combination of inverse-Compton and hadronic processes, see for e.g. [1].

Many radio surveys have been undertaken to study the synchrotron emission produced in the relativistic jets and how this relates to their dynamic morphology. For example, the MOJAVE program has studied the movement of superluminal components in AGN as well as the change in ejection direction and precession, over year-long time periods [2].

IOP Conf. Series: Journal of Physics: Conf. Series **905** (2017) 012001 doi:10.1088/1742-6596/905/1/012001

There are, however, still many unanswered questions about the processes that occur in jets and how they relate to the received emission. Numerical fluid dynamics is a powerful tool to study these dynamic structures. The macroscopic motion of relativistic jets can be described as a fluid and it must, therefore, adhere to the set of partial differential equations which describes the laws of conservation of mass, energy and momentum. By solving the conservation equations numerically, and evolving them with time, we can evolve the jet morphology to future states. This is a good method for investigating the variable nature of relativistic jets.

In order to compare the evolution of structures that are obtained using numerical simulations to the observed variability, we need to determine the synchrotron emission from the physical characteristics of the jet, such as internal energy, magnetic field strength, pressure and density.

In this proceedings we present a numerical simulation of an ideal relativistic jet. In this model the magnetic field was assumed to be dynamically negligible compared to the kinetic energy of the particles. A detailed discussion of the numerical simulation is given in section 2 while the emission modelling follows in section 3. Section 4 contains our results followed by a conclusion in section 5.

2. Numerical models

In hydrodynamical simulations a numerical jet environment is constructed and evolved with time according to the conservation equations of fluid dynamics which can be expressed as

$$\frac{\partial \vec{U}}{\partial t} = -\nabla \cdot \vec{\vec{T}}(\vec{U}) + \vec{\vec{S}}(\vec{U}), \tag{1}$$

where $\vec{U}$ is a column vector consisting of conserved quantities, $\vec{\vec{T}}(\vec{U})$ is a tensor containing the flux vectors as a function of $\vec{U}$, and $\vec{\vec{S}}(\vec{U})$ is a tensor containing source terms that can be used to introduce effects such as viscosity and gravitational forces [3]. Since AGN jets are have a high bulk Lorentz factor it is necessary to perform the simulations in the relativistic hydrodynamic (RHD) regime. In this case, the constituents of equation (1) have the following form:

$$\vec{U} = \begin{bmatrix} \rho\Gamma \\ \rho\Gamma^2 h\mathbf{v} \\ \rho\Gamma^2 h - P \end{bmatrix}, \quad \vec{\vec{T}}(\vec{U}) = \begin{bmatrix} \rho\Gamma\mathbf{v} \\ \rho\Gamma^2 h\mathbf{v}\mathbf{v} + P\mathbf{I} \\ \rho\Gamma^2 h\mathbf{v} \end{bmatrix}, \quad \vec{\vec{S}}(\vec{U}) = 0, \tag{2}$$

where ρ is the proper density, P is the pressure, h is the specific enthalpy, Γ is the Lorentz factor, $\mathbf{I}$ is a 3x3 unit matrix and $\mathbf{v}$ is the three velocity[3].

All other fluid properties, such as the pressure and the internal energy, can be determined from the above mentioned variables [4]. For example, the pressure can be solved by combining the ideal caloric equation of state,

$$e = \frac{P}{\rho(\gamma_{\rm ad} - 1)}, \tag{3}$$

where e is the internal energy density of the fluid, with the definition of the sound speed ($C_s = v/M_b$, M_b representing the Mach number), yielding

$$P = \frac{(\gamma_{\rm ad} - 1)\rho C_s^2}{\gamma_{\rm ad}(\gamma_{\rm ad} - 1 - C_s^2)}. \tag{4}$$

The complex nature of this set of equations (equation 1) makes it impossible to find analytical solutions for these types of environments. We, therefore, need to numerically solve these equations. For this study the numerical hydrodynamical code PLUTO was used to evolve the numerical simulation. PLUTO is a open-source relativistic magnetohydrodynamical code which

Table 1. Parameters used in the set up of the initial conditions.

Parameter		Value (arbitrary units)
Lorentz factor	Γ	10
Velocity	v	$0.995c$
Density ratio	η	10^{-6}
Jet density	ρ_b	0.1
Ambient density	ρ_{am}	10000
Mach number	M_b	7.8
Adiabatic index	γ_{ad}	5/3
Pressure	P	0.001

uses high resolution shock capturing algorithms to solve the conservation equations and evolve them with time [5].

For our simulation we set up a three dimensional Cartesian mesh grid consisting of $256 \times 256 \times 512$ computational cells. The grid was initially assigned a uniform background medium at rest. A nozzle with a radius of 4 computational cells, was created on the $z = 0$ axis at the bottom of the simulation and jet material was injected into the environment at a constant rate. For this simulation the density of the injected jet was chosen to be less than the ambient density ($\eta = \frac{\rho_b}{\rho_{am}} = 10^{-6}$). The simulation was performed using arbitrary units in order to avoid numerical errors. A pressure matched model was used for the simulation in order to help collimate the jet. The parameters used for the simulation were based on previous papers such as [4, 6]. All parameters used in the simulation are listed in table 1.

The simulation was run using the HLLC Riemann solver, piecewise parabolic interpolation and characteristic tracing time stepping [7]. The environment was evolved until the bow shock created by the injection of jet material reached the edge of the computational domain.

3. Synchrotron emission

In order to compare the numerical simulations with the observed emission in radio frequencies we need to translate the three dimensional environment of physical parameters onto a two dimensional image plane. In order to do this we first calculate the synchrotron emission and absorption coefficients for each cell in the simulation, using a post-processing code. In this code we assumed that the electrons inside the jet consist of particles with a power-law energy distribution with the number density, $n(\gamma)$, given by,

$$n(\gamma) = n_0 \gamma^{-p}, \tag{5}$$

where the normalization factor, n_0, is determined by [8]

$$n_0 = \left(\frac{e(p-2)}{1 - C_E^{2-p}} \right)^{p-1} \left(\frac{1 - C_E^{1-p}}{\frac{\rho}{m_p}(p-1)} \right)^{p-2}. \tag{6}$$

Here m_p is the mass of a proton, p is the power-law index and C_E is the ratio of the maximum and minimum energies. For our initial estimates we chose $p = 1.8$ and $C_E = 10^3$. The magnetic field energy density is assumed to be an equipartition fraction of the internal energy density,

$$u_B = \epsilon_B e \tag{7}$$

where $\epsilon_B = 10^{-3}$ based on values fitted by [9].

An analytical delta-approximation model was used to estimate the synchrotron emission (j_v^{sy}) and absorption (α_v^{sy}) coefficients for each cell [10], given by

$$j_\nu^{sy} = \frac{4}{9}\left(\frac{q^2}{mc^2}\right)^2 u_B \nu^{\frac{1}{2}} \nu_0^{\frac{-3}{2}} n\left(\sqrt{\frac{\nu}{\nu_0}}\right), \tag{8}$$

$$\alpha_\nu^{sy} = \frac{2}{9}\frac{p+2}{m\nu^2}\left(\frac{q^2}{mc^2}\right)^2 u_B \nu_0^{-1} n\left(\sqrt{\frac{\nu}{\nu_0}}\right), \tag{9}$$

where,

$$\nu_0 = \frac{3qB}{4\pi mc}, \tag{10}$$

represents the electron gyro-frequency. Here q is the charge and m is the mass of the non-thermal electrons in the jet, c is the speed of light, u_B is the magnetic field energy density given by equation 6 and ν is the frequency of emission in the co-moving frame [1].

Since the jet material is moving at relativistic velocities we need to apply Lorentz transformations to each cell. The synchrotron coefficients of each cell transform as,

$$j_\nu^{sy\prime} = \frac{j_\nu^{sy}}{(\Gamma[1-\beta\mu'])^2}, \tag{11}$$

$$\alpha_\nu^{sy\prime} = \alpha_\nu^{sy}(\Gamma[1-\beta\mu']) \tag{12}$$

where the primed quantities are in the observer frame and the unprimed quantities are in the co-moving frame, β is the magnitude of the velocity in units of c and μ' is the cosine of the angle between the observer and the velocity of the fluid in the observer frame. For our initial calculations additional effects such as the light travel time and the expansion of the universe were neglected.

The post-processing code finally integrates the coefficients along a line of sight, s, defined by an azimuth angle ϕ' and a polar angle θ',

$$I'_\nu = \int (j_\nu^{sy\prime} - \alpha_\nu^{sy\prime} I'_\nu) ds', \tag{13}$$

to create a 2D intensity map. This is illustrated in figure 1.

4. Results

The xz-slices of the density, pressure and velocity distributions for the simulation are shown in figure 1. From these figures we can note five distinct regions in the simulation. First we have the outer uniform ambient medium. A bow wave propagates through this medium due to the injection of the supersonic jet material through the nozzle on the initial z boundary. The bow wave increases the pressure of the medium as it propagates creating a high pressure region between the jet and the ambient medium.

The centre of the jet contains collimated jet material moving at a high Lorentz factor through the environment. Surrounding this collimated beam we find a turbulent backflow of material which is generated as the jet material interacts with the shock front, called the working surface, which forms between the jet material and the ambient medium.

Figure 2 shows 1 dimensional plots of the pressure and density in the z direction along the centre of the beam. We note initial oscillation in both pressure and density. The oscillations indicate the formation of stationary re-collimation shocks. The shocks are caused by the over-pressured cocoon acting on the central beam. These shocks are damped as the jet material

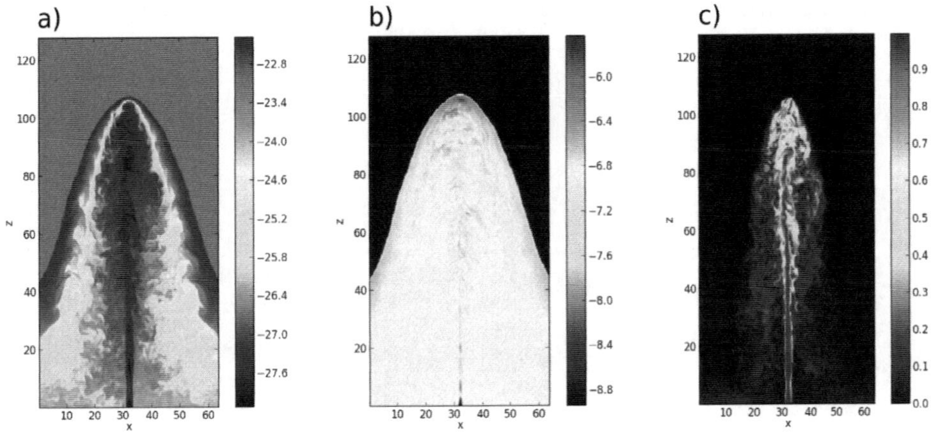

Figure 1. Two dimensional visualizations of the simulation through the xz-plane of the jet, showing a) the density, b) the pressure and c) the velocity component in the z direction. Logarithmic scales are shown for the density and pressure plots in arbitrary units, while the velocity plots have a linear scaling in units of c.

propagates along the z-direction. At larger distances from the injection site turbulence causes the beam of the jet to become unstable. Figure 2 shows this as a rapid change in the density and pressure of the beam. Finally we see a discontinuity in pressure and density at the working surface where the jet material collides with the ambient medium.

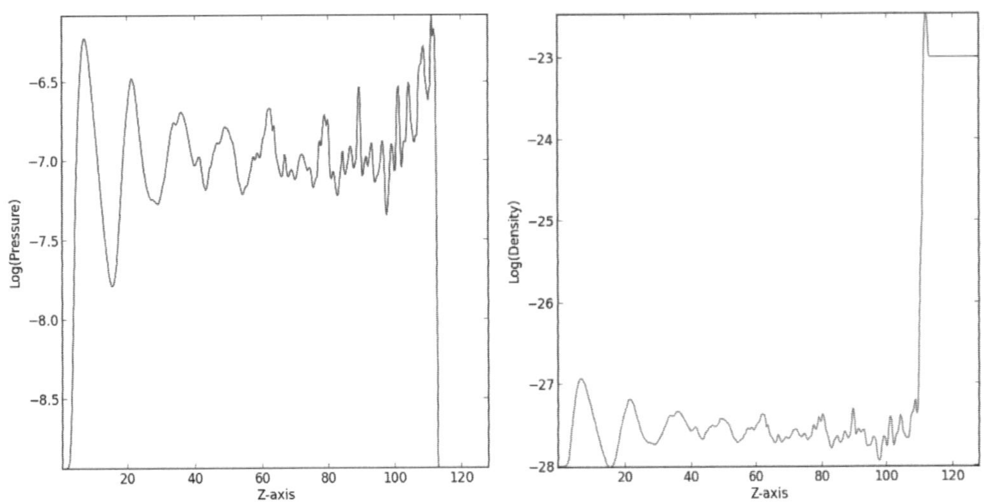

Figure 2. One dimensional plots of the pressure (left) and the density (right) along the central beam.

The emission modelling code was applied to the simulation with a viewing angle of $\psi = 90°$ and $\theta = 90°$. Figure 3 shows 6 sequential frames of the approximated synchrotron emission

IOP Conf. Series: Journal of Physics: Conf. Series **905** (2017) 012001 doi:10.1088/1742-6596/905/1/012001

at 15 GHz illustrating the time dependence of the main emission regions, or hot spots, in the emission. The hot spot at the head of the jet is shown to change its location, size and maximum intensity. As time progresses we note that the hotspot breaks apart to form several extended emission regions.

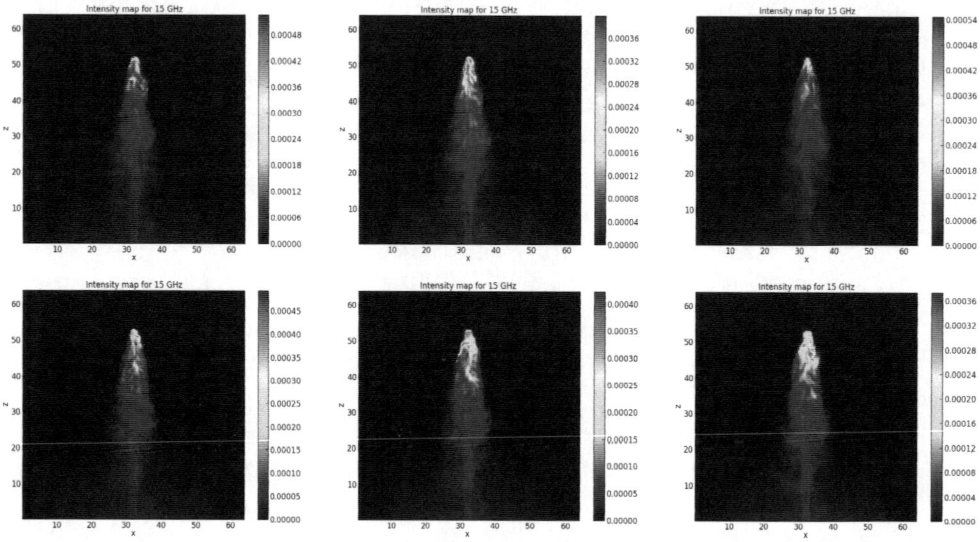

Figure 3. Intensity maps calculated at 15 GHz showing 6 sequential time steps in the numerical simulation.

5. Conclusion

In this paper a numerical model of a relativistic jet was set up and evolved with time using the PLUTO hydrodynamical code. The results showed the formation of a relativistic beam surrounded by a cocoon of backflowing material. Plots of the pressure and density along the central beam revealed the presence of periodic re-collimation shocks within 40 jet radii. Beyond this region the beam becomes turbulent with rapid variations in pressure and density, ending in a shock front at the head of the jet.

A delta-approximation synchrotron model was applied to the simulation to produce a 2D intensity map of an edge-on system. The resulting intensity maps showed time dependent emission regions varying in intensity, position and size. This result indicates that variations in the emission of AGN jets can occur even if the system is fed by a constant injection.

In order to make a precise comparison between the synthetic intensity maps and real observation, however, additional corrections to emission model must be applied for effects such as light travel time and expansion of the universe and radiative losses. The future aim for this project is to include an inverse Compton model into the emission code to investigate the correlation of variability at multi-wavelengths. Based on this we will attempt SED modelling of the simulations in order to produce a better qualitative comparison with observational data. We also plan to include a variable injection rate in the simulation to model how variable injection influences the morphology and stability of the resulting jet. Finally the project can be extended to a much larger scope, including many different astronomical sources such as X-ray binaries and gamma-ray bursts.

Acknowledgments

The financial assistance of the National Research Foundation (NRF) towards this research is hereby acknowledged. Opinions expressed and conclusions arrived at, are those of the author and are not necessarily to be attributed to the NRF.
This work is based on the research supported in part by the National Research Foundation of South Africa for the grant 87919. Any opinion, finding and conclusion or recommendation expressed in this material is that of the authors and the NRF does not accept any liability in this regard.

References

[1] Böttcher M 2011 Modeling the spectral energy distributions and variability of blazars *2011 Fermi & Jansky: Our Evolving Understanding of AGN* (*preprint arXiv:1205.0539*)
[2] Lister M L, *et al.* 2009 MOJAVE: Monitoring of jets in active galactic nuclei with VLBA experiments. V. Multi-epoch VLBA images *ApJ* **137** 37183729
[3] Toro E F 2009 *Riemann Solvers and Numerical Methods for Fluid Dynamics: A Practical Introduction Third Edition* (Berlin: Springer) Chapter 1 pp 1-40
[4] Marti J M, Müller E, Font J A, Ibáñez J M and Marquina 1997 Morphology and dynamics of relativistic jets *ApJ* **479** 151-163
[5] Mignone A, Bodo G, Massaglia S, Matsakos T, Tesileanu O, Zanni C and Ferrari A 2007 PLUTO: a numerical code for computational astrophysics *ApJ* **170** 228-42
[6] Rossi P, Mignone A, Bodo G, Massaglia S and Ferrari A 2008 Formation of dynamical structures in relativistic jets: the FRI case *A&A* **488** 795-806
[7] Mignone A and Bodo G 2006 An HLLC Riemann solver for relativistic flows - II. Magnetohydrodynamics *MNRAS* **368** 1040-54
[8] Gomez J L, Marti J M, Marscher A P, Ibáñez J M and Marcaide J M 1995 Parsec-scale synchrotron emission from hydrodynamic relativistic jets in active galactic nuclei *ApJ* **449** 19-21
[9] Böttcher M and Dermer C 2010 Timing signatures of the internal-shock model for blazars *ApJ* **711** 445-460
[10] Böttcher M, Harris D E and Krawczynski H, 2012 *Relativistic Jets from Active Galactic Nuclei*, Weinheim, Germany: Wiley-VCH Verlag GmbH & Co. KGaA Chapter 3 pp 39-80

IOP Conf. Series: Journal of Physics: Conf. Series **905** (2017) 012002 doi:10.1088/1742-6596/905/1/012002

Computational modelling of cosmic rays in the neighbourhood of the Sun

M S Potgieter and R D Strauss

Centre for Space Research, North-West University, Potchefstroom, South Africa

E-mail: marius.potgieter@nwu.ac.za

Abstract. The heliosphere is defined as the plasmatic influence sphere of the Sun and stretches far beyond the solar system. Cosmic rays, as charged particles with energy between about 1 MeV and millions of GeV, arriving from our own Galaxy and beyond, penetrate the heliosphere and encounter the solar wind and embedded magnetic field so that when observed they contain useful information about the basic features of the heliosphere. In order to interpret these observations, obtained on and near the Earth and farther away by several space missions, and to gain understanding of the underlying physics, called heliophysics, we need to simulate the heliosphere and the acceleration, propagation and transport of these astroparticles with numerical models. These types of models vary from magnetohydrodynamic based approaches for simulating the heliosphere to using standard finite-difference numerical schemes to solve transport-type partial differential equations with varying complexity. A large number of these models have been developed locally to do internationally competitive research and have become as such an important training tool for human capacity development in computational physics in South Africa. How these models are applied to various aspects of heliospheric space physics, with illustrative examples, is discussed in this overview.

1. Introduction

The Sun has a plasmatic atmosphere that is blowing away from it at supersonic speeds between 450 km/s and 800 km/s and is called the solar wind. Embedded in this solar wind is a magnetic field and together they defined the influence sphere of the Sun which is called the heliosphere. Between 70-90 AU from the Sun, the solar wind becomes subsonic and creates a termination shock (TS); 1 AU is the averaged distance between the Earth and the Sun. At around 120 AU a heliopause (HP), is formed and much further out a bow shock. This whole region constitutes the heliosphere, as will be further illustrated and discussed in Figure 1, and can expand easily to 500 AU in the direction in which the heliosphere is moving through the local interstellar medium (ISM). The Sun has an 11-year activity cycle and on top of this, its magnetic field changes polarity every 11 years to create a 22-year cycle.

Cosmic rays are charged particles with energies from about 1 MeV up to EeV. The are accelerated to these energies in our Galaxy, and beyond, and penetrate the heliosphere from all directions. Below 50 GeV these particles are reduced systematically in the heliosphere, progressively more the lower the energy, to the point where this reduction becomes significant, typically below about 10 GeV. Since solar activity has an 11-year cycle, the cosmic ray flux is also modulated by this cycle, that is, as a function of time, space and energy. This is known as the solar modulation of cosmic rays.

There are four primary populations of charged particles in the heliosphere: Galactic cosmic rays entering the heliosphere from interstellar space while solar energetic particles are accelerated during so-called solar flare or coronal mass ejection events; then there is the anomalous cosmic ray component considered to be accelerated in the outermost regions of the heliosphere, and lastly, charged particles that originate from planetary magnetospheres, for example, from Jupiter and Saturn. See the review by [1].

Numerous space missions have contributed significantly over several decades to the observation of cosmic rays in the heliosphere, from the Earth to the HP. Presently, the two Voyager spacecraft explore the outer regions of the heliosphere, with Voyager 1 already beyond the HP [2] at a distance of 135 AU from the Sun.

2. Transport of cosmic rays in the heliosphere

The propagation of cosmic rays as charged particles in the heliosphere is generally described by a transport equation (TPE) derived by [3] as a special case of the general Fokker-Planck equation. Within a heliocentric spherical coordinate system, this equation can be expressed as

$$\underbrace{\frac{\partial f}{\partial t}}_{\text{a}} = -(\underbrace{\mathbf{V}_{sw}}_{\text{b}} + \underbrace{\langle \mathbf{v}_D \rangle}_{\text{c}}) \cdot \nabla f + \underbrace{\nabla \cdot (\mathbf{K}_s \cdot \nabla f)}_{\text{d}} + \underbrace{\frac{1}{3}(\nabla \cdot \mathbf{V}_{sw}) \frac{\partial f}{\partial \ln p}}_{\text{e}} + \underbrace{Q}_{\text{f}}, \tag{1}$$

for the omnidirectional distribution function $f(\mathbf{r}, p, t)$ of cosmic rays, where f is a function of position $\mathbf{r}$, particle momentum p and time t. The transport and modulation mechanisms contained in this equation are: (a) time-dependent changes in the cosmic rays, (b) outward convection caused by the expanding solar wind velocity $\mathbf{V}_{sw}$, (c) gradient, curvature and current sheet drifts in the global heliospheric magnetic field (HMF) in terms of the average drift velocity $\langle \mathbf{v}_D \rangle$, (d) spatial diffusion caused by the irregular HMF through the symmetric diffusion tensor $\mathbf{K}_s$, (e) adiabatic energy changes (deceleration or acceleration) determined by the solar wind divergence $(\nabla \cdot \mathbf{V}_{sw})$, and (f) possible additional sources of cosmic rays within the heliosphere, such as the Jovian and Cronian electrons. The diffusion tensor contains a diffusion coefficient parallel to the mean HMF, $\kappa_{||}$, and in two perpendicular directions, $\kappa_{\perp r}$ and $\kappa_{\perp \theta}$. All three depend on position, time and on the energy of the particles so that the underlying physics can be very complicated. This TPE has been applied to all the mentioned populations of cosmic rays; see [4] and [5] for overviews of these transport processes and cosmic ray modulation in general.

Solving this equation using different numerical techniques, while keeping the physics as up to date as possible, is the essence of the numerical modelling of cosmic rays in the heliosphere. A similar approach is used to solve the TPE for cosmic rays in the Galaxy although the transport and other processes, such as energy losses, can be different; see [6] and [7].

3. Modelling the heliosphere

Magnetohydrodynamic (MHD) models are frequently used to describe the main features of the heliosphere, for example, its full geometrical extent in all directions, and important issues such as where are the TS and the HP located. Most of these models are quite comprehensive and as such complicated, sometimes with overwhelming detail [10], or they can be designed to be of more practical use for cosmic ray studies as done by [9]. The latter serves the purpose of defining the basic features of the heliosphere as relevant to the study of cosmic ray modulation. As the Sun moves through interstellar space, the heliosphere interacts with the ISM so that the heliosphere is shaped into distinctive narrow nose and extended tail regions, where the narrower nose region points in the direction in which the heliosphere is moving.

In Figure 1, a MHD simulated meridional cut of the heliosphere is shown. The colour scales in the panels labelled (a) to (d) respectively indicate particle density, plasma speed, magnitude

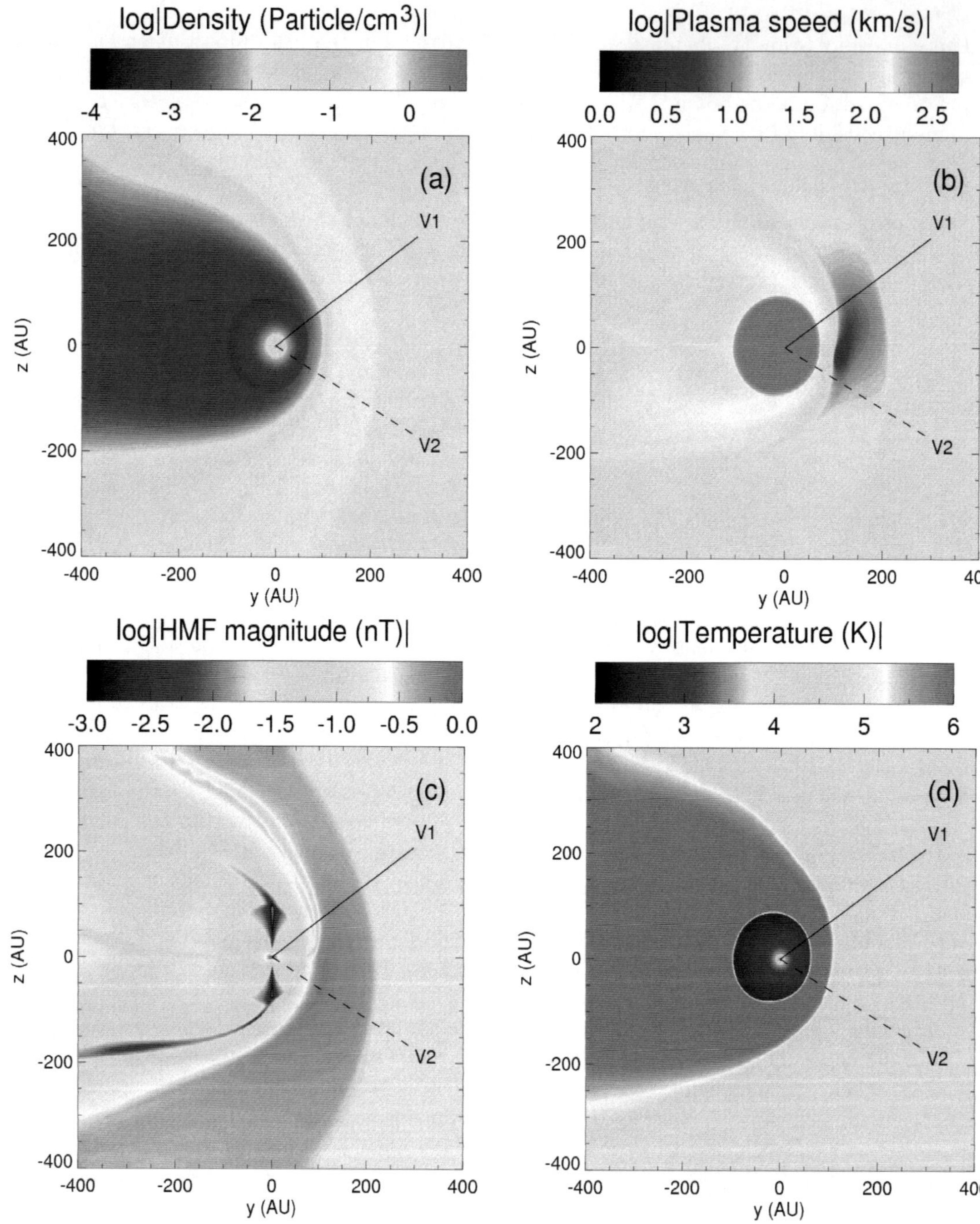

Figure 1. The MHD heliospheric environment in terms of particle density, plasma (solar wind) speed, magnitude of the heliospheric magnetic field, and temperature in the meridional plane as shown in panels labelled (a), (b), (c) and (d), respectively. This simulated heliosphere moves to the right through the interstellar medium. From right to left, notice the distinguishable structures in panel (c) particularly, first the bow shock, then the heliopause and termination shock, with the Sun at (0,0) AU. The approximate trajectories of the Voyager 1 and 2 spacecraft (projected onto the same plane) are indicated by the solid and dashed lines, respectively. Voyager 1 is now at about 135 AU, already beyond the heliopause, while Voyager 2 is at 112 AU. Figure is from [8], see also [9].

of the HMF and temperature. The ISM flow is from the right and the first notable structure to form is the bow shock (BS) at about 220 AU from the Sun in the nose direction along the solar equatorial plane. The assumed trajectories of the Voyager 1 and 2 spacecraft are also indicated, projected onto the same plane. According to this MHD approach, moving radially inward along the equatorial plane from the position of the BS, the HP is located at about 110 AU from the Sun. The HP separates the heliospheric medium from the ISM and as such serves as the effective boundary between the heliosphere and the local ISM. The region between the BS and the HP is known as the outer heliosheath. In this figure, at about 75 AU in the equatorial plane, the supersonic solar wind decreases to subsonic speeds so that the TS is formed. The region of the heliosphere between the HP and the TS is known as the inner heliosheath. The existence of the TS was confirmed when Voyager 1 and 2 crossed into the heliosheath at distances of 94 AU in 2004 and 84 AU in 2007, respectively. Voyager 1 crossed the HP at about 121 AU in August 2012, while Voyager 2 is still inside the inner heliosheath. (See also http://voyager.jpl.nasa.gov/).

A comprehensive approach to the numerical modelling of solar modulation consists of combining these MHD models with transport models for cosmic rays, the so-called hybrid models. For illustrative examples of such an approach and for several applications, see the recent modelling of [9] and [11].

4. Solar modulation models

The TPE in Eq. 1 written in terms of heliocentric spherical coordinates can be rearranged to emphasize the contributions of four major modulation processes which are: diffusion, particle drifts, convection and adiabatic energy losses, so that it becomes:

$$
\begin{aligned}
&\overbrace{\left[\frac{1}{r^2}\frac{\partial}{\partial r}\left(r^2 K_{rr}\right) + \frac{1}{r\sin\theta}\frac{\partial K_{\phi r}}{\partial \phi}\right]\frac{\partial f}{\partial r} + \left[\frac{1}{r^2\sin\theta}\frac{\partial}{\partial\theta}\left(K_{\theta\theta}\sin\theta\right)\right]\frac{\partial f}{\partial\theta}}^{\text{diffusion}} \\
+ &\overbrace{\left[\frac{1}{r^2\sin\theta}\frac{\partial}{\partial r}(rK_{r\phi}) + \frac{1}{r^2\sin^2\theta}\frac{\partial K_{\phi\phi}}{\partial\phi} - \Omega\right]\frac{\partial f}{\partial\phi}}^{\text{diffusion}} \\
+ &\overbrace{K_{rr}\frac{\partial^2 f}{\partial r^2} + \frac{K_{\theta\theta}}{r^2}\frac{\partial^2 f}{\partial\theta^2} + \frac{K_{\phi\phi}}{r^2\sin^2\theta}\frac{\partial^2 f}{\partial\phi^2} + \frac{2K_{r\phi}}{r\sin\theta}\frac{\partial^2 f}{\partial r\partial\phi}}^{\text{diffusion}} \\
+ &\overbrace{\left[-\langle \mathbf{v}_D\rangle_r\right]\frac{\partial f}{\partial r} + \left[-\frac{1}{r}\langle \mathbf{v}_D\rangle_\theta\right]\frac{\partial f}{\partial\theta} + \left[-\frac{1}{r\sin\theta}\langle \mathbf{v}_D\rangle_\phi\right]\frac{\partial f}{\partial\phi}}^{\text{drift}} \\
- &\overbrace{V_{sw}\frac{\partial f}{\partial r}}^{\text{convection}} + \overbrace{\frac{1}{3r^2}\frac{\partial}{\partial r}\left(r^2 V_{sw}\right)\frac{\partial f}{\partial \ln p}}^{\text{adiabatic energy losses}} = 0,
\end{aligned}
\tag{2}
$$

given here in terms of five different diffusion coefficients; Ω is the rotational speed of the Sun.

In order to solve this TPE it is written in a condensed form, in terms of P, the particle's rigidity (momentum/charge), as

$$
\begin{aligned}
&a_0(r,\theta,\phi,P)\frac{\partial f}{\partial r} + b_0(r,\theta,\phi,P)\frac{\partial f}{\partial\theta} + c_0(r,\theta,\phi,P)\frac{\partial f}{\partial\phi} + d_0(r,\theta,\phi,P)\frac{\partial^2 f}{\partial r\partial\phi} \\
+ \; &e_0(r,\theta,\phi,P)\frac{\partial^2 f}{\partial r^2} + l_0(r,\theta,\phi,P)\frac{\partial^2 f}{\partial\theta^2} + m_0(r,\theta,\phi,P)\frac{\partial^2 f}{\partial\phi^2} \\
+ \; &s_0(r,\theta,\phi,P)\frac{\partial f}{\partial \ln P} = 0.
\end{aligned}
\tag{3}
$$

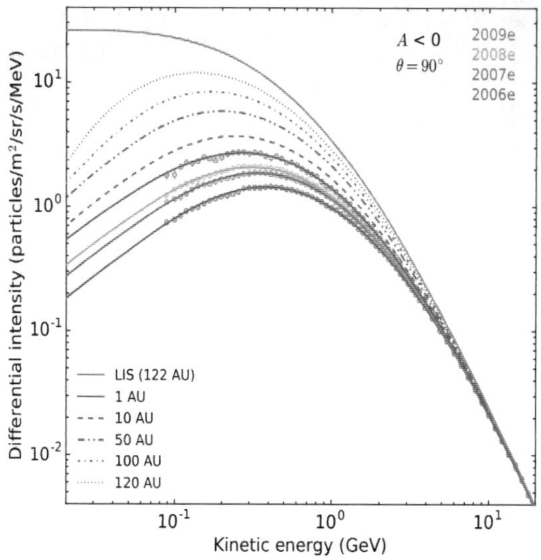

Figure 2. Computed proton spectra as a function of kinetic energy at the Earth (1 AU) are shown in comparison with observed spectra from the PAMELA space detector for the periods 2006e, 2007e, 2008e and 2009e. Additional computed spectra are shown in the heliospheric equatorial plane ($\theta = 90°$) at 10 AU, 50 AU, 100 AU and 120 AU (blue lines) with respect to a new very local interstellar spectrum (LIS; solid grey line) from [14] and [15], specified at 122 AU, assumed to be the position of the HP.

This is a parabolic differential equation which can be solved with a modified Crank-Nicholson finite difference method, called the Alternating Direction Implicit (ADI) method. For cosmic ray studies, this ADI method, as described by [12], is modified to include three spatial coordinates and energy (rigidity). Adding a fifth numerical dimension does not work well when using this ADI approach so that when time-dependent modulation is simulated, one of the spatial dimensions, typically the ϕ-dependence, is substituted by a time-dimension.

The very local interstellar spectra (LIS) of various cosmic ray species (from hydrogen to iron, including electrons, positrons and anti-particles) are taken as individual input spectrum at the outer boundary of the heliosphere, which typically is considered to be the HP for all values of θ and ϕ, while the inner boundary is taken close to the solar surface. At the heliospheric poles ($\theta = 0; \pi$), it usually is assumed that $\partial f/\partial\theta = 0$, so that a symmetry exists with respect to the heliospheric polar line, mostly done to same computational time.

For the ADI approach, an initial solution is obtained at one third of a rigidity step forward by solving Eq. 3 implicitly in the direction of the first spatial coordinate, i.e. radial distance r. A second solution is then calculated at another third of a rigidity step forward, using the first solution, by solving the differential equation implicitly in the direction of the second spatial coordinate, the polar angle θ. For a solution at the final third of the rigidity step, this process is repeated for the last spatial coordinate, the azimuthal angle ϕ, using the previous two solutions. As a result, a system of linear equations is obtained which can be solved using, for example, the Thomas algorithm. In the process, a 3D grid is constructed across which the TPE is solved. The radial grid points $r_i = (i-1)\Delta r + r_1$, for $i = 1, 2, \ldots, N_r$, run from $r = r_1$ at the inner boundary to $r = r_b$ at the outer boundary, with the radial increments $\Delta r = (r_b - r_1)/(N_r - 1)$. The polar grid points are $\theta_j = (j-1)\Delta\theta$, for $j = 1, 2, \ldots, N_\theta$, and run from $\theta = 0°$ at the north pole to $\theta = 180°$ at the south pole, with increments of $\Delta\theta = \pi/(N_\theta - 1)$. The azimuthal grid points $\phi_k = (k-1)\Delta\phi$, for $k = 1, 2, \ldots, N_\phi$, run from $\phi = 0°$ to $\phi = 360°$ for a full solar rotation, where the azimuthal increments are $\Delta\phi = 2\pi/(N_\phi - 1)$. Furthermore, the rigidity decreases logarithmically from an initial maximum value P_{max} to a minimum value P_{min}, so that $P_{n+1} = P_n/\exp(\Delta \ln P)$, for $n = 1, 2, \ldots$, with $\Delta \ln P > 0$ the rigidity increment. The solution, however, becomes unstable when the rigidity steps are large relative to the squares of the spatial increments, so that the value for $\Delta \ln P$ is chosen mostly in order to improve the stability of the solution.

In terms of the generalized grid points, the distribution function is

$$f(r_i, \theta_j, \phi_k, P_n) = f(i\Delta r, j\Delta\theta, k\Delta\phi, n\Delta \ln P) = f_{i,j,k,n}, \tag{4}$$

so that, from Taylor series expansions, standard central finite difference approximations for the first, second and mixed derivatives of f are obtained. The solutions for f at the different rigidity steps are indicated by

$$f_{i,j,k,n} = f_{i,j,k}; f_{i,j,k,n+\frac{1}{3}} = f^*_{i,j,k}; f_{i,j,k,n+\frac{2}{3}} = g_{i,j,k}; f_{i,j,k,n+1} = h_{i,j,k}; \tag{5}$$

where for the first solution, $f^*_{i,j,k}$, is calculated by solving Eq. 3 implicitly in the radial direction at one third of a rigidity step backward. This is achieved by evaluating half of the central difference equations in r at the current rigidity step and half of them at a third of a rigidity step backward. For the second solution, $g_{i,j,k}$, this process is repeated by once again solving Eq. (3) implicitly, in terms of $f^*_{i,j,k}$, in the θ-direction. Similarly, the third solution, $h_{i,j,k}$, in terms of $f^*_{i,j,k}$ and $g_{i,j,k}$, is obtained at one full rigidity step backward by solving Eq. 3 implicitly in the ϕ-direction. This process is repeated for all rigidity values in order to obtain a solution for the final distribution function f, which is related to the differential intensity $j = P^2 f$.

A typical solution of Eq. 2 is shown in Figure 2. It depicts computed proton spectra in the solar equatorial plane ($\theta = 90°$) at 1 AU, 10 AU, 50 AU, 100 AU and 120 AU, relative to the LIS for protons, specified at the HP located here at 122 AU [14]. Four computed spectra are compared to the observed proton spectra from the PAMELA space experiment (http://pamela.roma2.infn.it/index.php) for November 2006, December 2007, December 2008 and December 2009, indicated as 2006e, 2007e, 2008e and 2009e [14]. This shows that these types of numerical models are successful at reproducing major cosmic ray observations. For more elaborate computations of radial and latitudinal gradients for cosmic rays throughout the heliosphere, see [13] and [15].

5. SDE models

The calculus of stochastic differential equations (SDEs) is a well developed field and the reader is referred to the work of [16] and [17] in which the relevant references are given.

For the purpose of this overview, it suffices to define an SDE as one that can be cast into the general form

$$\frac{\partial x(t)}{\partial t} = \mu(x,t) + \sigma(x,t)\zeta(t), \tag{6}$$

for a one-dimensional (1D) case where $\mu(x,t)$ and $\sigma(x,t)$ are continuous functions, while $\zeta(t)$ represents a rapidly varying stochastic function, also referred to as the noise term. The first and second terms on the right hand side of this equation are referred to as the drift and diffusion terms, respectively, and should not to be confused with the physical drift and diffusive terms present in the TPE. Moreover, only SDEs of the Itō type are considered, so that this equation can be rewritten as

$$dx(t) = \mu(x,t)dt + \sigma(x,t)dW(t), \tag{7}$$

where $dW(t)$ represents the Wiener process. This is a time stationary stochastic Levy process where the time increments have a Normal distribution with a mean of zero (i.e. a Gaussian distribution) and a variance of dt; $dW(t) = W(t) - W(t-dt) \sim \mathcal{N}(0, dt)$. Eq. (7) can be integrated as

$$x(t) = x_0 + \int_0^t \mu(x, t')dt' + \int_0^t \sigma(x, t')dW(t'), \quad (8)$$

where the first integral is a normal (Riemann or Lebesgue) integral, while the second is an Itō type stochastic integral. Since analytical solutions for SDEs are only available for a limited few scenarios, this equation is usually integrated numerically, using the Euler-Maruyama numerical scheme [18], where a finite time step Δt is chosen so that

$$x^{t+\Delta t} = x^t + \mu(x, t)\Delta t + \sigma(x, t)\Delta W(\Delta t), \quad (9)$$

from an initial position $x = x_0$ at time $t = 0$ and continued until the boundary is reached at $x = x_\mathrm{e}$ at $t = t_\mathrm{e}$, or until a temporal integration limit is reached at $t = t^\mathrm{N}$. The Wiener process is discretized as

$$\Delta W(\Delta t) = \sqrt{\Delta t} \cdot \Lambda(t), \quad (10)$$

where $\Lambda(t)$ is a simulated Gaussian distributed pseudo-random number, which should be handled with great care. The temporal evolution of x forms a trajectory through phase space, which is generally referred to as the trajectory of a pseudo-particle. Integrating Eq. 8 for a single pseudo-particle has limited significance so that integration is carried out over all possible trajectories, usually done $N \gg 1$ times; this is also referred to as tracing N pseudo-particles.

A normalised (conditional) probability density $\rho'(x, t)$ can be calculated from the pseudo-particle trajectories, such that

$$\int_\Omega \rho'(x, t)d\Omega = 1, \quad (11)$$

where Ω denotes the entire integration space. Numerically this is done by dividing e.g. x into a series of bins, e.g. N' intervals, and calculating the number of pseudo-particles ending up in each bin, i.e. bin i has N_i particles, and dividing by N. The discretized version of this equation thus reads

$$\sum_{i=1}^{N'} \rho_i'(x, t) = \sum_{i=1}^{N'} \frac{N_i}{N} = 1. \quad (12)$$

The set of SDEs for cosmic ray transport in heliocentric spherical dimensions (r,θ,ϕ) and momentum (p) is

$$\begin{aligned} dr &= \mu_r \cdot ds + \sigma_{rr} \cdot dW_r + \sigma_{r\theta} \cdot dW_\theta + \sigma_{r\phi} \cdot dW_\phi \\ d\theta &= \mu_\theta \cdot ds + \sigma_{\theta\theta} \cdot dW_\theta + \sigma_{\theta\phi} \cdot dW_\phi \\ d\phi &= \mu_\phi \cdot ds + \sigma_{\phi\phi} \cdot dW_\phi \\ dp &= \mu_p \cdot ds, \end{aligned} \quad (13)$$

which can be integrated as mentioned above. The coefficients σ_{ij} are related to the elements of tensor in the TPE as in Eq. 1, with specifics as given in Eq. 2. The full set of SDEs to solve for the 3D transport of cosmic rays in the heliosphere is given in detail by [8] and [19]; see also [17] .

The momentum SDE can be rewritten in terms of the kinetic energy of particles as

$$dE = 1/3\,(\nabla \cdot \mathbf{V}_{sw})\,\Gamma E \cdot ds, \quad (14)$$

with

$$\Gamma \equiv (E + 2E_0)/(E + E_0), \quad (15)$$

where E_0 is the rest mass energy.

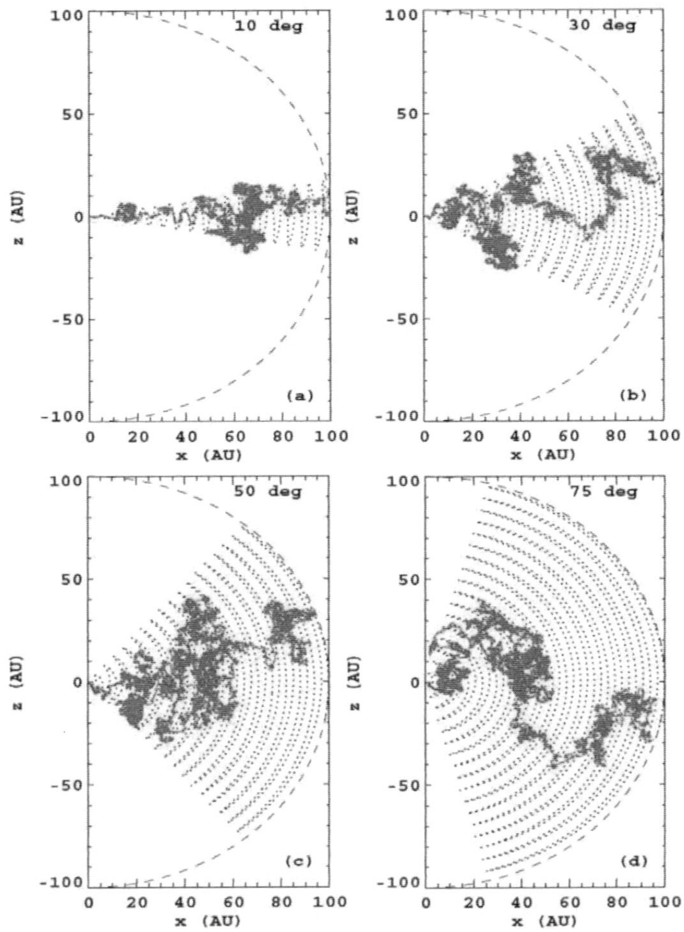

Figure 3. Pseudo-particle trajectories (red stochastic-like lines) for 100 MeV galactic protons propagating through the heliosphere towards the Earth at 1 AU, projected onto the meridional plane with the Sun at (0,0) AU. The HP position, in this example at 100 AU, is indicated by the circular dashed line, while the dotted line shows a projection of the wavy heliospheric current sheet onto the same plane, for four values of its tilt angle, shown as the dotted background lines. This figure is from [21].

This TPE usually is solved by means of SDEs in a so-called time-backward fashion. The time-forward and time-backward approaches are equivalent, but for cosmic ray modulation the latter approach is numerically more optimal and computationally much faster, with an appropriate handling of outer and inner boundary conditions. The use of SDE-models to solve the TPE, in order to overcome the characteristic problems and issues associated with finite-difference and finite-volume numerical schemes in higher dimensions, has been quite successful.

The validity and reliability of the SDE-based numerical model have been established in the past by extensive benchmarking. In the benchmarking and applications of [16], [20] and [21], the SDE-based model was utilised to model various aspects of the modulation of cosmic rays in the simulated heliosphere, serving as confirmation that the SDE approach is indeed reproducing the basic features of cosmic ray modulation. The SDE-based models also allow for the calculation of pseudo-particle trajectories, from which valuable insight into the solar modulation process is gained.

Pseudo-particle trajectories are shown in Figure 3 for galactic protons with kinetic energy of 100 MeV. Four different scenarios of the wavy heliospheric current sheet (HCS) are shown, indicated by different tilt angles, from a small value of 10 degrees (top, left panel) to a large value of 75 degrees (bottom right panel). The latter represents maximum solar activity and the former solar minimum activity when maximum cosmic ray intensity occur, as was observed in 2009 (see Figure 2). The dashed line shows the position of the HP, whereas the dotted line shows a projection of the wavy HCS onto the meridional plane. These traces are indicative of how cosmic ray protons propagate into the heliosphere through the equatorial regions, following the HCS to a large extent while its waviness is small but they get significantly scattered off the HCS with increasing solar activity, as the tilt angle increases. Evidently, under these conditions these particles propagate mainly in latitudes covered by this HCS; see [21] for further elaborations.

It is worthwhile to mention that negatively charged cosmic rays, such as galactic electrons or anti-protons, will propagate differently into the heliosphere depending on the polarity of the HMF. This creates what is known as charge-sign dependent modulation which is an important special feature in the 22-year cycles of the solar modulation of cosmic rays; see the review by [5].

6. Implementing GPUs

The computing power of graphics processing units (GPUs) has become readily accessible with the introduction of Nvidia's CUDA (Compute Unified Device Architecture) parallel computing platform and application programming interface (API) almost ten years ago. The primary advantage of utilizing the massive parallel design of the CUDA architecture, is the significant increase in runtime performance compared to central processing units (CPUs). Many fields of science have benefited from the gain in performance that GPUs and the CUDA API can offer, including space and heliospheric physics.

A GPU-based modulation model for which the CUDA API was used to solve the TPE given above and to simulate the transport of charged particles using SDEs was introduced by [22]. A comparison was made between the run-times of the GPU-based model and a CPU version of the same model. The latter model was also optimized with OpenMP (Open Multi-Processing), a multi-thread technique that facilitates parallel processing in CPUs. Because SDE modulation models calculate the trajectory of individual pseudo-particles, where the computation of each trajectory is independent of all other trajectories, the process can easily be done in parallel. This GPU-based model [22] showed a performance increase of a factor of between ~ 10 and ~ 60 over the usual CPU-model, depending on the choice of input parameters.

The modulation model based on a finite difference numerical scheme as described above give a solution at each rigidity-step which depends on the solution from the previous rigidity-step, which makes these type of numerical models not fully parallelizable. However, numerous variables and modulation parameters, such as the solar wind speed, HMF-values and the diffusion coefficients, have to be calculated at each of the grid points in the simulated heliosphere. Such calculations across large arrays are parallelizable and can be performed much faster by using the CUDA API as opposed to performing the calculations sequentially with a CPU, or even with CPU-based parallel systems like OpenMP or MPI (Message Passing Interface). This advantage that CUDA offers has been exploited in order to perform all of the time-expensive array-based calculations, resulting in a runtime performance increase of a factor of between ~ 10 and ~ 20. Even though CUDA is capable of achieving significant performance increases for calculations that can be performed independently, the application of this technology to traditional finite difference numerical schemes remains a challenge. The best approach for improving runtime performance in such models will be to use a combination of CPU and GPU computing resources; see [13] for details. For scaling and efficiency factors done for the SDE and CUDA approaches, see [22] and [23].

7. Summary

State-of-the-art numerical models have been developed over the years in the Centre for Space Research at the North-West University in South Africa. Because of the substantial computational skills required to develop these models, the process has become a good and productive training school for computational physics in South Africa.

The application of the described ADI-based models has made a significant impact on the study of the solar modulation of cosmic rays in the heliosphere as comprehensively reviewed by [4].

The SDE approach has proven to be remarkably successful in describing the heliospheric modulation of these particles. Numerically, SDE-based models are relatively easier to construct than a 3D finite difference model, and being unconditionally stable, makes it possible to study cosmic ray transport phenomena for a broader range of scenarios, while utilizing the power of modern day parallel computers. The CUDA version of the code is considerably faster than a multi-threaded CPU implementation and even small clusters. The GPU-approach can be of benefit when using modern desktop computers or if large CPU clusters are not accessible.

Acknowledgments

The authors thank Andreas Kopp, Xi Lou, Etienne Vos and Jan-Louis Raath for valuable discussions, and thank the South African National Research Foundation (NRF) for providing partial research funding and the Centre for High Performance Computing (CHPC) in Cape Town for allocating computational time.

References

[1] Strauss R D, Potgieter M S and Ferreira S E S 2012 *Adv. Space Res.* **49**, 392
[2] Gurnett D A, Kurth W S, Burlaga L F and Ness N F 2013 *Science* **341** 1489
[3] Parker E N 1965 *Planet. Space Sci.* **13** 9
[4] Potgieter M S 2013 *Living Rev. Solar Phys.* **10** 3
[5] Potgieter M S 2014 *Adv. Space Res.* **53** 1415
[6] Strong A W and Moskalenko I V 1998 *Astrophys. J.* **509** 212
[7] Kopp A, Büsching I, Potgieter M S, Strauss R D 2014 *New Astron.* **30** 32
[8] Strauss R D 2013 PhD thesis (North-West University, Potchefstroom, South Africa)
[9] Strauss R D, Potgieter M S, Ferreira S E S, Fichtner H and Scherer K 2013 *Astrophys. J. Lett.* **765** L18
[10] Opher M, Drake J F, Zieger B and Gombosi T I 2015 *Astrophys. J.* **800** L28
[11] Luo X, Zhang M, Potgieter M S, Feng X and Pogorelov P 2015 *Astrophys. J.* **808** 82
[12] Douglas J 1962 *Numerische Mathematik* **4** 41
[13] Vos E E 2016 PhD thesis (North-West University, Potchefstroom, South Africa)
[14] Vos E E and Potgieter M S 2015 *Astrophys. J.* **815** 119
[15] Vos E E and Potgieter M S 2016 *Solar Phys.* in press
[16] Strauss R D, Potgieter M S, Büsching I and Kopp A 2011a *Astrophys. J.* **735** 83
[17] Kopp A, Büsching I, Strauss R D and Potgieter M S 2012 *Comp. Phys. Comm.* **183** 530
[18] Maruyama G. 1955 *Rend. Circ. Mat. Palermo* **4** 48
[19] Raath J L, Potgieter M S, Strauss R D and Kopp A 2016 *Adv. Space Res.* **57** 1965
[20] Strauss R D, Potgieter M S, Kopp A and Büsching I 2011b *J. Geophys. Res.* **116** 105
[21] Strauss R D, Potgieter M S, Büsching I and Kopp A 2012 *Astrophys. Space Sci.* **339** 223
[22] Dunzlaff P, Strauss R D and Potgieter M S 2015 *Comp. Phys. Comm.* **192** 156
[23] Strauss R D and Moeketsi D M 2016 *Proc. Conf. South Afr. Inst. Phys.* in press

IOP Conf. Series: Journal of Physics: Conf. Series **905** (2017) 012003 doi:10.1088/1742-6596/905/1/012003

Computing of radiation parameters for atoms and multicharged ions within relativistic energy approach: Advanced Code

V V Buyadzhi, P A Zaichko, O A Antoshkina, T A Kulakli, G P Prepelitsa, V B Ternovsky and V F Mansarliysky

Odessa State Environmental University, L'vovskaya str.15, Odessa-16, 65016, Ukraine

E-mail: vbuyad@mail.ru

Abstract. The advanced numerical atomic code based on the combined relativistic energy approach and relativistic many-body perturbation theory with the optimized zeroth ab initio model potential approximation is presented and used in computing the radiation characteristics of the multielectron atoms and multicharged ions. There are presented the results of computing the energy levels and oscillator strengths (reduced dipole matrix elements) and lifetimes for heavy alkali atom of the caesium, in particular, there are listed data for the 7s1/2 – np1/2,3/2 (n=7-10) transitions and some Ne-like multicharged ions. The comparison of the calculated values with available theoretical and experimental (compillated) data is performed.

1. Introduction

It is well known that the correct data about different radiation, energetic and spectroscopic characteristics of the multielectron atoms and multicharged ions, namely, radiative decay widths, probabilities and oscillator strengths of atomic transitions, excitation and ionization cross-sections are needed in astrophysics and laboratory, thermonuclear plasma diagnostics and in fusion research. In this light, a special interest attracts studying the spectral characteristics of the alkali elements. There have been sufficiently many reports of calculations and compilation of energies and oscillator strengths for these atoms and corresponding ions (see, for example, [1–28]). In many papers the standard Hartree-Fock, Dirac-Fock methods, model potential approach, quantum defect approximation etc in the different realizations have been used for calculating energies and oscillator strengths. However, it should be stated that for the heavy alkali atoms (such as caesium and francium and corresponding ions) and particularly for their high-excited (Rydberg) states, there is not enough precise information available in literatures.

This work goes on our studying the spectroscopic properties of the neutral and highly ionized atoms, which has a fundamental importance in many fields of atomic physics (spectroscopy, spectral lines theory), astrophysics, plasma physics, laser physics and so on. In this paper the combined generalized relativistic energy approach [1-4,11-14] and relativistic many-body perturbation theory with the zeroth order ab initio model potential optimized one-particle representation [1,5,13-16] are used for computing the energy levels and oscillator strengths (reduced dipole matrix elements) of radiative transitions in some heavy multielectron atoms and multicharged ions, in particular, the alkali atoms and ions and the Ne-like multicharged ions. The comparison of the calculated dipole matrix elements with available theoretical and experimental (compillated) data is performed. A generalized relativistic energy approach to computing the radiative decay characteristics for atoms and multicharged is based on the Gell-Mann and Low adiabatic S- formalism. Originally the energy approach to radiative and autoionization processes in multielectron atoms and ions has been developed by Ivanova-Ivanov et al [22-27] (the PC code "Superatom-ISAN"). More advanced version of the relativistic energy approach has been developed in Refs. [28-32]. Different advanced computational generalizations have been considered in Refs. [33-50].

2. The theoretical method

In the relativistic energy approach (REA) [22-24,30] an imaginary part of the electron energy shift of an atom is directly connected with a radiation decay possibility (transition probability). An approach, is based on the Gell-Mann and Low formula with the QED scattering matrix. The total energy shift of the state in relativistic atom can be presented in the standard form:

$$\Delta E = \mathrm{Re}\Delta E + i\,\Gamma/2 \tag{1}$$

where Γ is interpreted as the level width, and the transition probability P = Γ. An imaginary part of electron energy can be defined in the lowest order of the perturbation theory as [22]:

$$\mathrm{Im}\Delta E(B) = -\frac{e^2}{4\pi} \sum_{\substack{\alpha>n>f \\ [\alpha<n\leq f]}} V^{|\omega_{\alpha n}|}_{\alpha n \alpha n}\ , \tag{2}$$

where (α>n>f) for electron and (α<n<f) for vacancy. The matrix element is determined as follows:

$$V^{|\omega|}_{ijkl} = \iint dr_1 dr_2 \Psi^*_i(r_1)\Psi^*_j(r_2)\frac{\sin|\omega| r_{12}}{r_{12}}(1-\alpha_1\alpha_2)\Psi^*_k(r_2)\Psi^*_l(r_1) \tag{3}$$

The separated terms of the sum (3) represent the contributions of different channels; for example, a probability of the dipole transition is as follows:

$$\Gamma_{\alpha_n} = \frac{1}{4\pi}\cdot V^{|\omega_{\alpha_n}|}_{\alpha_n\alpha_n} \tag{4}$$

When calculating the matrix elements (8), one should use the angle symmetry of the task and write the corresponding expansion for $\sin|\omega|r_{12}/r_{12}$ on spherical harmonics as follows [1,24]:

$$\frac{\sin|\omega| r_{12}}{r_{12}} = \frac{\pi}{2\sqrt{r_1 r_2}}\sum_{\lambda=0}^{\infty}(\lambda) J_{\lambda+1/2}(|\omega| r_1) J_{\lambda+1/2}(|\omega| r_2) P_\lambda(\cos \widehat{\mathbf{r}_1\mathbf{r}_2})\ , \tag{5}$$

where J is the Bessel function of first kind and (λ)= 2λ + 1. This expansion is corresponding to usual multipole one for probability of radiative decay. Substitution of the expansion (5) to matrix element of interaction gives as follows [24,27]:

$$V^{\omega}_{1234} = [(j_1)(j_2)(j_3)(j_4)]^{1/2}\sum_{\lambda\mu}(-1)^{\mu}\begin{pmatrix} j_1 j_3 & \lambda \\ m_1 - m_3 & \mu \end{pmatrix}\times \mathrm{Im}Q_{\lambda}(1234),$$

$$Q_{\lambda} = Q^{\mathrm{Qul}}_{\lambda} + Q^{\mathrm{Br}}_{\lambda}, \qquad (6)$$

where j_i is the total single electron momentums, m_i – the projections; Q^{Qul} is the Coulomb part of interaction, Q^{Br} - the Breit part. The imaginary part Q^{Cul}_{λ} contains the radial R_{λ} and angular S_{λ} integrals as follows:

$$\begin{aligned}\mathrm{Im}\,Q^{Cul}_{\lambda}(12;43) = Z^{-1}\,\mathrm{Im}\{&R_{\lambda}(12;43)S_{\lambda}(12;43) + R_{\lambda}(\tilde{1}2;4\tilde{3})S_{\lambda}(\tilde{1}2;4\tilde{3}) + \\ &+R_{\lambda}(1\tilde{2};\tilde{4}3)S_{\lambda}(1\tilde{2};\tilde{4}3) + R_{\lambda}(\tilde{1}\tilde{2};\tilde{4}\tilde{3})S_{\lambda}(\tilde{1}\tilde{2};\tilde{4}\tilde{3})\}.\end{aligned} \qquad (7)$$

The angular coefficient has only a real part:

$$S_{\lambda}(12;43) = S_{\lambda}(13)S_{\lambda}(24) \qquad S_{\lambda}(13) = \{\lambda l_1 l_3\}\begin{pmatrix} j_1 & j_3 & \lambda \\ \frac{1}{2} & -\frac{1}{2} & 0 \end{pmatrix} \qquad (8)$$

$\{\lambda l_1 l_3\}$ means that λ, l_1 and l_3 must satisfy the triangle rule and the sum $\lambda + l_1 + l_3$ must be an even number. The rest terms in (7) include the small components of the Dirac functions. The tilde designates that the large radial component f must be replaced by the small one g, and instead of $l_i, \tilde{l}_i = l_i - 1$ should be taken for $j_i < l_i$ and $\tilde{l}_i = l_i + 1$ for $j_i > l_i$. The Breit (magnetic) part can be expressed as follows [27]:

$$Q^{Br}_{\lambda} = Q^{Br}_{\lambda,\lambda-1} + Q^{Br}_{\lambda,\lambda} + Q^{Br}_{\lambda,\lambda+1} \qquad (9)$$

The corresponding imaginary part (17) is as follows:

$$\begin{aligned}\mathrm{Im}\,Q^{Br}_{\lambda,l} = \frac{1}{Z}\mathrm{Im}\{&R_{\lambda}(12;\tilde{4}\tilde{3})S^{l}_{\lambda}(12;\tilde{4}\tilde{3}) + R_{\lambda}(\tilde{1}\tilde{2};43)S^{l}_{\lambda}(\tilde{1}\tilde{2};43) + \\ &+R_{\lambda}(\tilde{1}2;\tilde{4}3)S^{l}_{\lambda}(\tilde{1}2;\tilde{4}3) + R_{\lambda}(\tilde{1}\tilde{2};\tilde{4}\tilde{3})S^{l}_{\lambda}(\tilde{1}\tilde{2};\tilde{4}\tilde{3})\}.\end{aligned} \qquad (10)$$

The detailed expressions for the Coulomb and Breit parts and the corresponding radial R_{λ} and angular S_{λ} integrals can be found in Refs. [22-32].The total probability of a λ - pole transition is usually represented as a sum of the electric P^{E}_{λ} and magnetic P^{M}_{λ} parts. The electric (or magnetic) λ - pole transition $\gamma \rightarrow \delta$ connects two states with parities which by λ (or λ +1) units. In our designations

$$\begin{aligned}&P^{E}_{\lambda}(\gamma\rightarrow\delta) = 2(2j+1)Q^{E}_{\lambda}(\gamma\delta;\gamma\delta) \qquad Q^{E}_{\lambda} = Q^{Cul}_{\lambda} + Q^{Br}_{\lambda,\lambda-1} + Q^{Br}_{\lambda,\lambda+1} \\ &P^{M}_{\lambda}(\gamma\rightarrow\delta) = 2(2j+1)Q^{M}_{\lambda}(\gamma\delta;\gamma\delta) \qquad Q^{M}_{\lambda} = Q^{Br}_{\lambda,\lambda}.\end{aligned} \qquad (11)$$

In our work the relativistic wave functions are determined by solution of the Dirac equation with the potential, which includes the Ivanova-Ivanov “outer electron- ionic core” potential [22] and polarization potential [32,38]. The calibration of the single model potential parameter has been performed on the basis of the special ab initio procedure [37,38] (see also [1,29,30]). In Refs. [29,30] the lowest order multielectron effects, in particular, the gauge dependent radiative contribution Im δE_{ninv} for the certain class of the photon propagator calibration has been treated. The value of this contribution allows to estimate an effectiveness of quantitative accounting for the multielectron correlation effects. The minimization of the gauge-non-invariant contribution Im δE_{ninv} provides a reasonable criterion in the construction of an optimized one-electron representation in the relativistic many-body perturbation theory. The minimization procedure leads to the system of the integral-differential equation that can be solved using one of the standard numerical codes. In result, this provides the construction of the optimized one-particle representation and improves an effectiveness of the numerical code. As it is known (c.f.[1,27]), an accuracy of computing the transition probabilities can be significantly increased by means of adding a multi-electron polarization exchange-correlation potential into the transition amplitude. The correct relativistic expression for the polarization operator has been presented in Refs. [2,36-40] and used by us in this work. All calculations are performed on the basis of the code Superatom-ISAN (version 93).

3. Illustration results and conclusions

Here we present some results of computing the oscillator strengths (reduced dipole matrix elements) for a number of transitions in spectra of the heavy alkali atoms and corresponding ions. As the first illustration, we present below the data for the caesium. In Table 1 there are listed the theoretical reduced dipole matrix elements for a number of transitions, computed within: i) relativistic Hartree-Fock (RHF) method [6], ii) the empirical relativistic model potential method (ERMP) [7], iii) the relativistic single-double (SD) method in which single and double excitations of the Dirac-Hartree-Fock (DHF) wave function are included to all orders of perturbation theory [8] and iv) our data. Let us note that the precise experimental data for the $6p_{1/2,3/2}$-6s transition are as follows: $6p_{1/2}$-6s=4.4890(7)and $6p_{3/2}$-6s=6.3238(7) [9]. The important feature of the approach used is using the optimized one-particle representation and an effective account for the exchange-correlation (including the core polarization) effects (see details in Refs. [1,27,38,50]). An estimate of the gauge-non-invariant contributions (the difference between the oscillator strengths values calculated with using the transition operator in the form of "length" G1 and "velocity" G2) is about 0.3%. The REA results, obtained with using the different photon propagator gauges (Coulomb, Babushkin, Landau) are practically equal.

Table 1. Theoretical reduced dipole matrix elements for a number of transitions of Cs (see text)

Transition	i: RHF	ii: ERMP	iii: SD-DHF	iv: Our data
$6p_{1/2}$-6s		4.489	4.482 4.535	4.487(G1) 4.488(G2)
$7p_{1/2}$-6s	0.282	0.283	0.297 0.279	0.282
$8p_{1/2}$-6s		0.088	0.091 0.081	0.086
$6p_{3/2}$-6s		6.323	6.382 6.304	6.320
$7p_{3/2}$-6s	0.582	0.583	0.601 0.576	0.584
$8p_{3/2}$-6s		0.228	0.232 0.218	0.229

In figure 1 we present experimental (squares) and theoretical (the empirical Coulomb approximation data by Feng et al – circles [17]; n<40; our work data – continuous line) lifetimes for the Rydberg states $nS_{1/2}$ in spectrum of Cs.

Note that the theoretical results obtained by Feng et al within the empirical Coulomb approximation (n <40), and our computing data are in a good agreement with experimental data. Resulting in the majority only in recent years (see Ref. [46]) experimental data mainly refer to states with n <40.

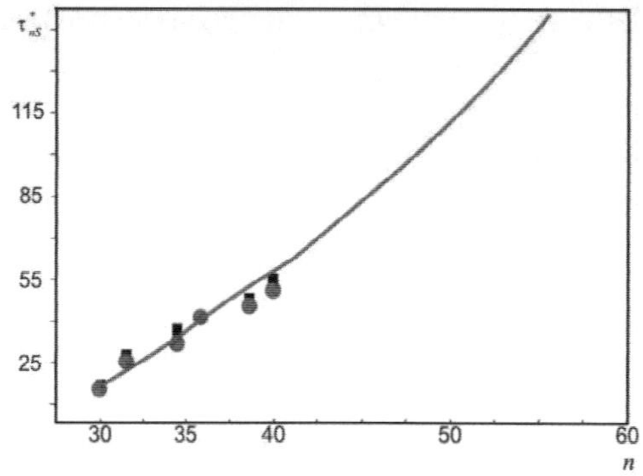

Figure 1. Experimental (squares) [17] and theoretical (the empirical Coulomb approximation data by Feng et al – circles; n<40; our work data – continuous line) lifetimes for the Rydberg states $nS_{1/2}$ in spectrum of Cs

Another illustration is computing the transition probabilities in the Ne-like multicharged ions. Let us note that the isoelectronic sequence of neon has been especially thoroughly investigated, but nevertheless remains of interest because of the spectra of Ne-like ions the source of the most important information for the solution of a wide variety of problems in the hot, dense, thermonuclear plasmas spectroscopy, physics of the shortwave lasers etc. The detailed reviews of work on spectroscopy of the Ne-like ions are presented in many papers (look, for example, [27,28,51] and Refs. therein). In tables 2 and 3 we present the values of probabilities of the transitions between levels of the configurations $2s^22p^53s$,3d,4s,4d and $2s2p^63p$,4p in the Ne-like ions of the Ni XIX, Br XXVI (in s^{-1}; total angle moment J=1): a – the MCDF method; b- relativistic PT with the empirical zeroth approximation (RPTMP); c1 – REA-PT data (without correlation corrections); c2 – REA-PT data (with an account for the correlation); exp.- experimental data (look [17-21,27,28] and Refs therein); This work -our data.

Table 2. Probabilities of radiation transitions between levels of the configurations $2s^22p^53s$,3d,4s,4d and $2s2p^63p$,4p in the Ne-like ion of Ni XIX (in s^{-1}; total angle moment J=1): a – the MCDF method; b- relativistic PT with the empirical zeroth approximation (RPTMP); c1,2 – REA PT data (without and with account for correlation corrections); exp.- experimental data; this work-our data (see text).

Level *J*=1	Exp.	a-MCDF	b-RPTMP	c1-REA PT	c2-REA PT	This work
$2p_{3/2}3s_{1/2}$	7.6+11	9.5+11	1.3+12	9.7+11	8.1+11	7.9+11
$2p_{1/2}3s_{1/2}$	6.0+11	1.8+12	1.0+12	7.6+11	6.2+11	6.1+11
$2p_{3/2}3d_{3/2}$	1.4+11	2.2+11	1.5+11	1.7+11	1.4+11	1.3+11
$2p_{3/2}3d_{5/2}$	1.2+13	2.1+13	1.2+13	1.5+13	1.2+13	1.1+13
$2p_{1/2}3d_{3/2}$	3.2+13	4.8+13	3.6+13	4.0+13	3.3+13	3.2+13
$2s_{1/2}\ 3p_{1/2}$			8.5+11	9.5+11	8.1+11	8.0+11
$2s_{1/2}\ 3p_{3/2}$			5.1+12	5.6+12	4.7+12	4.6+12
$2p_{3/2}4s_{1/2}$	3.3+11		3.6+11	4.1+11	3.4+11	3.3+11
$2p_{1/2}4s_{1/2}$	2.0+11		3.0+11	3.1+11	2.4+11	2.2+11
$2p_{3/2}4d_{3/2}$	4.5+10		5.2+10	5.4+10	4.8+10	4.6+10
$2p_{3/2}4d_{5/2}$	8.3+12		8.3+12	9.2+12	8.2+12	8.1+12
$2p_{1/2}4d_{3/2}$	8.1+12		7.9+12	8.9+12	8.0+12	8.0+12
$2s_{1/2}4p_{1/2}$				6.3+11	5.7+11	5.6+11
$2s_{1/2}4p_{3/2}$				2.7+12	2.4+12	2.3+12

Table 3. Probabilities of radiation transitions between levels of the configurations $2s^22p^53s,3d,4s,4d$ and $2s2p^63p,4p$ in the Ne-like ion of Br XXVI (in s^{-1}; total angle moment J=1): a – the DF method; b- RPTMP; c1,2 – REA PT data (without and with account for correlation corrections); exp.- experimental data; this work-our data (see text).

Level *J*=1	Exp.	a-MCDF	b-RPTMP	c1-QED PT	c2-QED PT	This work
$2p_{3/2}3s_{1/2}$	4.5+12	6.2+12	4.4+12	5.5+12	4.4+12	4.3+12
$2p_{1/2}3s_{1/2}$	3.1+12	4.8+12	2.8+12	3.6+12	2.7+12	2.6+12
$2p_{3/2}3d_{3/2}$	2.8+11	3.9+11	2.9+11	3.5+11	2.8+11	2.7+11
$2p_{3/2}3d_{5/2}$	6.1+13	8.0+13	6.3+13	7.5+13	6.1+13	6.1+13
$2p_{1/2}3d_{3/2}$	8.6+13	9.5+13	8.7+13	9.9+13	8.6+13	8.5+13
$2s_{1/2}\,3p_{1/2}$	3.9+12		4.2+12	4.7+12	4.0+12	3.9+12
$2s_{1/2}\,3p_{3/2}$	1.4+13		1.5+13	1.8+13	1.4+13	1.3+13
$2p_{3/2}4s_{1/2}$	1.1+12		1.2+12	1.5+12	1.1+12	1.1+12
$2p_{1/2}4s_{1/2}$	2.1+12		2.5+12	2.8+12	2.3+12	2.2+12
$2p_{3/2}4d_{3/2}$	2.8+10		7.3+10	6.9+10	6.3+10	6.0+10
$2p_{3/2}4d_{5/2}$			2.8+13	2.7+13	2.3+13	2.2+13
$2p_{1/2}4d_{3/2}$	2.0+13		2.2+13	2.3+13	2.0+13	1.9+13
$2s_{1/2}4p_{1/2}$	2.5+12			2.9+12	2.6+12	2.5+12
$2s_{1/2}4p_{3/2}$	7.1+12			8.9+12	8.0+12	7.8+12

Analysis of the data shows that the computational method used provides a physically reasonable agreement between the theoretical and experimental data. Let us note that the transition probabilities values in the different photon propagator gauges are practically equal. Besides, an account of the inter particle (electron) correlation effects is of a great importance.

Acknowledgements

The authors would like to thank Prof. Nithaya Chetty, Prof. Marius Potgieter and colleagues for invitation to make contributions to CCP-2016 (Gauteng, South Africa). The useful comments of the anonymous referees are very much acknowledged too.

References

[1] Glushkov A V 2012 *Advances in the Theory of Quantum Systems in Chemistry and Physics* (*Progress in Theor.Chemistry and Physics* vol 28) K Nishikawa, J Maruani, E Brandas, G Delgado-Barrio, P Piecuch (Berlin: Springer) chapter 8 pp 31-52
[2] Glushkov A V 1991 *Opt. Spectrosc.* **70** 555; 1990 *Soviet Phys. J* **33**(1) 1
[3] Malinovskaya S V and Glushkov A V 1992 *Russian Phys. J.* **35** 999
[4] Glushkov A V, Khetselius O Yu and Malinovskaya S V 2008 *Europ. Phys. J. ST* **160** 195
[5] Glushkov A V, Ambrosov S V, Loboda A V, Chernyakovca Y G, Khetselius O Y and Svinarenko A A 2004 *Nucl. Phys. A* **734** E21
[6] Bièmont E 1977 *Astr. & Astrophys. Suppl. Ser.* **27** 489
[7] Dzuba V A, Flambaum V V and Sushkov O P 1995 *Phys. Rev. A* **51** 3454
[8] Marinescu M, Vrinceanu D and Sadeghpour H R 1998 *Phys. Rev. A* **58** R4259
[9] Safronova U I, Johnson W and Derevianko A 1999 *Phys. Rev.A.* **60** 4476
[10] Simsarian J E, Orozco L A, Sprouse G D and Zhao W Z 1998 *Phys. Rev. A* **57** 2448
[11] Glushkov A V 2012 *J. Phys.: Conf. Ser.* **397** 012011
[12] Glushkov A V 2014 *J. Phys.: Conf. Ser.* **548** 012020
[13] Glushkov A V, Khetselius O Yu and Svinarenko A A 2013 *Phys. Scr.* **T153** 014029
[14] Khetselius O Yu 2009 *Int. J. Quant. Chem.* **109** 3330
[15] Khetselius O Yu 2009 *Phys. Scr.* **T135** 014023
[16] Khetselius O Yu 2012 *J. Phys.: Conf. Ser.* **397** 012012
[17] Zhi-Gang Feng, Lin-Jie Zhang, Jian-Ming Zhao, Chang-Yong Liand and Suo-Tang Jia 2009 *J.*

Phys. B: At. Mol. Opt. Phys. **42** 145303
[18] Martin W 2004 *NIST Spectra Database, version 2.0* (NIST: Washington)
[19] Moore C 1987 *NBS Spectra Database*, (NBS: Washington)
[20] Safranova U I, Safranova M S and Johnson W 2005 *Phys. Rev. A* **71** 052506
[21] Safranova M S, Johnson W, Safranova U I and Cowan T 2006 *Phys. Rev. A* **74** 022504.
[22] Ivanova E P and Ivanov L N 1979 *Atom. Data. Nucl. Data. Tabl.* **24** 95
[23] Ivanov L N and Letokhov V S 1975 *JETP* **68** 1748; 1985 *Com. Mod. Phys.* D**4** 169
[24] Ivanov L N, Ivanova E P and Aglitsky E V 1988 *Phys. Rep.* **166** 315
[25] Ivanov L N, Ivanova E P and Knight L 1993 *Phys. Rev. A* **48** 4365
[26] Ivanova E P and Grant I P 1998 *J. Phys.B: At. Mol. Opt. Phys.* **31** 2871
[27] Ivanova E P, Ivanov L N, Glushkov A V and Kramida A E 1985 *Phys. Scr.* **32** 512
[28] Ivanova E P and Glushkov A V 1986 *J. Quant. Spectr. Rad. Transfer.* **36** 127
[29] Glushkov A V, Ivanov L N and Ivanova E P 1986 *Autoionization Phenomena in Atoms*, (Moscow: Moscow University Press) p 58
[30] Glushkov A V, Ivanov L N 1992 *Phys. Lett. A.* **170** 33
[31] Glushkov A and Ivanov L N 1993 *J. Phys. B: At. Mol. Opt. Phys.* **26** L379
[32] Glushkov A V 2006 *Relativistic and correlation effects in spectra of atomic systems* (Odessa: Astroprint).
[33] Svinarenko A A 2014 *J. Phys.: Conf. Ser.* **548** 012039
[34] Glushkov A V, Malinovskaya S V, Prepelitsa G P and Ignatenko V M 2005 *J. Phys.: Conf. Ser.* **11** 199
[35] Glushkov A V, Malinovskaya S V, Gurnitskaya E P, Khetselius O Yu and Dubrovskaya Yu V 2006 *J. Phys.: Conf. Ser.* **35** 425
[36] Glushkov A V, Malinovskaya S V, Khetselius O Yu, Loboda A V, Sukharev D E and Lovett L 2009 *Int. J. Quant. Chem.* **109** 1717
[37] Malinovskaya S V, Glushkov A V, Khetselius O Yu, Svinarenko A A, Mischenko E V and Florko T A 2009 *Int. J. Quant. Chem.* **109** 3325
[38] Malinovskaya S V, Glushkov A V, Khetselius O Yu, Loboda A V, Lopatkin Yu M, Nikola L V, Svinarenko A A and Perelygina T B 2011 *Int. J. Quant. Chem.* **111** 288
[39] Khetselius O Yu, Florko T A, Svinarenko A A and Tkach T B 2013 *Phys. Scr.* **T153** 014037
[40] Florko T , Ambrosov S, Svinarenko A and Tkach T 2012 *J. Phys.: Conf. Ser.* **397** 012037
[41] Khetselius O Yu, Florko T A, Nikola L V, Svianrenko A A, Serga I N, Tkach T B and Mischenko E V 2010 *Quantum Theory: Reconsideration of Foundations-5 AIP Conf. Proc.* **1232** 243
[42] Khetselius, O.Yu. 2010 *AIP Conf. Proc.* **1290 29**
[43] Svinarenko A A, Nikola L V, Prepelitsa P, Tkach T B and Mischenko E V 2010 *AIP Conf. Proc.* **1290** 94
[44] Glushkov A V, Lovett L, Khetselius O Yu, Gurnitskaya E P, Dubrovskaya Yu V and Loboda A V 2009 *Int. Journ. Modern Phys. A* **24** 611
[45] Glushkov A V, Loboda A, Gurnitskaya E and Svinarenko A 2009 *Phys. Scr.* **T135** 014022
[46] Glushkov A V, Ambrosov S V, Loboda A V, Gurnitskaya E P and Prepelitsa G P 2005 *Int. J. Quant.Chem.* **104** 562
[47] Glushkov A V, Khetselius O Yu and Malinovskaya S V 2008 *Molec. Phys.* **106** 1257
[48] Malinovskaya S V, Dubrovskaya Yu V and Vitavetskaya L A 2005 *AIP Conf. Proc.* **796** 201
[49] Svinarenko A A, Ignatenko A V, Ternovsky V B, Nikola L V, Seredenko S S and Tkach T B 2014 *J. Phys.: Conf. Ser.* 548 012047
[50] Glushkov A V, Butenko Yu V, Serbov N G, Ambrosov S V, Orlova V E , Orlov S V, Balan A K and Dormostuchenko G M 1996 *J. Appl. Spectrosc.* **63** 28
[51] Andersson M, Grumer J, Brage T, Zou Y and Hutton R 2016 *Phys. Rev. A.* **93** 032506

Computational code in atomic and nuclear quantum optics: Advanced computing multiphoton resonance parameters for atoms in a strong laser field

A V Glushkov, M Yu Gurskaya, A V Ignatenko, A V Smirnov, I N Serga, A A Svinarenko and E V Ternovsky

Odessa State Environmental University, L'vovskaya str.15, Odessa-16, 65016, Ukraine

E-mail: dirac13@mail.ru

Abstract. The consistent relativistic energy approach to the finite Fermi-systems (atoms and nuclei) in a strong realistic laser field is presented and applied to computing the multiphoton resonances parameters in some atoms and nuclei. The approach is based on the Gell-Mann and Low S-matrix formalism, multiphoton resonance lines moments technique and advanced Ivanov-Ivanova algorithm of calculating the Green's function of the Dirac equation. The data for multiphoton resonance width and shift for the Cs atom and the ^{57}Fe nucleus in dependence upon the laser intensity are listed.

1. Introduction

At the present time physics of multiphoton phenomena in atoms, molecules and nuclei has reached a great progress that is mostly stimulated by a development of new laser technologies and perspectives of new impressive applications (see Refs. [1,2]). The appearance of the powerful laser sources allowing to obtain the radiation field amplitude of the order of atomic field in the wide range of wavelengths results to systematic investigations of the nonlinear interaction of radiation with atomic and nuclear systems [1-16]. At the same time a direct laser-nucleus interactions traditionally have been dismissed because of the well known effect of small interaction matrix elements. Some exceptions such as an interaction of x-ray laser fields with nuclei in relation to alpha, beta-decay and x-ray-driven gamma emission of nuclei have been earlier considered. With the advent of new coherent x-ray laser sources in the near future, however, these conclusions have to be reconsidered. From the design report (look table II in Ref.[2]) for SASE 1 at TESLA XFEL and parameters for current and future ion beam sources, the signal rate due to spontaneous emission after real excitations of the nuclei can be estimated. For nuclei accelerated with an energy resolution of 0.1% such that 12.4 keV photons produced by SASE 1 become resonant with the $E1$ transition in a whole number of nuclei (for example, ^{153}Sm, ^{181}Ta, ^{223}Ra, ^{225}Ac , ^{227}Th etc) [9]. It means that the resonance condition ($\omega \sim \Delta\varepsilon$, where $\Delta\varepsilon$ is a typical level spacing, ω is a laser frequency) is fulfilled [9-12]. The coherence of the laser light expected from new sources (TESLA XFEL at DESY) may allow to access the extended complex coherence or interference phenomena. In particular, in conjunction with moderate acceleration of the target nuclei it allows principally to achieve realization of multi-photon phenomena, nuclear Rabi oscillations or more advanced quantum optical schemes in nuclei [9,10].

The interaction of atoms with the external alternating fields, in particular, laser field, has been the subject of intensive experimental and theoretical studies (see, for example, Refs. [1-27]). A definition of the k-photon emission and absorption probabilities and atomic levels shifts, study of dynamical stabilization and field ionization etc are the most current problems to be solved. At the present time, a progress is achieved in the description of the processes for atoms interacting with the harmonic emission field [9, 14-17]. But in a realistic laser field the corresponding processes differ significantly from the ones in a harmonic field. A substantial complexity is provided by the photon-correlation effects and an influence of the laser pulse multi-mode features. Notably, a number of different theoretical approaches have been developed in order to give adequate description of the atoms in a strong laser field. The corresponding methods are the standard perturbation theory (surely for low laser filed intensities), the Green's function method, the density-matrix formalism, time-dependent density functional formalism, a direct numerical solution of the Schrödinger and/or Dirac equation, multi-body multi-photon approach, the time-independent Floquet formalism *etc.* (see [1,14-28] and references therein). The effects of the different laser line shape on an intensity and spectrum of resonance fluorescence from a two-level atom are studied in Refs. [1-5,15-17,19-23].

Earlier, a relativistic energy approach to studying the interaction of atoms with a realistic strong laser field, based on the Gell-Mann and Low S-matrix formalisms, has been developed. Originally, Ivanov has proposed an idea to describe quantitatively a behavior of an atom in a realistic laser field by means of studying the radiation emission and absorption lines and further the consistent theory of interaction of an atom with a Lorentzian laser pulse has been developed in Refs. [19-22]. The theory has been applied to computing the multi-photon resonance shifts and widths in hydrogen and caesium [20]. The consistent theory of interaction of an atom with the Gaussian and soliton-like laser pulses has been developed in Refs. [24-27], where the corresponding calculation results are listed.

Here we present an advanced generalized relativistic energy approach to computing multiphoton resonances parameters in the finite Fermi-systems (as example, atom of Cs and nucleus of Fe) in a strong laser field with using the Green's function technique for computing sums in expressions for a multiphoton resonance line moments.

2. Relativistic energy approach to atom in a strong laser field: Multiphoton resonances

The relativistic energy approach in the different realizations and the radiation lines moments technique are presented in Refs. [17-27] and here we are limited only by presenting the master moments. In the theory of the non-relativistic atom there is a convenient field procedure for computing the energy shifts δE of degenerate states, which is connected with the secular matrix M diagonalization. In constructing M, one could apply the known Gell-Mann and Low adiabatic formula for δE (look details in Refs. [19-22]. In relativistic theory, the Gell-Mann and Low formula δE is connected with electrodynamical scattering matrice, which includes interaction with as a laser field as a photon vacuum field. In relativistic theory the secular matrix elements are already complex in the second perturbation theory (PT) order and their imaginary parts are connected with radiation decay possibility. The total energy shift is usually presented in the form: $\delta E = \mathrm{Re}\delta E + i\,\mathrm{Im}\delta E$, $\mathrm{Im}\,\delta E = -P/2$, where P is the level width. Spectroscopy of an atom (nucleus) in a strong laser field is determined by position and shape of the radiation emission and absorption lines. The lines moments μ_n are strongly dependent upon the laser pulse quality: intensity and mode constitution. We describe the interaction "nucleus-laser field" by the Ivanov potential [20]:

$$V(r,t)= V(r) \int d\omega\, f(\omega - \omega_0)\ \sum_{n=-\infty}^{\infty} \cos(\,\omega_0 t + \omega_0\, n\tau). \qquad (1)$$

Here ω_0 is the central laser radiation frequency, n is the whole number. The potential V represents the infinite duration of laser pulses with known frequency τ. The function $f(\omega)$ is a Fourier component of the laser pulse. For example, we consider a pulse with the Gaussian shape. A case of the Lorentzian shape has been earlier studied [20,21]. A case of the Gaussian and soliton-like shape is studied in Refs. [24-27]. The master program results in the calculating an imaginary part of an energy shift $\mathrm{Im}\delta E_\alpha(\omega_0)$

for any atomic level as the function of the central laser frequency ω_0. An according function has the shape of the resonance, connected with the transition α-p with absorption (or emission) of the "k" number of photons. For the resonance we calculate the following values [19-21]:

$$\delta\omega(p\alpha|k) = \int' d\omega \, \text{Im}\, \delta E_\alpha(\omega) \, (\omega - \omega_{p\alpha} / k) / N, \qquad (2a)$$

$$\mu_m = \int' d\omega \, \text{Im}\, \delta E_\alpha(\omega) \, (\varpi - \omega_{p\alpha} / k)^m / N, \qquad (2b)$$

where $\int' d\omega \text{Im} E_\alpha$ is the normalizing multiplier; $\omega_{p\alpha}$ is position of the non-shifted line for transition α-p, $\delta\omega(p\alpha|k)$ is the line shift under k-photon absorption; $\varpi_{p\alpha}=\omega_{p\alpha}+ k\cdot\delta\omega(p\alpha|k)$. The first moments μ_1, μ_2 and μ_3 determine an atomic line centre shift, its dispersion and asymmetry. In order to compute the values μ_m, one should get an expansion of E_α to the PT series [20]: $E_\alpha = \sum E_\alpha^{(2k)}(\omega_0)$. Here one may use the known Gell-Mann and Low adiabatic formula for δE_α and the further representation of the S-matrix in the form of the PT [20]:

$$\delta E_\alpha(\omega_0) = \lim_{\gamma \to 0} \gamma \sum_{k_1 k_2 \ldots k_n} a\,(k_1, k_2, \ldots, k_n), \qquad (3)$$

$$I_\gamma(k_1, k_2, \ldots, k_n) = \prod_{j=1} (-1)^j \int_{-\infty}^{0} d\, t_1 \ldots \int_{-\infty}^{t_m - 1} d\, t_j \, \langle \Phi_\alpha | V_1 V_2 \ldots V_m | \Phi_\alpha \rangle,$$

$$V_j = \exp(iH_0 t_j)\, V(rt_j) \exp(-iH_0 t_j) \exp(\gamma t_j).$$

Here H is the atomic hamiltonian, a $(k_1, k_2,\ldots,k_n)$ are the numerical coefficients. The structure of matrix elements $S_\gamma^{(m)}$ and integral technique are in details described in [17-23]. The corresponding expressions for the Gaussian laser pulse are as follows [24,25]:

$$\delta\omega(p\alpha \,|\, k) = \{\pi\Delta / (k+1)\text{k}\} \, [\, E(p, \omega_{p\alpha}/k) - E(\alpha, \omega_{p\alpha}/k)], \qquad (4)$$

$$\mu_2 = \Delta^2/\text{k}$$

$$\mu_3 = \{4\pi\Delta^3 / [k\,(k+1)]\} \, [\, E(p, \omega_{p\alpha}/k) - E(\alpha, \omega_{p\alpha}/k)],$$

$$E(j, \omega_{p\alpha}/k) = 0{,}5 \sum_{p_i} V_{jpi} V_{pij} \left[\frac{1}{\omega_{jp_i} + \omega_{p\alpha}/k} + \frac{1}{\omega_{jp_i} - \omega_{p\alpha}/k} \right] \qquad (5)$$

where the summation in (5) is over all atomic states. Note that these formulas for the Gaussian pulse [24,25] differ from the expressions for Lorentzian shape laser pulse [19-22]. For the soliton-like pulse, it is necessary to fulfil the numerical calculation or use some approximations to simplify the expressions [27]. Earlier to determine the the PT second order sum in (5) in different calculations of atomic and meso-atomic characteristics [28-36], it has been applied the effective Ivanov-Ivanova technique [18]. The alternative approach can be provided by method of the Green's function of the Dirac equation (look below). The corresponding computational procedure includes solving an ordinary differential equations system for above mentioned functions and integrals. The scheme of computing electron c wave functions within the relativistic many-body PT with the Dirac-Kohn-Sham zeroth approximation, the corresponding technique of computing all matrix elements in the expressions (2,5) and other details are presented in Refs. [20-30]. In all numerical calculations the PC "Superatom-ISAN" package is used.

Further it is important to underline that above presented expressions remain the same for treating the multiphoton resonances in a nucleus in a strong laser field, however there is some difference in comparison with an atomic case. It is obviously connected with calculating the corresponding matrix elements in formulae for resonance parameters with substituting the nuclear wave functions and some other details. In a modern theory of a nucleus there is sufficiently great number of the different models for generating the proton and neutron wave functions (c.f. [37,38]). At present time, it is acceptable that a quite adequate description of the nuclear systems is provided by the relativistic mean-field (RMF) and other similar models [37,38]. As an alternative approach one could use the advanced shell models based on the effective Dirac-Wood-Saxon type Hamiltonian [26,39]. This model [39] has been used in our work.

3. Method of the Green's function of the Dirac equation with complex energy: Advanced Ivanov-Ivanova computational algorithm

The details of the method of computing the sum in (5) are presented in Refs. [18,20,23,27]. The fundamentals of the method have been developed by Ivanov-Ivanova et al (c.f. [18-20]). The advanced version of this approach has been presented in Ref. [23]. Here we are limited only by the principal points of an advanced computational algorithm [23]. As usually, the radial Dirac equations can be written as follows (in the Coulomb units):

$$f' = -(\chi+1)f/r - V^{-} g\tilde{\alpha};$$

$$g' = (\chi-1)g/r + V^{+} f\tilde{\alpha};$$

$$V^{\mp} = V(r) - i\xi \mp \tilde{\alpha}^{-2}; \tag{6}$$

where $V(r)$ is the potential of a nucleus, $\tilde{\alpha} = \alpha Z$ (α is the fine structure constant), f and g are the Dirac equations solutions. We are interested by the case, when the potential is regular for $r \to 0$. It is easy to show (c.f. [18]) that for such a potential the solutions of two types (regular and non-regular at $r \to 0$) exist for each value of ξ and χ:

for $\chi<0$

$$f \sim r^{|\chi|-1},\ g \sim r^{|\chi|},\ \tilde{f} \sim r^{-|\chi|},\ \tilde{g} \sim r^{-|\chi|-1},$$

for $\chi>0$

$$f \sim r^{|\chi|},\ g \sim r^{|\chi|-1},\ \tilde{f} \sim r^{-|\chi|-1},\ \tilde{g} \sim r^{-|\chi|}. \tag{7}$$

The regular solution (f, g) at $r \to 0$ is simply defined by the condition (6) with the accuracy to a normalization. At the same time the singular solutions are not defined by these conditions.

For large values of $|\chi|$ the functions (7) have a strong degree dependence at $r \to 0$ that is a reason of the known computational difficulties during the numerical integration of the Dirac equations. At large χ the radial functions F and G vary rapidly at the origin of co-ordinates (c.f.[19,28]):

$$F(r), G(r) \approx r^{\gamma-1}$$
$$\gamma = \sqrt{\chi^2 - \alpha^2 z^2} \tag{8}$$

and further it is convenient to introduce the new functions isolating the main power dependence:

$$(F, G) = (f, g) \cdot r^{1-|\chi|};$$

$$\left(\tilde{F},\tilde{G}\right)=\left(\tilde{f},\tilde{g}\right)\cdot r^{|\chi|+1}. \tag{9}$$

The Green's function is a combination of the Dirac equation modified (the power dependence is separated) fundamental solutions:

$$F' = -(\chi+|\chi|)F/r + V^{-}\tilde{\alpha}G;\quad G' = (\chi-|\chi|)G/r - V^{+}\tilde{\alpha}F; \tag{10a}$$

$$\tilde{F}' = -(\chi+|\chi|)\tilde{F}/r + V^{-}\tilde{\alpha}\tilde{G};\quad \tilde{G}' = (|\chi|+\chi)\tilde{G}/r - V^{+}\tilde{\alpha}\tilde{F}. \tag{10}$$

The functions (*F, G*) represent the first fundamental solution, which is regular for $r\to 0$ and singular for $r\to\infty$. Any combination $\left(\tilde{F},\tilde{G}\right)+Cr^{2|\chi|}(F,G)$ satisfies to above written equations for $\left(\tilde{F},\tilde{G}\right)$ and represents singular solution at zero [18,19,23]. The right chosen combination $\left(\hat{F},\hat{G}\right)$ for the single value of the mixing coefficient C (regular for $r\to\infty$) is second fundamental solution $\left(\hat{f},\hat{g}\right)$. Finally, the Green's function electron function is a four-component matrice with the functions (*F, G*) and $(\hat{F},\hat{G})$, which are the Dirac equations solutions with account of the corresponding asymptotic conditions. Further one can obtain from Eqs. (10) that for $r\to\infty$:

$$(F,G)\sim \exp Ar;\ \left(\hat{F},\hat{G}\right)\sim \exp(-Ar);\quad A=\left(\tilde{\alpha}^{-2}+\xi^{2}\tilde{\alpha}^{2}\right)^{1/2}. \tag{11}$$

As the bi-linear combinations of the Dirac function components are presented in the Green's function, only their relative normalization is important. It is determined by the known "Wronscian" condition ($W=F\hat{G}-\hat{F}G\equiv 1.$). The electron radial Green's function is the four component matrice as follows:

$$G(r_1\, r_2\mid E,\chi)=\begin{pmatrix}\hat{F}(r_>)F(r_<) & \hat{F}(r_>)G(r_<)\\ \hat{G}(r_>)F(r_<) & \hat{G}(r_>)G(r_<)\end{pmatrix}, \tag{12}$$

where $r_>$ $(r_<)$ is more (or less) value of r_1, r_2; the functions (F,G) and $\left(\hat{F},\hat{G}\right)$ satisfy the Dirac equations (10), their asymptotical conditions (11) and the "Wronscian" normalization condition. The corresponding computational procedure, which includes solving an ordinary differential equations system for relativistic wave functions, computing all matrix elements and other details can be found in Refs. [23-28,39].

4. Some results and conclusions

Below we present the calculation results for the three-photon transition 6S-6F (p-α transition) in the ^{133}Cs atom (wavelength 1,059 μm). The detailed experimental studying the multi-photon processes in Cs atom has been fulfilled in Ref. [15]. In this paper it has been experimentally studied a statistics of the laser radiation and there have been measured the characteristics of the multi-photon ionization. According to Ref.[15], the multiphoton line shift is linear to respect to a laser intensity (the laser intensity is increased from 1.4 to $5.7\cdot 10^{7}$ W/cm^2) and is equal (a case of the Gaussian multi-mode laser pulse): $\delta\omega(p\alpha\,|k)=bI$ with b=(5.6±0.3) cm^{-1}/GW·cm^{-2} (b is expressed in terms of energy of the three-photon transition 6S-6F). The corresponding shift for the coherent (one-mode) laser pulse is equal as follows [15]: $\delta\omega_0(p\alpha|k)=aI$, a=2 cm^{-1}/GW·cm^{-2}.The theoretical values, obtained within the relativistic energy approach [22,24,25] are as follows: i). $\delta\omega(p\alpha|k)=bI$ with b=5.8 cm^{-1}/GW·cm^{-2} for the Gaussian multi-mode pulse (chaotic light); ii). $\delta\omega_0(p\alpha\ |k)=aI$, a=2.1 cm^{-1}/GW·cm^{-2} for the coherent one-mode pulse. The corresponding theoretical values, obtained in this work, are as follows: i). $\delta\omega(p\alpha \mid k) = bI$, b=5.62 cm^{-1}/GW·cm^{-2} for the Gaussian multi-mode pulse; ii). $\delta\omega_0(p\alpha\ |k)=aI$, a=2.03 cm^{-1}/GW·cm^{-2} for the coherent one-mode pulse. One should note that the radiation line shift for the multi-mode pulse is significantly larger (in ~ 3 times) than the corresponding shift for the single-mode pulse. In fact the resonance line shift is enhanced by the photon-correlation effects. In figure 1 we present the results of computing the multi-photon resonance width for transition 6S-6F in the atom of Cs (wavelength 1059nm) in dependence upon a laser intensity. We use the following denotations: experimental data- o (Grance ; Lompre etal, 1981); theory – by Glushkov-Ivanov (the energy approach plus method of the differential equations; continuous lines) Γ_{L2} (1-mode Lorentzian pulse), Γ_2, Γ_4 - multimode Gaussian laser pulse (the linewidth 0.03, 0.08 cm^{-1}); this work: the dotted lines Γ_{L1} (the Lorentzian pulse), Γ_1, Γ_3, Γ_5, Γ_6, Γ_7 - multimode Gaussian laser pulse (the linewidth 0.03, 0.055, 0.08, 0.115, 0.15 cm^{-1} correspondingly). In Ref. [15] there are presented experimental data for laser pulse of the Gaussian shape with line band respectively 0.03cm^{-1}, 0.08cm^{-1}, 0.15cm^{-1}.

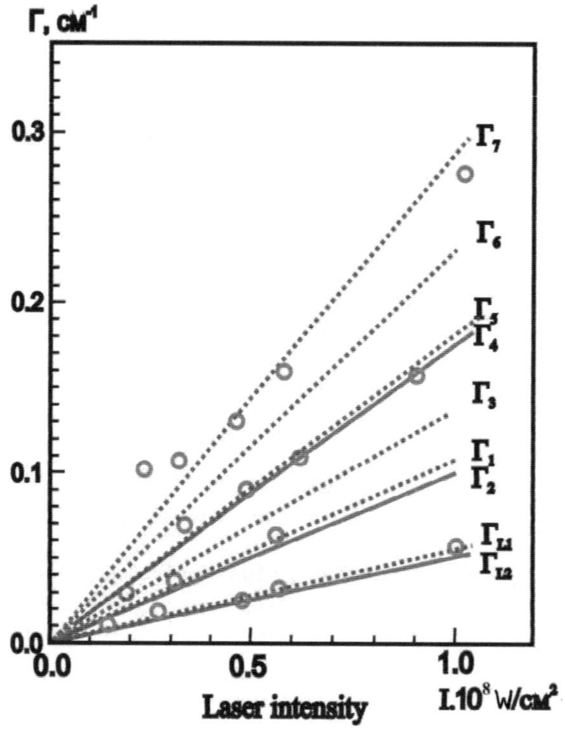

Figure 1. The results of computing the multi-photon resonance width for transition 6S-6F in the atom of Cs (wavelength 1059nm) in dependence upon laser intensity: experimental data- o (Grance, Lompre et al [15]); theory – Glushkov-Ivanov (the energy approach plus method of the differential equations; continuous lines) Γ_{L2} (1-mode Lorentzian pulse), Γ_2, Γ_4 - multimode Gaussian laser pulse (the linewidth 0.03, 0.08 cm^{-1}) [20,24]; This work: the dotted lines Γ_{L1} (the Lorentzin pulse), Γ_1, Γ_3, Γ_5, Γ_6, Γ_7 - multimode Gaussian laser pulse (the linewidth 0.03, 0.055, 0.08, 0.115, 0.15 cm^{-1} correspondingly).

In general there is a physically reasonable agreement between theoretical and experimental data. The analysis shows that the shift and width of the multi-photon resonance line in atomic system for the multimode laser pulse are greater than the corresponding resonance shift and width for the one-mode laser pulse. This fact is entirely corresponding to the experimental result [15]. From physical point of view it is provided by action of the photon-correlation effects and influence of the laser pulse multimodity (c.f.[14,15,20-27]).

Further let us present the first results of estimating the multiphoton resonance parameters in the ^{57}Fe nucleus in a super strong laser field. In order to estimate the corresponding matrix elements on the nuclear wave functions it was used the shell model based on the effective Dirac-Wood-Saxon type Hamiltonian, which contains the Bloumkvist-Wahlborn nuclear potential [39]. According to Ref. [40], the nuclear multiphoton transitions take a place in ^{57}Fe nucleus subjected to radio-frequency electromagnetic field ω_0=30MHz. This picture was experimentally observed in the Mössbauer spectra of ^{57}Fe nuclei in Permalloy by Tittonen et al [41]. Really, the eight transitions are possible between the four hyperfine substates of the 14.4 keV excited level e and the two substates of the ground state g in the radio-frequency magnetic field [16]. If the static magnetic hyperfine splitting of the ground and excited states are respectively $\omega_g>0$ and $\omega_e>0$, the transition frequencies corresponding to forbidden γ-ray transitions are $(E_e-E_g)/h\pm3\omega_e/2\pm\omega_g/2$, where E_e, E_g are respectively the energies of the 14.4-keV and ground states of the ^{57}Fe nucleus in an absence of any external field. The estimates of the ^{57}Fe resonance shift in dependence upon the mean laser intensity are as follows: $I=10^{25}$W/cm^2, $\delta\omega=5\cdot10^{-12}$ keV, $I=10^{28}$W/cm^2, $\delta\omega=10^{-9}$ keV. The corresponding asymmetric nature of the multiphoton resonance line in ^{57}Fe nucleus is naturally explained within the presented theoretical approach.

References

[1] Batani D and Joachain C ed 2006 *Matter in super-intense laser fields* (New York: AIP)
[2] Ivanova E P 2011 *Phys. Rev. A* **84** 043829
[3] Khetselius O Yu, Florko T A, Svinarenko A A and Tkach T B 2013 *Phys. Scr.* **T153** 014037
[4] Florko T A, Ambrosov S V, Svinarenko A A and Tkach T B 2012 *J. of Phys.: Conf. Ser.* **397** 012037
[5] Svinarenko A A 2014 *J. Phys.: Conf. Ser.* **548** 012039
[6] Mercouris T and Nikolaides C A 2003 *Phys. Rev. A* **67** 063403
[7] Brandas E and Floelich P 1977 *Phys. Rev. A* **16** 2207
[8] Wangt K, Hot T-S and Chu S-I 2006 *Phys. Rev. A* **73** 023403
[9] Bürvenich T, Evers J and Keitel C 2006 *Phys. Rev. Lett.* **96** 142501; 2006 Phys.*Rev.C* **74** 044601
[10] Glushkov A V, Ivanov L N and Letokhov V S 1991 Nuclear quantum optics *Preprint AS-N4 of Institute for Spectroscopy of the USSR Academy of Sciences* (Moscow-Troitsk: ISAN) p 16; Goldanskii V I and Letokhov V S 1974 *Sov. Phys. JETP.* **67** 513
[11] Khetselius O Yu 2009 *Int. J. Quant. Chem.* **109** 3330
[12] Belyaev V S, Kraynov V P, Lisitsa V S and Matafonov A P *Phys.-Uspekhi* **178** 823
[13] Malinovskaya S V, Dubrovskaya Yu V and Vitavetskaya L A 2005 *AIP Conf. Proc.* **796** 201
[14] Zoller P 1982 *J. Phys. B: At. Mol. Opt. Phys.* **15** 2911
[15] Lompre L, Mainfrau G, Manus C and Marinier J 1981 *J. Phys. B: At. Mol. Opt. Phys.* **14** 4307
[16] Landen O L, Perry M D and Campbell E M 1987 *Phys. Rev. Lett.* **59** 2558
[17] Ivanov L N and Letokhov V S 1975 *JETP* **68** 1748; 1985 *Com. Mod. Phys.* D**4** 169
[18] Ivanov L N and Ivanova E P 1974 *Theor. Math. Phys.* **20** 282
[19] Ivanov L N, Ivanova E P and Aglitsky E V 1988 *Phys. Rep.* **166** 315
[20] Glushkov A V, Ivanov L N and Ivanova E P 1986 *Autoionization Phenomena in Atoms*, (Moscow: Moscow University Press) p 58; Glushkov A V and Ivanov L N 1992 *Proc. of 3rd USSR Seminar on Atomic Spectroscopy* (Chernogolovka: Nauka) p 113
[21] Glushkov A V and Ivanov L N 1992 *Preprint AS-N3 of Institute for Spectroscopy of the USSR Academy of Sciences* (Troitsk: ISAN) p 12; 1992 *Phys. Lett. A* **170** 33
[22] Glushkov A and Ivanov L N 1992 *Preprint AS-N1 of Institute for Spectroscopy of the USSR Academy of Sciences* (Troitsk: ISAN) p 16; Glushkov A and Ivanov L N 1993 *J. Phys. B: At. Mol. Opt. Phys.* **26** L179

[23] Glushkov A V, Malinovskaya S V, Sukharev D E, Khetselius O Yu, Loboda A V and Lovett L 2009 *Int. J. Quant. Chem.* **109** 1717
[24] Glushkov A V, Loboda A, Gurnitskaya E and Svinarenko A 2009 *Phys. Scr.* **T135** 014022
[25] Glushkov A V 2014 *J. Phys.: Conf. Ser.* **548** 012020
[26] Glushkov A V 2012 *J. of Phys.: Conf. Ser.* **397** 012011
[27] Glushkov A V 2006 *Relativistic and correlation effects in spectra of atomic systems* (Odessa: Astroprint)
[28] Ivanova E P , Ivanov L N, Kramida A E and Glushkov A V 1985 *Phys.Scr.* **32** 512
[29] Glushkov A V, Malinovskaya S V, Prepelitsa G P and Ignatenko V M 2005 *J. Phys.: Conf. Ser.* **11** 199
[30] Glushkov A V, Kondratenko P A, Buyadgi V V, Kvasikova A S, Shakhman A S and Sakun T N 2014 *J. Phys.: Conf. Ser.* **548** 012025
[31] Khetselius O Yu 2008 *Spectral Line Shapes (*vol 15*) AIP Conf. Proc.* **1058** 363
[32] Khetselius O Yu 2010 *Spectral Line Shapes AIP Conf. Proc.* **1290** 29
[33] Khetselius O Yu 2012 *J. Phys.: Conf. Ser.* **397** 012012
[34] Khetselius O Yu 2009 *Phys.Scr.* **T135** 014023
[35] Svinarenko A A, Khetselius O Yu, Buyadzhi V V, Florko T A, Zaichko P A and Ponomarenko E L 2014 *J. Phys.: Conf. Ser.* **548** 012048
[36] Glushkov A V, Kondratenko P A, Buyadgi V V, Kvasikova A S, Sakun T N and Shakhman A S 2014 *J. Phys.: Conf. Ser.* **548** 012025
[37] Serot B and Walecka J 1986 *Adv. Nucl. Phys.* **16** 1; Reinhard P G 1989 *Rep. Prog. Phys.* **52** 439
[38] Glushkov A V , Rusov V D , Ambrosov S V and Loboda A V 2003 *New Projects and New Lines of Research in Nuclear Physics (*vol 1*)* ed Fazio and F Hanappe (Singapore: World Scientific) pp 126-132
[39] Malinovskaya S V, Glushkov A V, Dubrovskaya Yu V and Vitavetskaya L A, Quantum calculation of cooperative muon-nuclear processes: discharge of metastable nuclei during negative muon capture *Recent Advances in the Theory of Chemical and Physical Systems (Progress in Theoretical Chemistry and Physics* vol 15*)* ed J-P Julien, J Maruani, D Mayou, S Wilson and G Delgado-Barrio (Dordrecht: Springer) pp 301-308
[40] Olariu S, Sinor T W and Collins C B 1994 *Phys. Rev. B* **50** 616
[41] Tittonen I, Lippmaa M, Ikonen E, Linden J and Katila T 1992 *Phys. Rev. Lett.* **69** 281

IOP Conf. Series: Journal of Physics: Conf. Series **905** (2017) 012005 doi:10.1088/1742-6596/905/1/012005

On the non-linear spectroscopy including saturated absorption and four-wave mixing in two and multi-level atoms: a computational study

M Patel[1], G De Jager[2], Z Nkosi[3], A Wyngaard[4] and K Govender[5,6]
[1,3,4,5] Department of Electrical, Electronic and Computer Engineering, Cape Peninsula University of Technology, Symphony Way, Bellville, Cape Town, 7535, South Africa
[2] Department of Electrical Engineering, University of Cape Town, Rondebosch, 7701, South Africa
[6] School of Chemistry and Physics, University of KwaZulu-Natal, South Africa

E-mail: [1]210230177@mycput.ac.za

Abstract. In this paper we report on the study of two and multi-level atoms interacting with multiple laser beams. The semi-classical approach is used to describe the system in which the atoms are treated quantum mechanically via the density matrix operator, while the laser beams are treated classically using Maxwells equations. We present results of a two level atom interacting with single and multiple laser beams and demonstrate Rabi oscillations between the levels. The effects of laser modulation on the dynamics of the atom (atomic populations and coherences) are examined by solving the optical Bloch equations. Plots of the density matrix elements as a function of time are presented for various parameters such as laser intensity, detuning, modulation etc. In addition, phase-space plots and Fourier analysis of the density matrix elements are provided. The atomic polarization, estimated from the coherence terms of the density matrix elements, is used in the numerical solution of Maxwells equations to determine the behaviour of the laser beams as they propagate through the atomic ensemble. The effects of saturation and hole-burning are demonstrated in the case of two counter propagating beams with one being a strong beam and the other being very weak. The above work is extended to include four-wave mixing in four level atoms in a diamond configuration. Two co-propagating beams of different wavelengths drive the atoms from a ground state $|1\rangle$ to an excited state $|3\rangle$ via an intermediate state $|2\rangle$. The atoms then move back to the ground state via another intermediate state $|4\rangle$, resulting in the generation of two additional correlated photon beams. The characteristics of these additional photons are studied.

1. Background

Quantum properties of light, besides being a subject of fundamental interest, is being used in emerging technologies such as quantum cryptography and quantum computing. There are a variety of ways of generating quantum light, however here we aim to investigate the generation and properties of quantum light using non-linear processes in cold atoms. We investigate the generation and subsequent properties of coherent light by means of four-wave mixing in an ensemble of cold rubidium atoms.

The motivation for this study is the fact that there is growing evidence that the next generation of computing and communication is going to be quantum based, i.e. these systems operate using the principles of quantum physics. These systems will operate using fundamentally new principles that involve light-atom interactions on a single photon level and ambitious schemes such as quantum teleportation. For long distance communications these schemes have to be implemented using entangled photons.

Entangled photons produced by four-wave mixing in a cold atomic ensemble have been shown to have desirable features. These photons have a narrow bandwidth, which is determined by the dispersion of the vapour and can be narrow enough to match the absorption profile of the atoms. These biphoton sources have properties such as long coherence time, long coherence length, high spectral brightness and high conversion efficiency. There have been a number of books and papers describing laser-atom interactions (e.g. see [1, 2, 3, 4, 5, 6]), and entangled photon pair generation using four-wave mixing in rubidium atomic ensembles [7, 8, 9, 10, 11].

We report on the computational study of the interaction between laser beams and, two and multi-level atoms. We use a semi-classical approach in which the dynamics of the atoms (described by the density matrix elements) are governed by the Liouville-von Neumann equation,

$$\frac{\partial \hat{\rho}}{\partial t} = -\frac{\imath}{\hbar}\left[\hat{H}, \hat{\rho}\right] + L_{damp}(\hat{\rho}) \qquad (1)$$

while the laser beam is described by the wave equation derived from Maxwell's equations. The Hamiltonian of the total system is $\hat{H} = \hat{H}_0 + \hat{H}_I$, where $\hat{H}_0$ is the unperturbed Hamiltonian and $\hat{H}_I$ is the interaction term. The effects of dissipation are contained in $L_{damp}(\hat{\rho})$. We first discuss a two level atom interacting with a single laser beam and examine the dynamics of the population and coherence terms of the density matrix elements. Thereafter we examine the behaviour of the laser beam as it propagates first in two level atoms and later in multi-level atoms.

The overview of the paper is as follows: Section 2 describes the interaction of a two level atom with a single laser beam. Results showing the effects of a uniform and modulated laser beam are given. Saturated absorption spectroscopy is discussed in Section 3. Results are given for various parameters followed by a discussion on non-linear mixing and entangled photon generation.

2. Laser-atom interactions in two level atoms

Consider an atom having two energy states represented by a ground state $|1\rangle$ having energy E_1 and an excited state $|2\rangle$ having energy E_2. Assume these states are separated in frequency by ω_0. Now let a laser beam with an electric field

$$\vec{E}(t) = \vec{E}_0 \cos \omega t$$

interact with a sample of these atoms, which are stationary. In the above, $\vec{E}_0$ is the amplitude of the electric field of the laser beam, $\cos \omega t$ represents the time variation of the electric field of the laser beam and ω is the laser frequency.

Then the following Hamiltonian describes the interaction:

$$\hat{H}_I = -\hat{\mu} \cdot \vec{E}_0 \cos \omega t \qquad (2)$$

where $-\hat{\mu}$ represents the dipole moment operator of the atom. The Liouville-von Neumann equation makes use of $\hat{H}_I$ to describe the time evolution of the density matrix elements of the system $\hat{\rho} = \sum_i P_i |i\rangle\langle i|$, where P_i is the probability of the atom being in state $|i\rangle$. It is used to derive the optical Bloch equations [3](below) for a two level atom:

$$\frac{\partial \rho_{11}}{\partial t} = \frac{\imath}{2}\Omega(\rho_{12} - \rho_{21}) + 2\gamma_{sp}\rho_{22} \qquad (3)$$

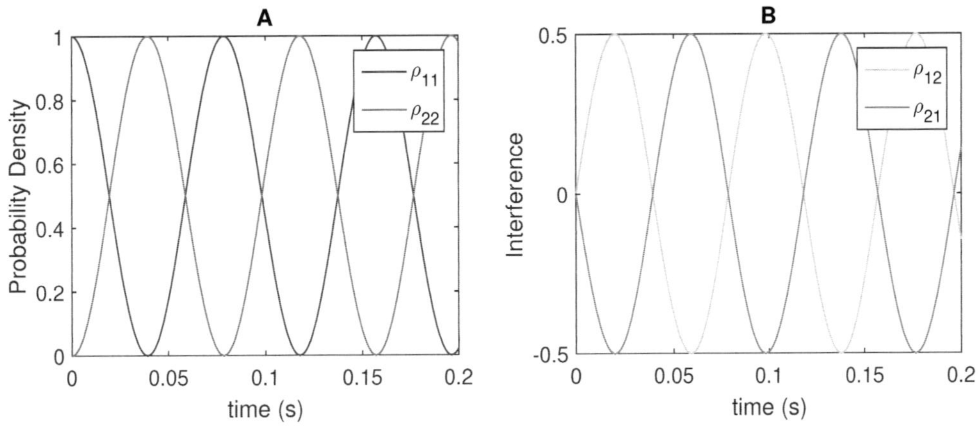

Figure 1. Rabi oscillations of a two level atom interacting with a uniform laser beam. Plots are: (**A**) - probability density and (**B**) - interference terms.

$$\frac{\partial \rho_{12}}{\partial t} = \frac{\imath}{2}\Omega(\rho_{11} - \rho_{22}) + \Big[\imath(\omega_0 - \omega) - \gamma_{sp}\Big]\rho_{12} \tag{4}$$

$$\frac{\partial \rho_{21}}{\partial t} = -\frac{\imath}{2}\Omega(\rho_{11} - \rho_{22}) + \Big[-\imath(\omega_0 - \omega) - \gamma_{sp}\Big]\rho_{21} \tag{5}$$

$$\frac{\partial \rho_{22}}{\partial t} = -\frac{\imath}{2}\Omega(\rho_{12} - \rho_{21}) - 2\gamma_{sp}\rho_{22} \tag{6}$$

where $\Omega = \langle\hat{\mu}\rangle E_0/\hbar$ is known as the Rabi frequency and γ_{sp} represents a decay/de-coherence coefficient. ρ_{11} and ρ_{22} tell us the probability of the atom being in the ground and excited states respectively. ρ_{12} and ρ_{21} are the interference terms of the atom indicative of superposition. We solve the above numerically first for a uniform laser beam and then for a frequency modulated laser beam.

2.1. Results

Results of the numerical solution of Equations (3-6) are given in Figures 1-3 which correspond to a uniform laser beam while Figures 4-6 deal with a frequency modulated laser beam. In all the plots labelled **A**, the blue curve represents ρ_{11} and the red curve ρ_{22}. The green curve represents ρ_{12} and the magenta curve ρ_{21} in Figures 1-3**B**. The laser is switched on at time $t = 0\ s$ with $\rho_{11} = 100\%$ and $\rho_{22} = 0\%$. Rabi oscillations are demonstrated in Figure 1 where $\omega_0 - \omega = 0$. When $\rho_{11}, \rho_{22} = 50\%$ maximum superposition is achieved. In Figure 2 the effects of laser detuning can be seen i.e. $\omega_0 - \omega \neq 0$. The interference between Ψ_1 and Ψ_2 is decreased, where Ψ_i is the wave function corresponding to level i. When a small dissipation is introduced in the form of γ_{sp}, the effects can be seen in Figure 3. The interference between Ψ_1 and Ψ_2 decays to the point where the probability of superposition is close to 0%.

Laser beam modulation is introduced by adding a sinusoidal variation to the existing laser frequency

$$\omega_{laser} = \omega + D\sin(2\pi\omega_m t) \tag{7}$$

where ω is a fixed value, ω_m is the modulation frequency and D represents the magnitude of variation in ω_{laser}. The effects of laser modulation can be seen in Figures 4-6. In Figure 4 the modulation frequency is chosen to be greater than the Rabi frequency $\Omega = 12.73\ s^{-1}$. From Figure 4**A** the time series shows a clear pattern for the modulation. Figure 4**B** is the phase space plot indicating that the atom returns to the ground state after each oscillation. The

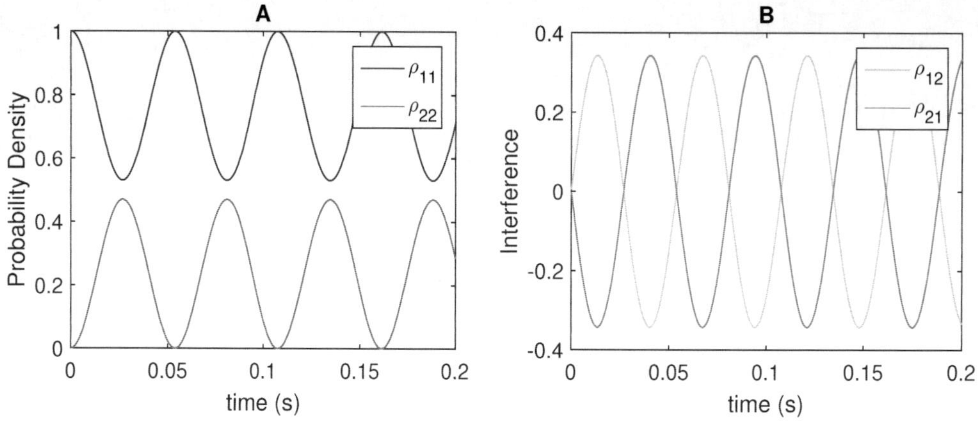

Figure 2. A two level atom interacting with a uniform laser beam that is detuned. Plots are: (**A**) - probability density and (**B**) - interference terms.

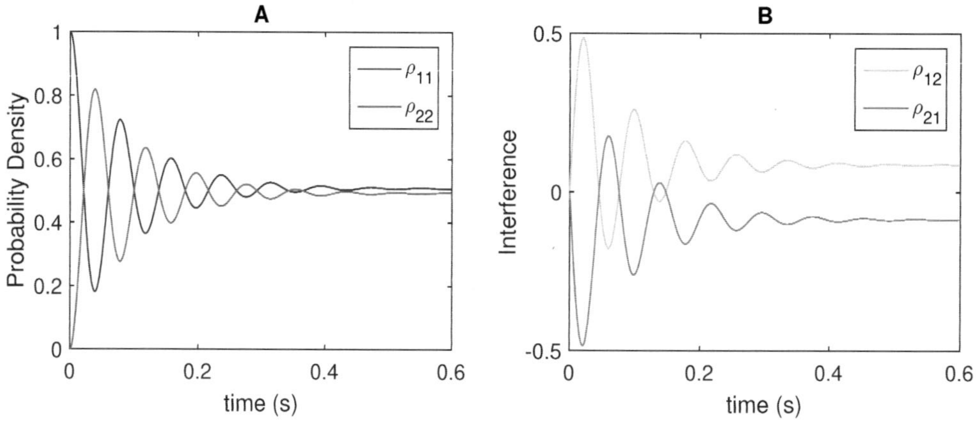

Figure 3. A two level atom interacting with a uniform laser beam where the dissipation coefficient is non-zero. Plots are: (**A**) - probability density and (**B**) - interference terms.

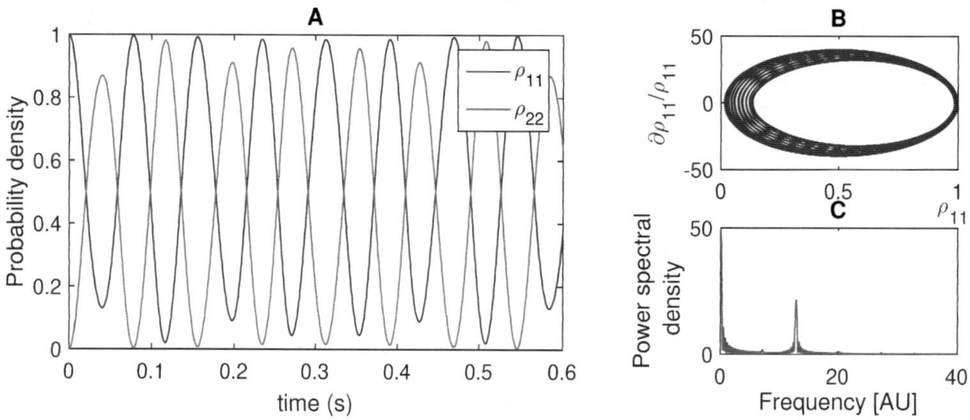

Figure 4. A two level atom interacting with a modulated laser beam where the modulation frequency is greater than the Rabi frequency. Plots are: (**A**) - probability density; (**B**) - phase space; and (**C**) - power spectral density. NB: In (**A**) only a few cycles are shown.

IOP Conf. Series: Journal of Physics: Conf. Series **905** (2017) 012005 doi:10.1088/1742-6596/905/1/012005

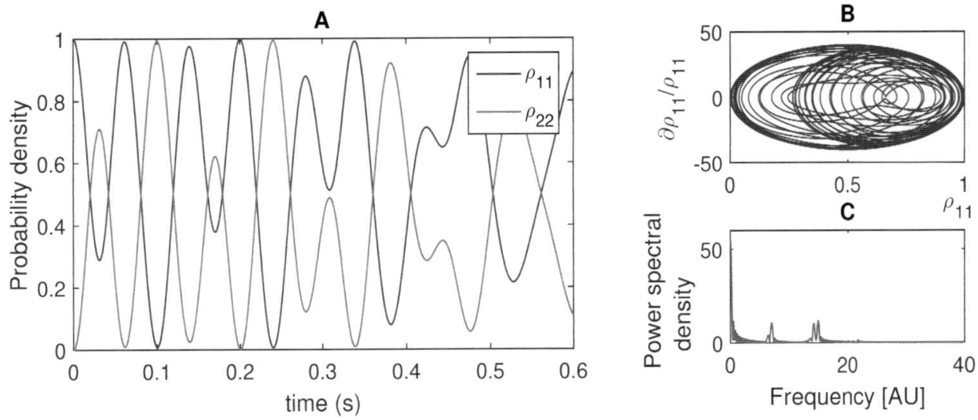

Figure 5. A two level atom interacting with a modulated laser beam where the modulation frequency is less than the Rabi frequency. Plots are: (**A**) - probability density; (**B**) - phase space; and (**C**) - power spectral density. NB: In (**A**) only a few cycles are shown.

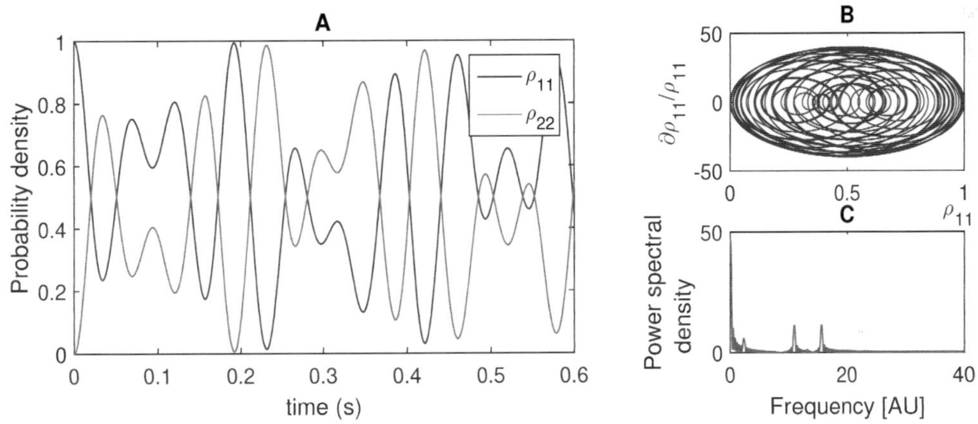

Figure 6. A two level atom interacting with a modulated laser beam where the modulation frequency is similar to the Rabi frequency. Plots are: (**A**) - probability density; (**B**) - phase space; and (**C**) - power spectral density. NB: In (**A**) only a few cycles are shown.

power spectral density, Figure 4**C**, shows a peak (around $13\,s^{-1}$) indicating the most prominent frequency besides the average value at the origin. A frequency that is less than the Rabi frequency is used to generate Figure 5. The time series, Figure 5**A**, depicts an irregular pattern for the modulation. The phase space plot, Figure 5**B** confirms this by indicating that the atom does not always transition to an expected state. Multiple peaks of similar amplitude are seen in the power spectral density, Figure 5**C** where the first peak is located at the modulated frequency ω_m and the second is located at $2\omega_m$. Finally, the modulation frequency is chosen to be similar to the Rabi frequency and is depicted in Figure 6. From Figure 6**A** it can once again be seen that an irregular pattern occurs which is also confirmed by the phase space plot, Figure 6**B**. In Figure 6**C** the power spectral density shows two peaks of the same amplitude located on either side of the Rabi frequency ($12.73\,s^{-1}$). These results show that the atom displays chaotic behaviour. Similar results have been seen by Pisipati et al [12].

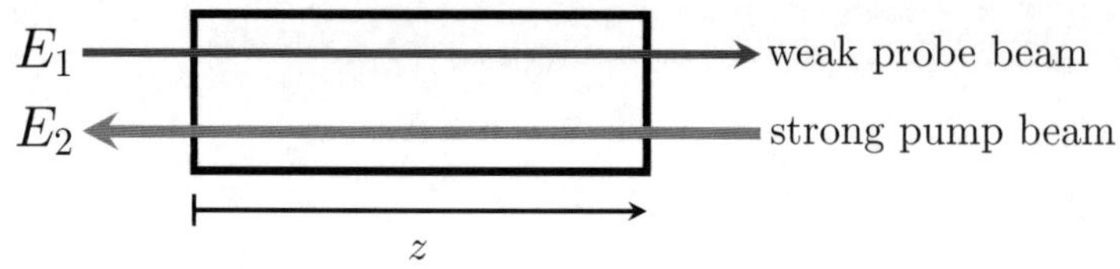

Figure 7. Basic arrangement for saturated absorption spectroscopy.

3. Saturated absorption spectroscopy

Next we look at the effect atoms have on the propagation of the laser beams which is done by making use of the basic arrangement depicted in Figure 7. Counter-propagating beams, E_1 and E_2, having the same frequency but different electric fields are passed through a gaseous sample after which the probe beam intensity is measured. The total electric field (in the laboratory frame) is given by

$$\vec{E} = (\vec{E}_1 e^{\imath kz} + \vec{E}_2 e^{-\imath kz}) e^{-\imath \omega t} \tag{8}$$

Assuming all the waves propagate in the z-direction we substitute Equation (8) in the wave equation

$$\nabla^2 \vec{E} - \frac{1}{c^2}\frac{\partial^2}{\partial t^2}\left(\vec{E} + \frac{\vec{P}}{\varepsilon_0}\right) = 0 \tag{9}$$

where $\vec{P}$ represents the polarization due to the applied electric filed, and we make use of the slowly varying approximation to obtain the following equation for the amplitude E_1 [2]

$$\frac{\partial E_1(z)}{\partial z} = \imath \frac{\omega}{2\epsilon_0 nc} P \tag{10}$$

Equation (10) is used to predict the behaviour of the electric field as it propagates through the sample where P is the total polarization given by

$$P = \int dv \underbrace{\left[\frac{1}{\sqrt{\pi} v_0} e^{-\frac{v^2}{v_0^2}}\right]}_{\mathbf{a}} \underbrace{\left[1 - \frac{\Omega_2^2}{\Gamma_{sp}} \frac{\gamma}{(\omega_0 - \omega_2)^2 + \gamma^2 + \Omega_2^2 \frac{\gamma}{\Gamma_{sp}}}\right]}_{\mathbf{b}} \cdots$$

$$\cdots \underbrace{\left[\frac{N}{V}\frac{|\mu_{12}|^2}{2\hbar} \frac{\imath\gamma + (\omega_0 - \omega_1)}{(\omega_0 - \omega_1)^2 + \gamma^2 + \Omega_1^2 \frac{\gamma}{\Gamma_{sp}}} E_1 e^{-\imath\omega t}\right]}_{\mathbf{c}} \tag{11}$$

where μ_{12} is the matrix element of the dipole moment and Ω_1 and Ω_2 are the Rabi frequencies due to the electric fields E_1 and E_2 respectively. The velocity of the atoms is denoted by v and v_0 is the rms velocity, N/V is the density of atoms, Γ_{sp} is the decay of the population of the upper level and γ is the decay of the coherence [2, 13]. Equation (11) is a sum of all the polarizations due to the atoms and takes into account the various thermal velocities as well. When doing this, the atoms see the laser frequency as being Doppler shifted, this results in a broadening of the absorption profile. The square bracket labelled **a** in Equation (11) caters for the atomic velocity distribution and has a Gaussian profile. The fraction of atoms that part-take in the polarization is given by the square bracket labelled **b** in Equation (11) and is unity for all ω except close to $\omega_2 = \omega_0$ where ω_2 is the frequency of the pump beam as seen by the atoms moving with velocity v. At and close to $\omega_2 = \omega_0$ a dip indicates a reduction in the number of atoms in the ground state, this effect is referred to as hole burning. The square bracket labelled

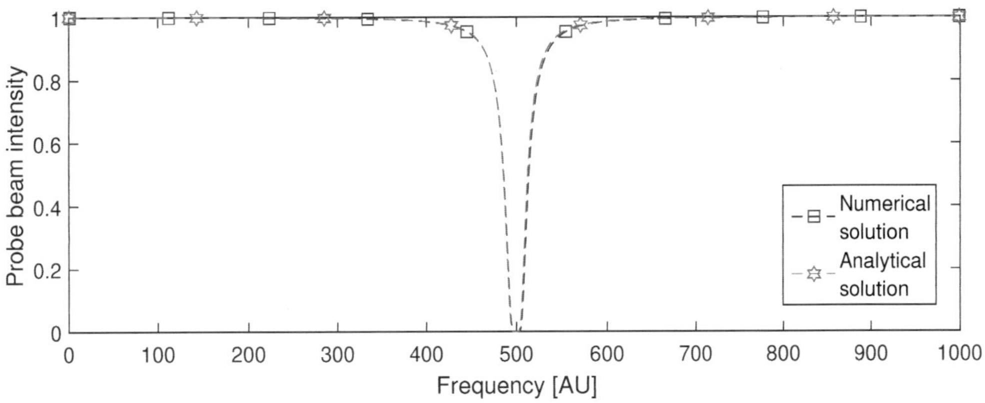

Figure 8. A comparison between the analytical and numerical solution of the probe beam intensity for frequencies above and below resonance for a particular spatial point in the sample.

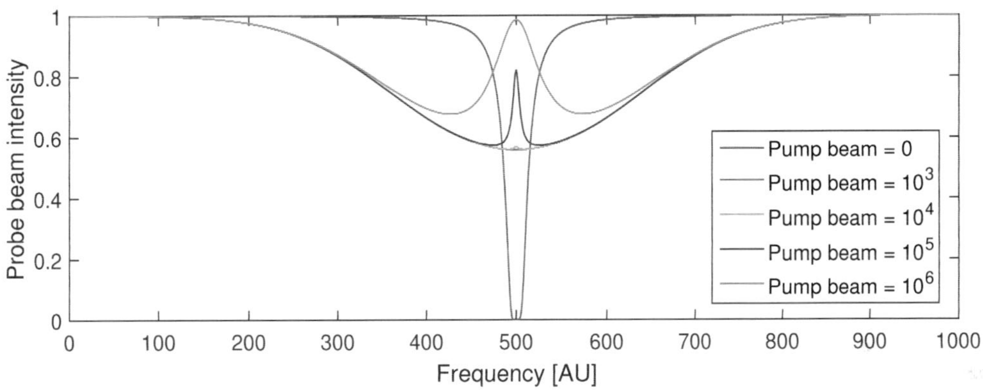

Figure 9. The probe beam intensity with increasing values of pump beam field, E_2.

c in Equation (11) is the term representing the polarization due to the probe beam which has a frequency ω_1. These terms are derived theoretically from the optical Bloch equations [2, 13].

3.1. Results

First we analyse the system with all atoms having zero velocity and interacting with the probe beam. Using Equation (10) results in the analytical solution for the intensity of the probe beam

$$|I| = I_0 e^{-2\alpha z} \tag{12}$$

where

$$\alpha = \frac{\omega}{2\epsilon_0 n c}\frac{N}{V}\frac{|\mu_{12}|^2}{2\hbar}\frac{\gamma}{(\omega_0-\omega)^2+\gamma^2} \tag{13}$$

This demonstrates absorption and is depicted by Figure 8 which shows a comparison between the analytical solution, where I is calculated using Equation (12), and the numerical solution, where I is obtained by solving Equation (10). Figure 8 shows satisfactory agreement between the two solutions.

The ground and excited state populations of the atoms are altered by the pump beam. The electric field of the pump beam is made strong enough to saturate the sample, therefore changing the measured absorption. Figure 9 demonstrates this change where the blue curve depicts the probe beam intensity for atoms having zero velocity and the pump beam electric field being zero. The thermal velocity of the atoms are then taken into account and the pump beam electric field is gradually increased. From Figure 9 the orange, yellow, purple and green curves depict the effects the pump beam has on the sample. It can be seen that when the thermal velocities are taken into account the absorption profile is broadened whilst the absorption at and close to resonance is decreased (showing the true Lorentzian profile) resulting in the intensity to increase. We assume in the above that the pump beam is strong enough so that it remains unchanged as it propagates through the sample.

4. Four-wave Mixing

We now examine a non-linear mixing process using a diamond energy level structure to simulate four-wave mixing in rubidium gas. We examine four-wave mixing using the energy level structure shown in Figure 10. These states could represent those of ^{87}Rb as shown in brackets.

E_1 to E_4 are the corresponding energies of each level, ω_1 and ω_2 are frequencies of two pump

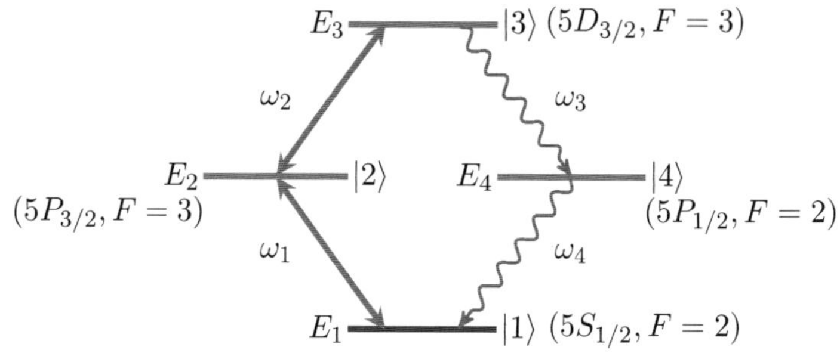

Figure 10. Four-wave mixing geometry.

lasers and ω_3 and ω_4 are frequencies of two internally generated photons.

We will assume that the pump laser beams are strong enough that they do not deplete as they propagate through the atomic ensemble. We will also assume that all photons propagate in the positive z-direction. The atomic levels are assumed to be such that the photons of frequency ω_1, ω_2, ω_3 and ω_4 couple only transitions between $|1\rangle \leftrightarrow |2\rangle$, $|2\rangle \leftrightarrow |3\rangle$, $|3\rangle \leftrightarrow |4\rangle$ and $|4\rangle \leftrightarrow |1\rangle$, respectively, that is they are very much detuned compared to the other transitions. The total electric field is then

$$E = \sum_{i=1}^{4} \left(\tilde{E}_i e^{-\imath\omega_i t} + c.c. \right) \tag{14}$$

where

$$\tilde{E}_i = E_i(z) e^{\imath k_i z} \tag{15}$$

The electric field amplitudes E_i should not be confused with the energy levels. The behaviour of the electric field E_3 and E_4 due to the frequencies ω_3 and ω_4 are described by Maxwell's equations:

$$\frac{\partial}{\partial z} E_3(z) = \imath \frac{\omega_3}{2\epsilon c} \frac{N}{V} \mu_{34} \rho_{34}^{(3)} \tag{16}$$

$$\frac{\partial}{\partial z} E_4(z) = \imath \frac{\omega_4}{2\epsilon c} \frac{N}{V} \mu_{41} \rho_{41}^{(3)} \tag{17}$$

where $\rho_{ij}^{(3)}$ are the third order density matrix elements and are obtained from the master equation as described below.

The total Hamiltonian is

$$H = H_0 + H_I \tag{18}$$

where $H_I = -\hat{\mu} \cdot E$ and $-\hat{\mu}$ is the dipole moment of the atom. We will use the following as a basis:

$$|\Psi_1\rangle = e^{-\imath E_1 t/\hbar}|1\rangle \tag{19}$$
$$|\Psi_2\rangle = e^{-\imath E_2 t/\hbar}|2\rangle \tag{20}$$
$$|\Psi_3\rangle = e^{-\imath E_3 t/\hbar}|3\rangle \tag{21}$$
$$|\Psi_4\rangle = e^{-\imath E_4 t/\hbar}|4\rangle \tag{22}$$

In the above basis

$$H_0 = \begin{pmatrix} E_1 & 0 & 0 & 0 \\ 0 & E_2 & 0 & 0 \\ 0 & 0 & E_3 & 0 \\ 0 & 0 & 0 & E_4 \end{pmatrix} \tag{23}$$

and $H_I' = e^{-\imath H_0 t/\hbar}\, H_I\, e^{\imath H_0 t/\hbar}$ becomes

$$E \begin{pmatrix} 0 & -\mu_{12}e^{\imath\omega_{12}t} & -\mu_{13}e^{\imath\omega_{13}t} & -\mu_{14}e^{\imath\omega_{14}t} \\ -\mu_{21}e^{\imath\omega_{21}t} & 0 & -\mu_{23}e^{\imath\omega_{23}t} & 0 \\ -\mu_{31}e^{\imath\omega_{31}t} & -\mu_{32}e^{\imath\omega_{32}t} & 0 & -\mu_{34}e^{\imath\omega_{34}t} \\ -\mu_{41}e^{\imath\omega_{41}t} & 0 & -\mu_{43}e^{\imath\omega_{43}t} & 0 \end{pmatrix} \tag{24}$$

where $\omega_{ij} = (E_i - E_j)/\hbar$ and $\mu_{ij} = \langle i|\hat{\mu}|j\rangle$. Then the Liouville-von Neumann equation (also known as the master equation) that we solve is

$$\dot{\rho} = -\frac{\imath}{\hbar}\left[H_I', \rho\right] + relaxation\ terms \tag{25}$$

Using perturbation theory we let

$$\rho = \rho^{(0)} + \lambda\rho^{(1)} + \lambda^2\rho^{(2)} + ... \tag{26}$$

We also expand each $\rho^{(i)}$ in frequency as well and also we replace H_I' by $\lambda H_I'$. We assume at zero order $\rho_{11}^{(0)} = 1$ with all other terms being zero, i.e.

$$\rho^{(0)} = \begin{pmatrix} 1 & 0 & 0 & 0 \\ 0 & 0 & 0 & 0 \\ 0 & 0 & 0 & 0 \\ 0 & 0 & 0 & 0 \end{pmatrix} \tag{27}$$

In the rotating wave approximation, the first order solutions we find that are non-zero are

$$\tilde{\rho}_{21}^{(1)} \simeq \frac{1}{\hbar}\mu_{21}\frac{E_1}{\Delta_{21} - \imath\Gamma_{21}}e^{-\imath\omega_1 t} \tag{28}$$

and

$$\tilde{\rho}_{41}^{(1)} \simeq \frac{1}{\hbar}\mu_{41}\frac{E_4}{\Delta_{41} - \imath\Gamma_{41}}e^{-\imath\omega_4 t} \tag{29}$$

plus their complex conjugates where $\tilde{\rho}_{ij}^{(1)} = \rho_{ij}^{(1)} e^{-\imath\omega_{ij}t}$. All the other terms are zero. The third order terms that are important for solving Maxwell's equation for the ω_3 and ω_4 terms are

$$\begin{aligned}\tilde{\rho}_{34}^{(3)} \simeq \frac{1}{\Delta_{34} - \imath\Gamma_{34}} \Bigg[\frac{2}{\gamma_{41}} \frac{|\mu_{14}|^2 \mu_{34}}{\hbar^3} \frac{|E_4|^2 E_3 \Gamma_{41}}{\Delta_{41}^2 + \Gamma_{41}^2} + \frac{\mu_{14}\mu_{32}\mu_{21}}{\hbar^3} \frac{E_1 E_2 E_4^*}{(\Delta_{42} + \imath\Gamma_{41})} \left(\frac{1}{(\Delta_{21} - \imath\Gamma_{21})} \right. \\ \left. - \frac{1}{(\Delta_{41} + \imath\Gamma_{41})} \right) - \frac{\mu_{14}\mu_{32}\mu_{21}}{\hbar^3} \frac{E_1 E_2 E_4}{(\Delta_{31} - \imath\Gamma_{31})(\Delta_{21} - \imath\Gamma_{21})} \\ - \frac{\mu_{14}\mu_{34}\mu_{41}}{\hbar^2} \frac{E_3 |E_4|^2}{(\Delta_{31} - \imath\Gamma_{31})(\Delta_{41} - \imath\Gamma_{41})} \Bigg] e^{-\imath\omega_3 t}\end{aligned} \tag{30}$$

and

$$\begin{aligned}\tilde{\rho}_{41}^{(3)} \simeq \frac{1}{\Delta_{41} - \imath\Gamma_{41}} \Bigg[-\frac{2}{\gamma_{21}} \frac{|\mu_{12}|^2 \mu_{41}}{\hbar^3} \frac{|E_1|^2 E_4 \Gamma_{21}}{\Delta_{21}^2 + \Gamma_{21}^2} - \frac{4}{\gamma_{41}} \frac{|\mu_{14}|^2 \mu_{41}}{\hbar^3} \frac{|E_4|^2 E_4 \Gamma_{41}}{(\Delta_{41}^2 + \imath\Gamma_{41}^2)} \\ + \frac{\mu_{43}\mu_{32}\mu_{21}}{\hbar^3} \frac{E_1 E_2 E_3^*}{(\Delta_{31} - \imath\Gamma_{31})(\Delta_{21} - \imath\Gamma_{21})} + \frac{\mu_{43}\mu_{34}\mu_{41}}{\hbar^3} \frac{|E_3|^2 E_4}{(\Delta_{31} - \imath\Gamma_{31})(\Delta_{41} - \imath\Gamma_{41})} \Bigg] e^{-\imath\omega_4 t}\end{aligned} \tag{31}$$

where Δ_{ij} represents the detuning between levels $|i\rangle$ and $|j\rangle$ and the corresponding photon, Γ_{ij} represents the decay rate of the corresponding coherence ρ_{ij} and γ_{ij} represents the decay of the population ρ_{ii}. Using the following definition of Rabi frequencies

$$\begin{aligned}\Omega_1 &= \mu_{12} E_1 / \hbar \\ \Omega_2 &= \mu_{23} E_2 / \hbar \\ \Omega_3 &= \mu_{34} E_3 / \hbar \\ \Omega_4 &= \mu_{41} E_4 / \hbar\end{aligned} \tag{32}$$

we rewrite Equations (30) and (31) in terms of these Rabi frequencies and solve numerically self consistently together with Maxwell's equations, Equations (16) and (17), for various values of z along the atomic ensemble and various values of detuning. The above approach is used for a different geometry of four-wave mixing by Boyd et al [14].

4.1. Results

The variation of the intensities of ω_3 and ω_4 photons as a function of distance along the sample for various detuning of pump 1 and pump 2 are shown in Figures 11 and 12. Figure 13 shows a colour intensity plot of the intensity of ω_3 for both values of detuning. In order to use Equations (30) and (31) it was necessary to start the calculations with very small values for Ω_3 and Ω_4.

From Figures 11 and 12 we can see an exponential growth in the intensities of the internally generated photons where the intensities depend heavily on the detuning of the lasers. The highest intensity occurs when both lasers have zero detuning. The intensity is also more sensitive to the detuning of the ω_1 laser.

5. Summary and Conclusion

We have investigated laser-atom interactions by first examining a two level atom interacting with a single laser beam where Rabi oscillations have been demonstrated. Dissipation effects show up as decay in the populations and de-coherence terms in the density matrix elements. Chaotic behaviour was also seen to occur when a modulated laser is used.

The analysis was extended to include saturated absorption spectroscopy and non-linear mixing processes in multi-level atoms. We showed that two additional coherent beams can be generated by four-wave mixing using two pump laser beams. The intensities of these beams depend on the detuning of the pump lasers.

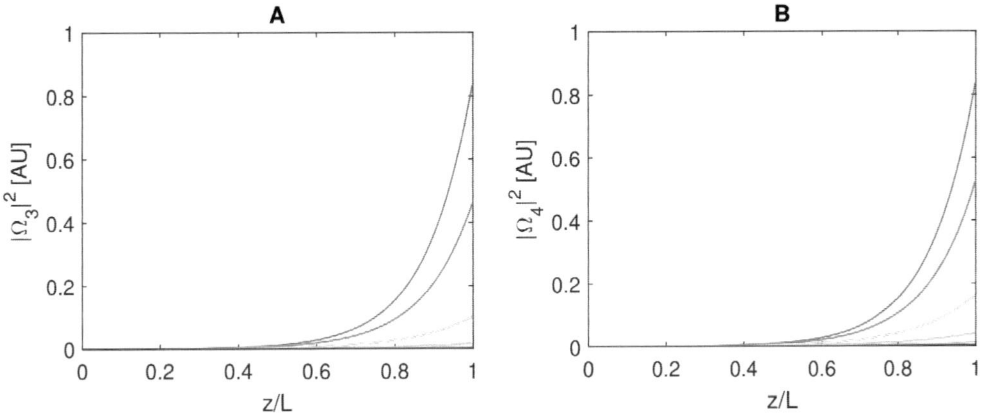

Figure 11. Intensities of the ω_3 (**A**) and ω_4 (**B**) photons for detuning values(Δ_{12}/Γ_{12}) (from top to bottom): 0 (red), 0.2 (orange), 0.4 (yellow), 0.6 (green), 0.8 (blue) and 1 (purple). These values are relative to the decoherence rate.

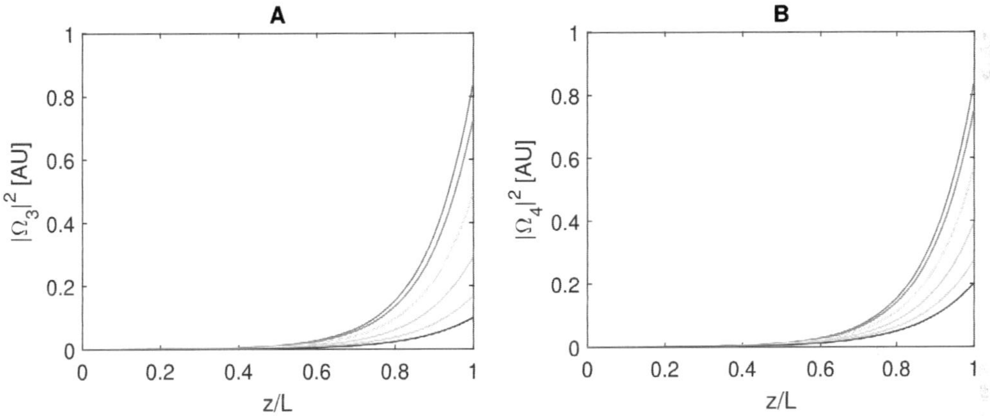

Figure 12. Intensities of the ω_3 (**A**) and ω_4 (**B**) photons for detuning values(Δ_{23}/Γ_{23}) (from top to bottom): 0 (red), 0.2 (orange), 0.4 (yellow), 0.6 (green), 0.8 (blue) and 1 (purple). These values are relative to the decoherence rate.

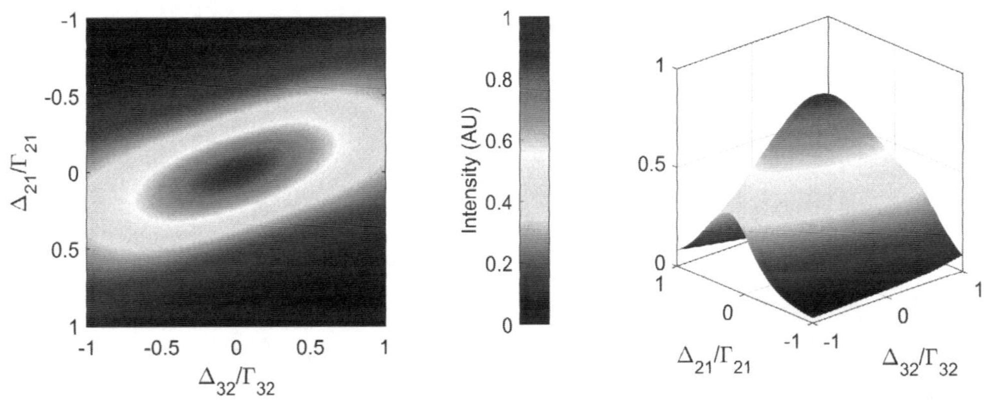

Figure 13. Colour plot of intensity of ω_3 photons as a function of detuning of pump 1 (Δ_{21}/Γ_{21}) and pump 2 (Δ_{32}/Γ_{32}).

IOP Conf. Series: Journal of Physics: Conf. Series **905** (2017) 012005 doi:10.1088/1742-6596/905/1/012005

References

[1] Allen L and Eberly J 1987 *Optical Resonance and Two-level Atoms* (New York: Dover)
[2] Grynberg G, Aspect A, Fabre C and Cohen-Tannoudji C 2010 *Introduction to Quantum Optics: From the Semi-classical Approach to Quantized Light* (Cambridge: Cambridge University Press)
[3] Loudon R 2000 *The Quantum Theory of Light* (Oxford: Oxford University Press)
[4] Jakowski J and Morokuma K 2009 *J. Chem. Phys.* **130** 224106
[5] Glushkov A V 2014 *JPCS* **548** 012020
[6] Glushkov A, Khetselius O Y, Svinarenko A A and Prepelitsa G P 2011 *Coherence and Ultrashort Pulse Laser Emission* (Croatia: Intech)
[7] Srivathsan B, Gulati G K, Brenda C M Y, Maslennikov G, Matsukevich D and Kurtsiefer C 2013 *Phys. Rev. Lett.* **111**(12) 123602
[8] de Melo N R and Vianna S S 2014 *J. Opt. Soc. Am. B* **31** 1735–1740
[9] Jen H H and Chen Y C 2016 *Phys. Rev. A* **93**(1) 013811
[10] Becerra F, Willis R, Rolston S and Orozco L 2011 *Rev. Mex. Fis.* **57** 23–28
[11] Vianna S S, Nussenzveig P, Magno W C and Tabosa J W R 1998 *Phys. Rev. A* **58**(4) 3000–3003
[12] Pisipati U, Almakrami I M, Joshi A and Serna J D 2012 *Am. J. Phys.* **80** 612–620
[13] Levenson M 1982 *Introduction to Nonlinear Laser Spectroscopy* (New York: Academic Press)
[14] Boyd R W, Malcuit M S, Gauthier D J and Rzążewski K 1987 *Phys. Rev. A* **35**(4) 1648–1658

IOP Conf. Series: Journal of Physics: Conf. Series **905** (2017) 012006 doi:10.1088/1742-6596/905/1/012006

Improved Maximum Entropy Method applied to Real-time Time-Dependent Density Functional Theory

M. Toogoshi, S. S. Kano and Y. Zempo

Computer and Information Sciences, Hosei University, 3-7-2 Kajino, Koganei 184-8584 Japan

E-mail: 16t9001@cis.k.hosei.ac.jp

Abstract. The maximum entropy method (MEM) is one of the key techniques for spectral analysis. The main feature is to describe spectra in low frequency with short time-series data. We adopted MEM to analyze the spectrum from the dipole moment obtained by the time-dependent density functional theory (TDDFT) calculation in real time, which is intensively studied and applied to computing optical properties. In the MEM analysis, we proposed that we use the concatenated data set made from several-times repeated raw data together with the phase. We have applied this technique to spectral analysis of the TDDFT dipole moment of oligo-fluorene with $n = 8$. As a result, the higher resolution can be obtained without any peak shift due to the phase jump. The peak position is in good agreement to that of FT with just raw data. This paper presents the efficiency and characteristic features of this technique.

1. Introduction

Time-dependent density functional theory (TDDFT) is a powerful tool for analyzing optical properties of medium-to-large sized molecule. We employ a real-time and real-space technique to solve the time-dependent Kohn-Sham equations. Within the framework of this approach, the wave functions are calculated by the finite difference method on real spatial grids [1] without using explicit bases such as plane waves or Gaussian. In our procedure to calculate optical properties, we use the time-series data, namely the dynamic dipole moment, from whose Fourier transform (FT) optical properties are calculated in a usual technique. The spectral resolution depends on the length of the dipole moment. To obtain the good resolution, the computational cost is quite expensive.

To solve this difficulty, we focus on Maximum Entropy Method (MEM)[2], which is one of the key techniques of spectrum analysis. The principles of MEM are methods of the information theory for estimating unknown probability distributions based on the information about their expected values. MEM is widely used to the solution of a variety of problems related to spectral estimation such as earth and planetary science[3], spectroscopy[4]. The main feature is to obtain a fairy high resolution and accuracy with a relatively small number of time-series data. MEM is based on the Fourier pair of the autocorrelation and power spectrum. We applied this technique to spectral analysis of time-dependent dipole moments of molecules, which is calculated by real-time TDDFT. In the comparison of MEM and FT spectrum for several molecules, we realize that MEM needs less time step than FT to obtain the optical spectrum with same resolution[6]. As a new improved MEM, we proposed to use the concatenated data set made from several-times

IOP Conf. Series: Journal of Physics: Conf. Series **905** (2017) 012006 doi:10.1088/1742-6596/905/1/012006

repeated raw data[7] together with the phase to avoid the side effect of the artificial periodicity. With this improvements, we have successively obtained the much better spectral resolution of the target peak. Thus, this improvements will make a further advantage of MEM.

We applied this technique to spectral analysis of the TDDFT dipole moment of typical some molecules such as benzene, oligo-fluorene and other materials. In the analysis of optical properties, we are interested in the lower energy peak, corresponding to the band gap area. The results show the higher resolution and the emphasized peak near the band gap without being affected by the artificial periodicity. In this paper, we present characteristic features of this technique, comparing with the FT and the conventional MEM.

The paper is organized as follows. In section 2, we briefly describe a procedure of our TDDFT calculation. Then, we recall a basis of MEM and our new techniques. In section 3, results of absorption spectra of molecules are presented and discussed. Finally, we summarize the efficiency and characteristic features of our technique.

2. Method

2.1. Time-dependent density functional theory

In this section, we briefly describe the procedure of our TDDFT calculation[8]. The basis is the density functional theory (DFT) [9] with the local density approximation (LDA). The total energy of the ground state can be derived from the Kohn-Sham equation (KS) [10]. DFT is much less successful in describing optical responses and absorption spectra where electronic excited states are involved. However, this difficulty is, in principle, solved by the extension of DFT to its time-dependent version, TDDFT, which was established by Runge and Gross [11].

In analogy to the time-independent case, the TDDFT equation of motion coupled with pseudopotentials is given by

$$\Big\{-\frac{1}{2}\nabla^2 + V_{ion}^{ps}(\mathbf{r}) + V_H(\mathbf{r},t) + V_{XC}[\rho(\mathbf{r},t)] + V_{ext}(\mathbf{r},t)\Big\}\psi_j(\mathbf{r},t) = i\frac{\partial}{\partial t}\psi_j(\mathbf{r},t)\ , \tag{1}$$

where V_{ion}^{ps} is an ionic pseudopotential, V_H is the Hartree potential, and V_{XC} is the exchange correlation potential. Since the exact time-dependent xc kernel is not known, the originally non-local time-dependent xc kernel is replaced with a time-independent local one. This is reasonable when the density varies slowly with time. This approximation allows the use of a standard local ground-state xc functional in the TDDFT frame work. The Hartree and exchange-correlation potentials can be determined from the electronic charge density,

$$\rho(\mathbf{r},t) = \sum\nolimits_j |\psi_j(\mathbf{r},t)|^2. \tag{2}$$

The summation is over all occupied states j. The Hartree potential is determined by $\nabla^2 V_H = -4\pi\rho$, and as the xc potential V_{XC}, the usual local density approximation (LDA) is used in our study. For the ionic potential, we employed the pseudopotential V_{ion}^{ps} in the separable form so that the only valence electrons were considered[12]. Prior to the calculation of optical responses, we first solved the usual, time-independent formulation of the pseudopotential-DFT method[13] to obtain the optimized electronic structure[14]. Then, we applied an external field V_{ext} to the system as a perturbation and followed linear responses of the system in real time. Although the xc functional we adopted is originally for electronic ground states, we also employed it in our practical calculations. This method of TDDFT has been effectively used, for the cases where the potential was the time-dependent, to study behaviors of electrons in oscillating electric and magnetic fields and hence excited-state reactions[15]. In our calculations, the real-time and real-space technique was adopted in solving Eq.(1) by means of the finite difference approach[1]. The uniform grid was used in our study for the simplicity.

The time dependent wave function is given by $\psi(t) = \exp[-iHt]\tilde{\psi}(0)$, with the initial wave function $\tilde{\psi}$ at $t = 0$,

$$\tilde{\psi}|_{t=0} = e^{ik_z z}\psi(0), \tag{3}$$

where H is the Hamiltonian of the system, and k is a small wave number corresponding to the external perturbation in the z direction. In the linear response, the time-dependent polarizability is proportional to the dipole matrix element $\mu_\xi(t) = \langle\psi(t)|\xi|\psi(t)\rangle$, where ξ represents x, y and z. The polarizability $\alpha_\xi(\omega)$ is numerically calculated using Fourier transformation (FT) of $\mu_\xi(t)$ and averaged as $\alpha = (\alpha_x + \alpha_y + \alpha_z)/3$. The oscillator strength distribution $S(\omega)$ is related to the imaginary part of the polarizability,

$$S(\omega) = \frac{2\omega}{\pi}\mathrm{Im}\alpha(\omega). \tag{4}$$

2.2. Maximum entropy method

In this section, we briefly describe the procedure and derivation of MEM. The entropy in information theory has been recognized as a measure of uncertainty[16, 17]. Any inferences made from incomplete information should be with the probability distribution which maximizes the entropy under the constraints of available information[18].

A framework of the present study is based on Burg's method[2]. The time-series data μ to be analyzed is given by real-time TDDFT, $\mu(n) = \mu(n\Delta t)$. If samples obey Gaussian distribution, the entropy h of a Gaussian process is given by

$$h \propto \int_{-\pi/\Delta t}^{\pi/\Delta t} \log P(\omega) d\omega, \tag{5}$$

where Δt is a sampling interval, and $P(\omega)$ is the power spectrum. We find $\hat{P}(\omega)$ which Eq.(5) is maximized under the constraints of the Wiener-Khintchin's theorem,

$$P(\omega) = \Delta t \sum_{m=-\infty}^{\infty} C_m \exp(-i\omega m\Delta t) \tag{6}$$

$$C_m = \frac{1}{2\pi}\int_{-\pi/\Delta t}^{\pi/\Delta t} P(\omega)\exp(i\omega m\Delta t)d\omega. \tag{7}$$

The solution is

$$\hat{P}(\omega)/\Delta t = \left|\frac{\beta}{1+\sum_{m=1}^{M} a_m \exp(-i\omega m\Delta t)}\right|^2, \tag{8}$$

where M is the maximum length of autocorrelation C_M. The parameters a_k and β are the Lagrange multipliers and they are the solution of Yule-Walker equation[19],

$$\begin{pmatrix} C_0 & C_1 & \cdots & C_M \\ C_1 & C_0 & \cdots & C_{M-1} \\ \vdots & \vdots & & \vdots \\ C_M & C_{M-1} & \cdots & C_0 \end{pmatrix} \begin{pmatrix} 1 \\ a_1 \\ \vdots \\ a_M \end{pmatrix} = \begin{pmatrix} |\beta|^2 \\ 0 \\ \vdots \\ 0 \end{pmatrix}. \tag{9}$$

We adopt the Levinson-Durbin algorithm to solve Eq.(9) efficiently. To solve Eq.(9), the autocorrelation C_m at the lag m is

$$C_m = \frac{1}{N}\sum_{n=0}^{N-m-1} \mu^*(n)\mu(n+m), \tag{10}$$

IOP Conf. Series: Journal of Physics: Conf. Series **905** (2017) 012006 doi:10.1088/1742-6596/905/1/012006

where N is the length of the time-series data, and a symbol $*$ is a complex conjugate. Generally, the autocorrelation shows the periodicity of the signal based on a lag. Given the large lag, it makes the estimation of lower frequency components more reliable. However, the values of high-order autocorrelations are almost close to zero with the too large lag M. The effective value of M is restricted by the number of the data N. Thus, we need a large enough N for a reasonable spectral resolution of the low frequency part.

2.3. Improvements of MEM
As a new improved MEM, we propose to use the concatenated data set made from several-times repeated raw data together with the phase to avoid the side effect of the artificial periodicity.

The information of low energy parts is important for the optical spectrum analysis. To emphasize this parts, we need a large M which is the maximum lag of autocorrelation. However, the unphysical results such as peak splits and false peaks are appeared by selecting too large value of M from our experience. Also, we cannot take the larger M where M is limited by the total number of time step N. To solve this difficulty, we duplicate the raw data and repeat it many times as shown in Fig. 1 (b). The lower energy resolution is attributed to the maximum lag M. With the repeated signal, the value of M may be selected large enough without causing additional computations, which save our calculation cost.

At the connection point, there may be a phase jump for each frequency component. This phase jump is considered as a noise. The side effect of phase jump may be added, for example the peak shifts. To reduce the side effect fo the phase jump, we propose that an appropriate phase for target frequency is introduced together with the repeated data

$$\tilde{\mu}(n) = \mu'(n)\exp(ik\phi), \tag{11}$$

where $\mu'(n)$ is the several repeated raw data, and $\tilde{\mu}(n)$ is the concatenated data $\mu'(n)$ with an appropriate phase ϕ for the target frequency$(-\pi \leq \phi \leq \pi)$. Even if we introduce this phase, the total power of the spectrum doesn't change. Then an absolute value of exponential in Eq. (11) is equal to 1.

Figure 1(c) shows the repeated data $N = 5000 \times 4$ with the phase $\phi = 0.25\pi$ in the case of the oligomer of fluorene with $n = 8$. In this case, k in Eq.(11) is equal to zero, then the value of a x-axis is in the area of $[0, 5)$; one in the area of $[5, 10)$; two in the area of $[10, 15)$; three in the area of $[15, 20)$. Since the phase ϕ is a constant value, it is necessary to choose the appropriate phase for the target frequency to reduce the side effect of the phase jump. We decide this phase to match the bandgap peak in our study.

3. Results and Discussion
In this section, we show the calculated absorption spectrum of the total oscillator strength of improved MEM. Then, we discuss characteristics of our method compared to the result of conventional MEM, and FT.

3.1. Benzene, Naphthalene, Anthracene and Tetracen
As a simple example to confirm the efficiency of our proposed method, we applied improved MEM to benzene, naphthalene, anthracene, and tetracene. The molecular structure is based on the ground state. The time evolution is carried out according to the real-time TDDFT [14, 8]. The spectrum $S(E)$ is calculated from the time-series data in our MEM technique. Results are shown in Fig.3.1, where the solid line is the result of benzene, the dashed line is that of naphthalene, the dash-dot line is that of anthracene, and the dotted line is that of tetracene. These spectra are normalized by the intensity of first peak. Figure 3.1(a) shows results of improved MEM. The parameters(M and ϕ) of our MEM are $M = 1800$ and $\phi = -0.25\pi$ at

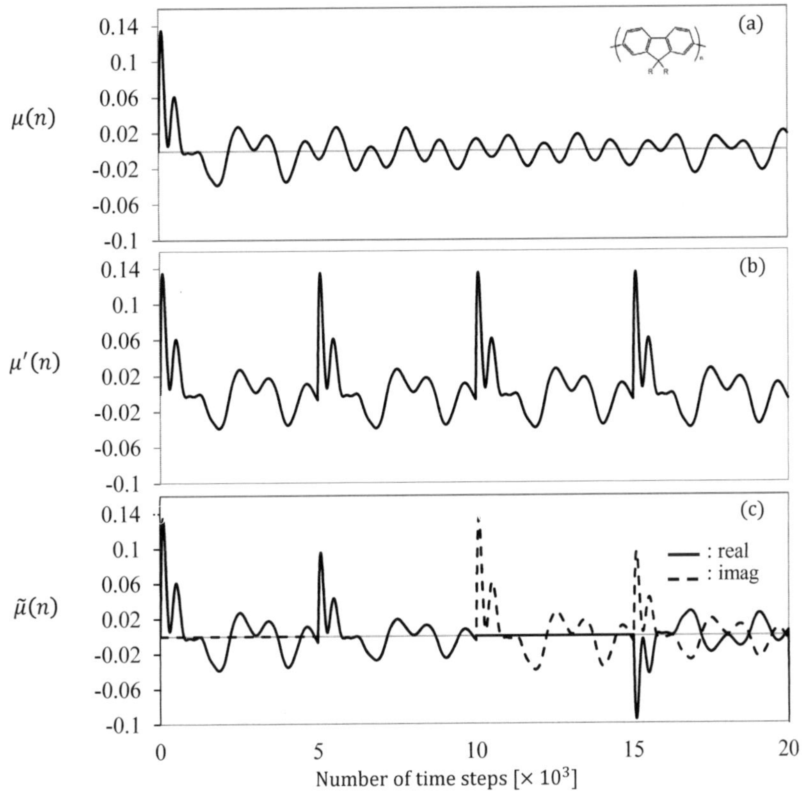

Figure 1. Comparison of the time-series data for oligo-fluorene with $n = 8$, which is made by real-time TDDFT. (a) Real time evolution up to $N = 20000$ for FT. (b) Four times repeated data of the the dipole moment with $N = 5000 \times 4$. (c) The data (b) together with the phase $\phi = 0.25\pi$. The solid line is the real part of the data, and the dashed line is the imaginary part of the data.

benzene; $M = 2500$ and $\phi = -0.20\pi$ at naphthalene; $M = 2000$ and $\phi = -0.23\pi$ at anthracene; $M = 3000$ and $\phi = -0.21\pi$ at tetracene. For both MEM calculations, the same number of the time step $N = 10000 \times 100$ is adopted. Figure 3.1(b) shows results of FT with $N = 10000$. Our method provides the good resolution compared with that of FT. Also, we can see the clear spectrum in comparison of FT. This is a one of the feature of our MEM.

3.2. Fluorene

Poly(9,9-dialkyl-fluorene) and their substituted derivatives are used for one of the Organic LED materials, and expected to be basic materials for a blue emission. Their electronic structures have been extensively studied [20, 21, 22, 23, 24]. We employed the oligomer of fluorene (oligo-FL) with $n = 8$ and carried out the MEM calculation. The molecule size is fairly big, then the calculation for each step is quite expensive. To obtain the low energy part of the spectrum, of course, it takes a long time to perform the calculation. If we can save the calculation cost to figure out the lower energy part, it is quite effective. To this signal analysis, we applied our improved MEM. Figure 3 shows the comparison of the absorption spectrum with four methods. The solid line shows the result of improved MEM with $N = 5000 \times 100$, $M = 2500$ and $\phi = 0.25\pi$. The dashed line is that of FT for $N = 20000$, the dash-dot line is our conventional

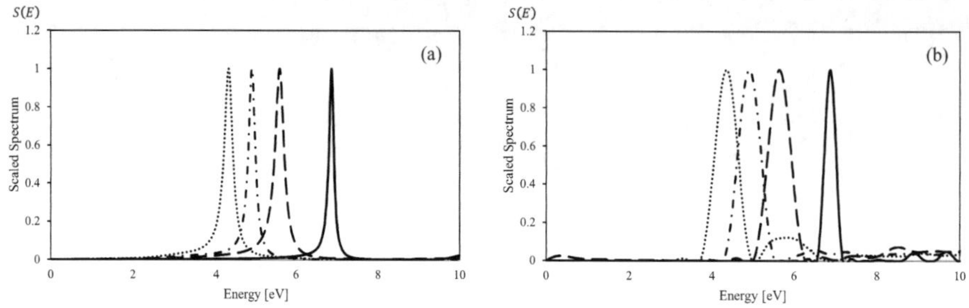

Figure 2. Comparison of improved MEM and FT spectrum using benzene(solid line), naphthalene(dashed line), anthracene(dash-dot line) and tetracene(dotted line). (a) is results of improved MEM using $N = 10000 \times 100$ together with $M = 1800$ and $\phi = -0.25\pi$ at benzene; $M = 2500$ and $\phi = -0.20\pi$ at naphthalene; $M = 2000$ and $\phi = -0.23\pi$ at anthracene; $M = 3000$ and $\phi = -0.21\pi$ at tetracene. (b) is results of FT with $N = 10000$. All the spectra are normalized at the first low energy peak.

MEM spectrum with just the real time steps $N = 5000$ and $M = 2500$. The dotted line is that of MEM with $N = 5000 \times 100$ and $M = 2500$. These spectra are normalized at the second low energy peak.

FT(dashed line) and MEM(dash-dot line) provide the broad peak at the bandgap area. With only the repeated signal, the strength of the first peak emphasizes, and the peak position shifts for the side effect of the phase jump[7]. On the other hand, we can obtain the clear first peak of our new method. In this case, we decide the matched phase corresponding to the first peak. Thus, the position of the second peak differs from that of other methods. If we choose the mismatched phase for the target peak, the position of that peak shifts, and the strength decreases. The bandgap peak at 2.53 eV for improved MEM may correspond to the experimentally observed one at 3.56 eV for poly-FL. This discrepancy is due to the inherent problem to DFT[8].

3.3. C60 fullerene

Our new MEM is not always works well for the target frequency. In this section, we show the case of our MEM does not work well.

Figure.4 shows that the result of C60 fullerene as the bad example of our MEM calculation. The solid line is the result of our MEM with hundred repeated signals $N = 10000 \times 100$ together with the phase $\phi = -0.125\pi$ and $M = 6000$. The dashed line shows that of FT with $N = 10000$. The result of FT has better resolution for first peak in comparison of that of MEM.

If the length of data N is not enough, we obtain the quite broad peak like Lorentz function in our MEM calculation. In particular, we can see that the first peak is not separated due to the lack of the information of that peak. We cannot obtain the low energy component which is not included in the data N. Thus, we have to calculate a certain time evolution until the information of the low energy part is obtained.

4. Concluding remarks

We employed MEM to obtain optical spectra of molecules, and the spectral resolution was quite improved, which is better than that of FT with the same number or total time steps. As a new aspect, we introduced the use of repeated the same signal together with the phase to make apparently long data, in which larger time lag was involved. In particular, it is efficient to find the low energy peak. We obtained quite better resolutions for target peak without adding practical machine time. This means that we can save our calculation cost efficiently. However,

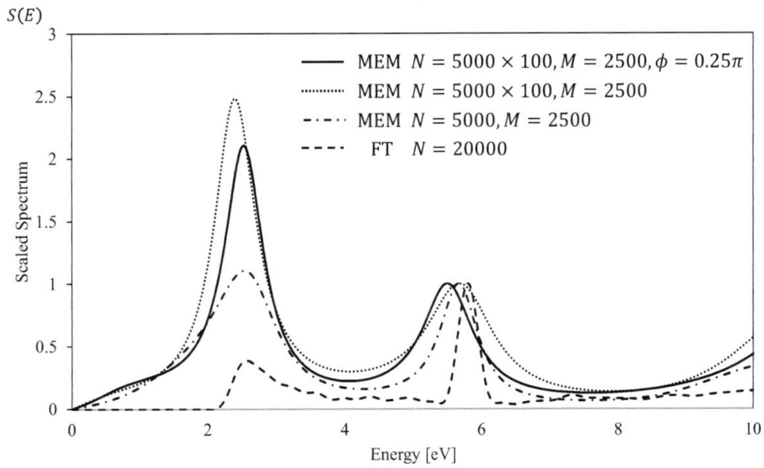

Figure 3. Comparison of the absorption spectrum for oligo-fluorene with $n = 8$. The dashed line is the result of FT with $N = 20000$, the dash-dot line is that of MEM with $N = 5000$, the dotted line is the that of MEM with repeated repeated data $N = 5000 \times 100$, and the solid line is that of improved MEM with $N = 5000 \times 100$, $M = 2500$ and $\phi = 0.25\pi$. Spectra are normalized at the second low energy peak.

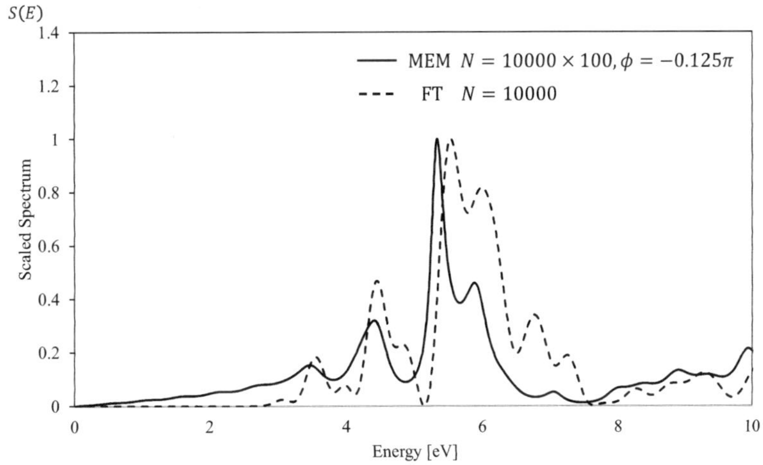

Figure 4. Absorption spectrum for C60 fullerene. The dashed line is the result of FT with $N = 10000$, and the solid line is that of improved MEM with $N = 10000 \times 100$, $M = 6000$ and $\phi = -0.125\pi$. Spectra are normalized at the highest peak in the area of 0-10 eV. The experimental absorption peaks at 3.8, 4.8 and 5.8 eV [14].

our new MEM is not always works well for target peak. With this method for the spectrum analysis, we cannot obtain the low energy component which is not included in the data. We have to calculate a certain level of the time evolution until the information of the low energy part is obtained.

Acknowledgments
Authors would like to express their acknowledgements to Professor H. Hanaizumi for useful discussions and encouragements. This work was partially supported by JSPS Grants-in-Aid for Scientific Research (C) Grant number 25390158, Sumitomo Chemical, Co. Ltd. and Simulatio Corporation.

References

[1] Chelikowsky J, Troullier N, Wu K and Saad Y 1994 *Phys. Rev. B***50**, 11355
[2] Burg J P 1968 *Advanced Study Institute on Signal Processing*, 15-0, NATO, Erschede, Netherlands
[3] Sakurai T 2002 *Earth Planets Space* **54**, 153
[4] Vartiainen E M, Peiponen K E, Kishida H and Koda T 1996 *J. Opt. Soc. Am.* B**13**, 2106.
[5] Smith C R and Erickson G J 1987 *"Maximum-Entropy and Bayesian Spectral Analysis and Estimation Problems"*, ed Smith C R and Erickson G J, D. Reidel Publishing Company(Holland)
[6] Toogoshi M, Kato M, Kano S S, Zempo Y 2014 *J. Phys. Conf. Ser.* **510**, 012027
[7] Toogoshi M, Kano S S, Zempo Y 2015 *J. Phys. Conf. Ser.* **640**, 012069
[8] Zempo Y, Akino, Ishida M, Ishitobi M and Kurita Y 2008 *J. Phys. Condens. Matter* **20**, 064231
[9] Hohenberg P and Kohn W 1964 *Phys. Rev.* **136**, B864
[10] Kohn W and Sham L. J 1965 *Phys. Rev.* **140**, A1133
[11] Runge E and Gross E K U 1984 *Phys. Rev. Lett.* **52**, 997
[12] Troullier N and Martins J L 1991 *Phys. Rev.* B**43**, 1993
[13] e.g. Pickett W E 1989 *Comput. Phys. Reports* **9** , 115
[14] Yabana K and Bertisch G F 1996 *Phys. Rev.* B**54**, 4484
[15] Tateyama Y, Oyama N, Ohno T and Miyamoto Y 2006 *J. Chem. Phys.* **124**, 124507
[16] Shannon C 1948 *Bell Syst. Tech. J.*,**27**, 379 and 623
[17] Jaynes E T 1957 *Phys. Rev.* **106**, 620
[18] Jaynes E T 1982 *Proc. IEEE* **70**, 939
[19] Haykin S and Kesler S 1979 *"Non linear Methods of Spectral Analysis"*, ed Haykin S, Springer(Berlin)
[20] e.g. Hummer K, Puschnig P, Sagmeister S and Ambrosch-Draxl C 2006 *Mod. Phys. Lett.* B**20**, 261 , and the references therein.
[21] Yu W L, Cao Y, Pei J, Huang W and Heeger A J 1990 *Appl. Phys. Lett.* **75** 3270
[22] Wang J F, Feng J K, Ren A M, Liu X D, Ma Y G, Lu P and Zhang H X 2004 *Macromolecules* **37**, 3451
[23] Brière J F and Côté M 2004 *J. Chem. Phys.* B**108**, 3123
[24] Grice A W, Bradley D D C, Bernius, M T, Inbasekaran M, Woo EP and Wu W W 1998 *Appl. Phys. Lett.* **73**, 629

Nonlinear dynamics of laser systems with elements of a chaos: Advanced computational code

V V Buyadzhi, A V Glushkov, O Yu Khetselius, A A Kuznetsova, A A Buyadzhi G P Prepelitsa and V B Ternovsky

Odessa State Environmental University, L'vovskaya str.15, Odessa-16, 65016, Ukraine

E-mail: vbuyad@mail.ru

Abstract. A general, uniform chaos-geometric computational approach to analysis, modelling and prediction of the non-linear dynamics of quantum and laser systems (laser and quantum generators system etc) with elements of the deterministic chaos is briefly presented. The approach is based on using the advanced generalized techniques such as the wavelet analysis, multi-fractal formalism, mutual information approach, correlation integral analysis, false nearest neighbour algorithm, the Lyapunov's exponents analysis, and surrogate data method, prediction models etc There are firstly presented the numerical data on the topological and dynamical invariants (in particular, the correlation, embedding, Kaplan-York dimensions, the Lyapunov's exponents, Kolmogorov's entropy and other parameters) for laser system (the semiconductor GaAs/GaAlAs laser with a retarded feedback) dynamics in a chaotic and hyperchaotic regimes.

1. Introduction

In a modern computational quantum and laser physics, electronics and others there are studied various systems and devices (such as atomic and molecular systems in an electromagnetic field, multi-element semiconductors and gas lasers etc), dynamics of which can exhibit a chaotic behaviour. These systems can be considered in the first approximation as a grid of autogenerators (quantum generators), coupled by different way [1-10]. It is easily to understand that a quantitative studying of the chaos phenomenon features is of a great interest and importance for many scientific and technical applications. At the present time it became one of the most actual and important problems of computational physics of the complex non-linear systems.

In this work we firstly applied a general, uniform chaos-geometric formalism to analysis and modelling of non-linear dynamics of the laser systems with elements of a chaos. The formalism is based on using the advanced generalized techniques such as the wavelet analysis, multi-fractal formalism, mutual information approach, correlation integral analysis, false nearest neighbour algorithm, the Lyapunov's exponents analysis, and surrogate data method, prediction models etc (see details in Refs. [6-19]). There are firstly presented the numerical data on topological and dynamical invariants of chaotic systems, in particular, the correlation, embedding, Kaplan-York dimensions, the Lyapunov's exponents, Kolmogorov's entropy etc for laser (the semiconductor GaAs/GaAlAs laser with retarded feedback) systems dynamics in chaotic and hyperchaotic regimes.

2. Chaos–geometric approach to modelling chaotic dynamics of the laser systems: General formalism

The important step of the quantitative studying chaotic dynamics of different dynamical systems is an numerical analysis of the characteristic time series, i.e. the time series of the key dynamical characteristics. Let us formally consider scalar measurements $s(n) = s(t_0 + n\Delta t) = s(n)$, where t_0 is the start time, Δt is the time step, and is n the number of the measurements. In a general case, $s(n)$ is any time series, particularly the amplitude level. Since processes resulting in the chaotic behaviour are fundamentally multivariate (look [18-23]), it is necessary to reconstruct phase space using as well as possible information contained in the $s(n)$. Such a reconstruction results in a certain set of d-dimensional vectors $\mathbf{y}(n)$ replacing the scalar measurements. Packard et al. [21] introduced the method of using time-delay coordinates to reconstruct the phase space of an observed dynamical system. The direct use of the lagged variables $s(n + \tau)$, where τ is some integer to be determined, results in a coordinate system in which the structure of orbits in phase space can be captured. Then using a collection of time lags to create a vector in d dimensions,

$$\mathbf{y}(n) = [s(n), s(n + \tau), s(n + 2\tau), \ldots, s(n + (d-1)\tau)], \tag{1}$$

the required coordinates are provided. In a nonlinear system, the $s(n + j\tau)$ are some unknown nonlinear combination of the actual physical variables that comprise the source of the measurements. The dimension d is called the embedding dimension, d_E.

According to Mañé and Takens [22,23], any time lag will be acceptable is not terribly useful for extracting physics from data. If τ is chosen too small, then the coordinates $s(n + j\tau)$ and $s(n + (j + 1)\tau)$ are so close to each other in numerical value that they cannot be distinguished from each other. Similarly, if τ is too large, then $s(n + j\tau)$ and $s(n + (j + 1)\tau)$ are completely independent of each other in a statistical sense. Also, if τ is too small or too large, then the correlation dimension of attractor can be under- or overestimated respectively. It is therefore necessary to choose some intermediate (and more appropriate) position between above cases. First approach is based on computing the linear autocorrelation function. Another approach is based on using the average mutual information.

The next principal step is to determine an embedding dimension determination and to reconstruct a Euclidean space R^d large enough so that the set of points d_A can be unfolded without ambiguity. In accordance with the embedding theorem, the embedding dimension, d_E, must be greater, or at least equal, than a dimension of attractor, d_A, i.e. $d_E > d_A$. In other words, we can choose a fortiori large dimension d_E, e.g. 10 or 15, since the previous analysis provides us prospects that the dynamics of our system is probably chaotic. However, two problems arise with working in dimensions larger than really required by the data and time-delay embedding [24]. Firstly, many of computations for extracting interesting properties from the data require search and other operations in R^d whose computational cost rises exponentially with d. Secondly, but more significant from the physical viewpoint, in the presence of noise or other high dimensional contamination of the observations, the extra dimensions are not populated by dynamics, already captured by a smaller dimension, but entirely by the contaminating signal. In too large an embedding space one is unnecessarily spending time working around aspects of a bad representation of the observations which are solely filled with noise. Further it is necessary to determine the dimension d_A. There are a few standard approaches to reconstruct an attractor dimension (see, e.g., [1-8]), but usually there are applied only two methods. The first correlation integral analysis uses the correlation integral, $C(r)$, to distinguish between chaotic and stochastic systems. To compute the correlation integral, the algorithm of Grassberger and Procaccia [25] is the most commonly used approach, where the correlation integral is

$$C(r) = \lim_{N\to\infty} \frac{2}{N(n-1)} \sum_{\substack{i,j \\ (1\le i<j\le N)}} H(r - \| \mathbf{y}_i - \mathbf{y}_j \|) \tag{2}$$

where H is the Heaviside step function with $H(u) = 1$ for $u > 0$ and $H(u) = 0$ for $u \leq 0$, r is the radius of sphere centered on $\mathbf{y}_i$ or $\mathbf{y}_j$, and N is the number of data measurements. If the time series is characterized by an attractor, then the integral $C(r)$ is related to the radius r given by

$$d = \lim_{\substack{r \to 0 \\ N \to \infty}} \frac{\log C(r)}{\log r}, \quad (3)$$

where d is correlation exponent that can be determined as the slop of line in the coordinates log $C(r)$ versus log r by a least-squares fit of a straight line over a certain range of r, called the scaling region.

If the correlation exponent attains saturation with an increase in the embedding dimension, then the system exhibits a chaotic dynamics. The saturation value of the correlation exponent is defined as the correlation dimension (d_2) of an attractor. The nearest integer above the saturation value provides the minimum or optimal embedding dimension for reconstructing the phase-space or the number of variables necessary to model the dynamics of the system.

To verify the results obtained by the correlation integral analysis, one can use surrogate data method. This method (look [6,7,26-29]) is an approach that makes use of the substitute data generated in accordance to the probabilistic structure underlying the original data. This means that the surrogate data possess some of the properties, such as the mean, the standard deviation, the cumulative distribution function, the power spectrum, etc., but are otherwise postulated as random, generated according to a specific null hypothesis. Here, the null hypothesis consists of a candidate linear process, and the goal is to reject the hypothesis that the original data have come from a linear stochastic process. Often, a significant difference in the estimates of the correlation exponents, between the original and surrogate data sets, can be observed. In the case of the original data, a saturation of the correlation exponent is observed after a certain embedding dimension value (i.e., 7), whereas the correlation exponents computed for the surrogate data sets continue increasing with the increasing embedding dimension. The high significance values of the statistic indicate that the null hypothesis (the data arise from a linear stochastic process) can be rejected and hence the original data might have come from a nonlinear process.

The next important step is computing the Lyapunov's exponents, which are the dynamical invariants of a nonlinear system. In a general case, the orbits of chaotic attractors are unpredictable, but there is the limited predictability of chaotic physical system, which is defined by the global and local Lyapunov's exponents. In a chaos theory, the spectrum of the Lyapunov's exponents is considered a measure of the effect of perturbing the initial conditions of a dynamical system. Note that both positive and negative Lyapunov's exponents can coexist in a dissipative system, which is then chaotic. In fact, if one manages to derive the whole spectrum of the Lyapunov's exponents, other invariants of the system, i.e. Kolmogorov entropy and attractor's dimension can be found. The Kolmogorov entropy, K, measures the average rate at which information about the state is lost with time. An estimate of this measure is the sum of the positive Lyapunov's exponents. The inverse of the Kolmogorov entropy is equal to the average predictability. The estimate of the dimension of the attractor is provided by the Kaplan and Yorke conjecture:

$$d_L = j + \frac{\sum_{\alpha=1}^{j} \lambda_\alpha}{|\lambda_{j+1}|}, \quad (4)$$

where j is such that $\sum_{\alpha=1}^{j} \lambda_\alpha > 0$ and $\sum_{\alpha=1}^{j+1} \lambda_\alpha < 0$, and the Lyapunov's exponents λ_α are taken in descending order. There are a few approaches to computing the Lyapunov's exponents. One of them is based on the Jacobi matrix of system [30]. In the case where only observations are given and the system function is unknown, the matrix has to be estimated from the data. In this case, all the suggested

IOP Conf. Series: Journal of Physics: Conf. Series **905** (2017) 012007 doi:10.1088/1742-6596/905/1/012007

methods approximate the matrix by fitting a local map to a sufficient number of nearby points. In our work we use the method with the linear fitted map proposed by Sano and Sawada [30], although the maps with higher order polynomials can be also used. To calculate the spectrum of the Lyapunov's exponents from the amplitude level data, one could determine the time delay τ and embed the data in the four-dimensional space. In this point it is very important to determine the Kaplan-York dimension and compare it with the correlation dimension, determined by the Grassberger-Procaccia algorithm. The estimations of the Kolmogorov entropy and average predictability can further show a limit, up to which the amplitude level data can be on average predicted. Surely, the important moment is a check of the statistical significance of results. It is worth to remind that results of state-space reconstruction are highly sensitive to the length of data set (i.e. it must be sufficiently large) as well as to the time lag and embedding dimension determined. Indeed, there are limitations on the applicability of chaos theory for observed (finite) time series arising from the basic assumptions that the time series must be infinite [31]. A finite and small data set may probably results in an underestimation of the actual dimension of the dynamical process. The statistical convergence tests together with surrogate data algorithm can provide the satisfactory significance of the investigated data regarding the state-space reconstruction. In Table 1 we present the main blocks of our universal chaos-geometric approach to computational studying a non-linear dynamics of the chaotic systems [6-19].

Table 1. A chaos-geometric approach to nonlinear analysis and prediction of chaotic dynamics of the complex systems

I. Preliminary study and assessment of the presence of chaos: 1. Test by Hottvod-Melben: $K \to 1$ – chaos; 2. Fourier decompositions, irregular nature of change – chaos; 3. Spectral analysis, Energy spectra statistics, the Wigner distribution, the spectrum of power, "Spectral rigidity";
⇓
II. The geometry of the phase space. Fractal Geometry: 4. Computation time delay τ using autocorrelation function or mutual information; 5. Determining embedding dimension d_E by the method of correlation dimension or algorithm of false nearest neighbouring points; 6. Calculation multi-fractal spectra. Wavelet analysis;
⇓
III. Prediction: 7. Computing global Lyapynov dimensions LE: λ_α; Kaplan-York dimension d_L, KE, average predictability measure Pr_{max}; 8. Determining the number of nearest neighbour points NN for the best prediction results; 9. Methods of nonlinear prediction. Neural network algorithm, the algorithm optimized trajectories, ...;

3. Results and conclusions

As illustration, below we present the results of computational studying the low- and high dimensional dynamics of a chaos generation in the semiconductor GaAs/GaAlAs laser with the retarded feedback. Fischer et al [32] have carried out the excellent experimental studying dynamics of a chaos generation in the semiconductor GaAs / GaAlAs Hitachi HLP1400 laser; an instability is generated by means of the retarded feedback during changing the control parameter such as the feedback strength μ (or in fact an injection current). Of course, depending on the system μ there is appeared a multi-stability of different states with the modulation period: $T_n=2\tau/(2n+1)$, n=0, 1,2,... The state of n = 0 is called as a

IOP Conf. Series: Journal of Physics: Conf. Series **905** (2017) 012007 doi:10.1088/1742-6596/905/1/012007

ground one. With respect to the frequency modulation, other states are called as the third harmonic, fifth harmonic and so on. In the figure 1 we list the measured data on the time-dependent intensities for a semiconductor laser device with feedback: a) the time series, which illustrates a chaotic wandering between the ground state and the state of the third harmonic; b) the time series for a system in a state of the global chaotic attractor.

In Table 2 we present our original data on the correlation dimension d_2, the embedding dimension, computed on the basis of the false nearest neighboring points algorithm (d_N) with percentage of false neighbors (%) which are calculated for different lag times τ. The data are presented for two interesting regimes: I. chaos and II. hyperchaos. In Table 3 our computational data on the Lyapunov's exponents, Kaplan-York attractor dimensions, the Kolmogorov entropy K_{entr} are listed.

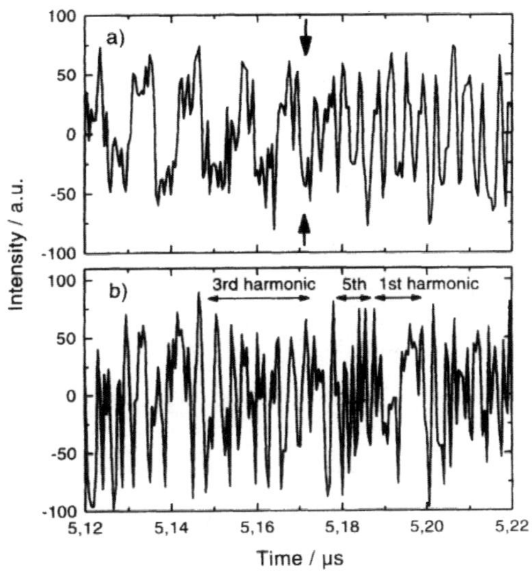

Figure 1. The time series of intensity in the GaAs/GaAlAs Hitachi HLP1400 laser (the measured data, from Ref. [32]).

Table 2. The correlation dimension d_2, the embedding dimension, computed on the basis of the false nearest neighboring points algorithm (d_N) with percentage of false neighbors (%) which are calculated for different lag times τ

Chaos regime (I)			Hyperchaos regime (II)		
τ	d_2	(d_N)	τ	d_2	(d_N)
58	3.4	5 (8.1)	67	8.4	11 (15)
6	2.2	4 (1.05)	10	7.4	8 (3.4)
8	2.2	4 (1.05)	12	7.4	8 (3.4)

Table 3. The Lyapunov's exponents: λ_1–λ_4, the Kaplan-York attractor dimension d_L and the Kolmogorov entropy K_{entr}

Regime	λ_1	λ_2	λ_3	d_L	K_{entr}
Chaos (I)	0.151	0.00001	-0.188	1.8	0.15
Hyperchaos (II)	0.517	0.192	-0.139	7.1	0.71

One can see that there are the Lyapunov's exponents positive and negative values. The resulting Kaplan-York dimensions in both cases are very similar to the correlation dimension, which is computed using the Grassberger-Procaccia algorithm [25]. The Kaplan-York dimension is less than the embedding dimension that confirms the correct choice of the latter. A scenario of chaos generation is in converting initially periodic states into individual chaotic states with increasing the parameter μ through a sequence of the period doubling bifurcations. Further there is appeared a global chaotic attractor after merging an individual chaotic attractors according to a few complicated scenario (see details in Refs. [13,32,33]).

To conclude, in this paper we have presented the results of computing nonlinear chaotic dynamics characteristics for the semiconductor GaAs/GaAlAs laser with retarded feedback system. The corresponding data have been obtained on the basis of using the advanced non-linear-analysis techniques such as a wavelet analysis, multi-fractal formalism, mutual information approach, correlation integral analysis and other methods. The correlation dimension method provided a low (or high-) fractal-dimensional attractor thus suggesting a possibility of an existence of the chaotic behaviour. The method of surrogate data, for detecting nonlinearity, provided significant differences in the correlation exponents between the original data series and the surrogate data sets. This finding indicates that the null hypothesis (linear stochastic process) can be rejected. It has been finally confirmed that the studied laser system dynamics exhibit a nonlinear behaviour with elements of the low-and high-dimensional chaos.

Acknowledgements

The authors would like to thank Prof. Nithaya Chetty, Prof. Marius Potgieter and colleagues for invitation to make contributions to CCP-2016 (Gauteng, South Africa). The useful comments of the anonymous referees are very much acknowledged too.

References

[1] Kenneth F 2003 *Fractal Geometry: Mathematical Foundations and Applications* (Chichester: John Wiley & Sons)

[2] Abarbanel H, Brown R, Sidorowich J and Tsimring L 1993 *Rev. Mod. Phys.* **65** 1331

[3] Cheng T, Liu J, Chen S and Guo H 2000 *Phys. Lett. A* **265** 384390

[4] Cecconi F, Cencini M, Falcioni M and Vulpiani A 2012 *Amer. J. Phys.* **80** 1001

[5] Gallager R 1986 *Information theory and reliable communication* (N.-Y.: Wiley)

[6] Glushkov A V 2008 *Modern theory of a chaos* (Odessa: OSENU)

[7] Glushkov A V, Khetselius O, Brusentseva S, Zaichko P and Ternovsky V 2014 Studying interaction dynamics of chaotic systems within a non-linear prediction method: application to neurophysiology *Advances in Neural Networks, Fuzzy Systems and Artificial Intelligence (Recent Advances in Computer Engineering*, vol 21*)* ed J.Balicki (Gdansk: WSEAS Press) pp 69-75

[8] Glushkov A V, Svinarenko A, Buyadzhi V, Zaichko P and Ternovsky V 2014 Chaos-geometric attractor and quantum neural networks approach to simulation chaotic evolutionary dynamics during perception process *Advances in Neural Networks, Fuzzy Systems and Artificial Intelligence (Recent Advances in Computer Engineering*, vol 21*)* ed J.Balicki (Gdansk: WSEAS Press) pp 143-150

[9] Glushkov A V, Khetselius O Yu, Svinarenko A A and Prepelitsa G P 2011 Energy approach to atoms in a laser field and quantum dynamics with laser pulses of different shape *Coherence and Ultrashort Pulsed Emission (*vol 1*)* ed F J Duarte (Rijeka: InTech) chapter 8 pp 159-186

[10] Glushkov A V, Buyadzhi V V and Ponomarenko E L 2014 *Proc. of Int. Geometry Center* **7**(1) 24

[11] Glushkov A V, Bunyakova Yu Ya, Fedchuk A P, Serbov N G, Svinarenko A A and Tsenenko I A 2007 *Sensor Electr. and Microsyst. Techn.* **3** 14

[12] Glushkov A V, Kuzakon V, Ternovsky V B and Buyadzhi V V 2013 Dynamics of laser systems with absorbing cell and backward-wave tubes with elements of a chaos *Dynamical Systems Theory (*vol T1*)* ed J. Awrejcewicz, M. Kazmierczak, P. Olejnik, and J. Mrozowski (Lodz:

Wyd. Politechniki Łódzkiej), pp 461-466

[13] Glushkov A V, Prepelitsa G P, Svinarenko A A and Zaichko P A 2013 Studying interaction dynamics of the non-linear vibrational systems within non-linear prediction method (application to quantum autogenerators) *Dynamical Systems Theory (*vol T1*)* ed J. Awrejcewicz, M. Kazmierczak, P. Olejnik, and J. Mrozowski (Lodz: Wyd. Politechniki Łódzkiej) pp 467-477

[14] Glushkov A V, Prepelitsa G P, Dan'kov S V, Polischuk V N and Efimov A V 1997 *J. Tech. Phys.* **38** 219

[15] Prepelitsa G P, Glushkov A V, Lepikh Ya I, Buyadzhi V V, Ternovsky V B and Zaichko P A 2014 *Sensor Electr. and Microsyst.Techn.* **11**(4) 43

[16] Glushkov A V and Ivanov L N 1993 *J. Phys. B: At. Mol. Opt. Phys.* **26** L379

[17] Glushkov A V, Malinovskaya S V, Shpinareva I M, Kozlovskaya V P and Gura V I 2005 *Int. J. Quant. Chem.* **104** 512

[18] Svinarenko A A 2014 *J. Phys.: Conf. Ser.* **548** 012039

[19] Florko T A, Ambrosov S V, Svinarenko A A and Tkach T B 2012 *J. Phys.: Conf. Ser.* **397** 012037

[20] Bezruchko B P, Kuznetsov S P and Seleznev Y P 2000 *Phys. Rev. E.* **62** 7828

[21] Packard N, Crutchfield J, Farmer J and Shaw R 1988 *Phys. Rev. Lett.* **45** 712

[22] Takens F 1981 Detecting strange attractors in turbulence *Dynamical systems and turbulence* ed D. Rand and L. Young (Springer: Berlin) pp 366-381

[23] Mañé R 1981 On the dimension of the compact invariant sets of certain non-linear maps *Dynamical systems and turbulence* ed D. Rand and L. Young (Springer: Berlin) pp 230-242

[24] Paluš M, Pelikán E, Eben K, Krejčíř P and Juruš P 2001 Nonlinearity and prediction of air pollution *Artificial Neural Nets and Genetic Algorithms* ed V Kurkova, N C Steel, R Neruda and M Neruda (Springer: Wien) pp 473-476

[25] Grassberger P and Procaccia I 1983 *Physica D.* **9** 189

[26] Kennel M, Brown R and Abarbanel H 1992 *Phys. Rev. A.* **45** 3403

[27] Theiler J, Eubank S, Longtin A, Galdrikian B and Farmer J 1992 *Phys. D.* **58** 77

[28] Fraser A and Swinney H 1986 *Phys. Rev. A.* **33** 1134

[29] Havstad J and Ehlers C 1989 *Phys. Rev. A.* **39** 845

[30] Sano M and Sawada Y 1995 *Phys. Rev. Lett.* **55** 1082

[31] Schreiber T 1999 *Phys. Rep.* **308** 1

[32] Fischer I, Hess O, Elsaber W and Gobel E 1994 *Phys. Rev. Lett.* **73** 2188

[33] Jalabert R A and Pastawski H M 2001 *Phys. Rev. Lett.* **86** 2490

IOP Conf. Series: Journal of Physics: Conf. Series **905** (2017) 012008 doi:10.1088/1742-6596/905/1/012008

Phase transition approach to bursting in neuronal cultures: quorum percolation models

P Monceau[1,2], R Renault[1], S Métens[1], S Bottani[1] and T Fardet[1]
[1]Laboratoire Matière et Systèmes Complexes, UMR 7057 CNRS, Université Denis Diderot-Paris 7, 10 rue A. Domon et L. Duquet, 75013 Paris Cedex, France.
[2]Université d'Evry-Val d'Essonne, France.

E-mail: Pascal.Monceau@univ-paris-diderot.fr

Abstract. The Quorum Percolation model has been designed in the context of neurobiology to describe bursts of activity occurring in neuronal cultures from the point of view of statistical physics rather than from a dynamical synchronization approach. It is based upon information propagation on a directed graph with a threshold activation rule; this leads to a phase diagram which exhibits a giant percolation cluster below some critical value m_C of the excitability. We describe the main characteristics of the original model and derive extensions according to additional relevant biological features. Firstly, we investigate the effects of an excitability variability on the phase diagram and show that the percolation transition can be destroyed by a sufficient amount of such a disorder; we stress the weakly averaging character of the order parameter and show that connectivity and excitability can be seen as two overlapping aspects of the same reality. Secondly, we elaborate a discrete time stochastic model taking into account the decay originating from ionic leakage through the membrane of neurons and synaptic depression; we give evidence that the decay softens and shifts the transition, and conjecture than decay destroys the transition in the thermodynamical limit. We were able to develop mean-field theories associated with each of the two effects; we discuss the framework of their agreement with Monte Carlo simulations. It turns out that the the critical point m_C from which information on the connectivity of the network can be inferred is affected by each of these additional effects. Lastly, we show how dynamical simulations of bursts with an adaptive exponential integrate-and-fire model can be interpreted in terms of Quorum Percolation. Moreover, the usefulness of the percolation model including the set of sophistication we investigated can be extended to many scientific fields involving information propagation, such as the spread of rumors in sociology, ethology, ecology.

Keywords: Phase transitions general studies, neural networks, percolation, complex networks.

1. Introduction

In vitro cultures of dissociated neurons have turned out to be a powerful tool in investigating fundamental questions in several scientific fields. From a conceptual point of view, compelling issues comprise the nature of biological computations, the mechanisms of information spreading throughout networks, and the understanding of collective behaviors [1], as these questions could bring new elements to the understanging of neuronal computation and to the field of artificial intelligence. Moreover, neuronal cultures are suitable for pharmaceutical drugs experimentation and have been useful in studying neurodegenerative diseases [2]. Lastly, populations of neurons could be the basic units in designing computational devices involving real living cells [3]. Experimentally, such cultures can be obtained by seeding dissociated neurons extracted from rodent embryos on a suitable substrate; axons and dendrites grow in such a way that neurons self-organize after a few days into a quasi two-dimensional network [4]. The cells can be maintained alive several weeks and display sustained electric activity. Let us recall that there are about 10^{11} neurons in the human brain, each of them being connected to 7 000 others on average through synapses; hence, it is a very complex network where neurons are organized in localized computational units connected according to a well defined hierarchical structure. However, *in vitro* cultures establish very different connectivity patterns during their growth [3] since they are characterized by a higher level of randomness. The neuronal cultures we are interested in hold between 10^3 and 10^5 neurons with typical densities between 500 and 5 000 neurons per mm^2, each of them connected via a number of synapses falling between 20 and 200. These changes in connectivity and scale which could at first glance appear as a loss from a neurobiologic point of view are largely compensated by the benefits associated with *in vitro* experimentation. The recent development of techniques such as micro-electrode arrays (MEA), [5] optogenetics and calcium imaging [6, 7] enables the experimentalists to carry out quantitative measurements inaccessible *in vivo*. Furthermore, more precise control over the system can now be attained: physicochemical parameters such as extracellular ionic concentrations can be modified [8], drugs can be injected [4], neurons can be electrically or optically excited. Microfabrication techniques are now also used to structure the connectivity between sub-populations by constraining mechanically the axon growth with obstacles or designed channels [3] in order to build *in silico* models of brain structures or build neuronal devices designed for specific functions [9]. Both *in vitro* and *in vivo*, neuronal rhythms are a widespread phenomenon observed at many temporal and spatial scales. Synchronized periodic bursts of spiking activity emerge spontaneously in cultures of dissociated neurons from rodent hippocampus and cortex [8, 10], depending on their density and age. Furthermore, bursts can be triggered by initially activating a fraction of neurons. Rather than describing collective behaviors observed in living neuronal networks grown *in vitro* in terms of synchronization [11, 12], the Quorum Percolation model (QP) tackles the issue of population wide activation from the point of view of statistical physics. The Quorum Percolation model, derived from bootstrap percolation, has been specifically designed to describe activity bursts observed in such cultures [13]. Under its original form, it is a discrete time dynamics model of information propagation on a directed graph, built up according to a simplification of the most relevant biological features: the neurons, located at the nodes of the graph, are two state systems whose activation is governed by a threshold (Quorum) rule. A burst is seen as a discontinuity in the activity of the network, interpreted as the occurrence of a giant excited cluster. We further refined the model by introducing the following biological relevant developments:

(i) Modulation of the neurons excitability. As a matter of fact, neuronal cultures exhibit some variability in the neuron excitability; we study the modifications induced in the behavior of a Quorum Percolation model by taking into account an uncorrelated Gaussian variability of the neuronal thresholds.

(ii) Decay of the subthreshold neuron voltage. The decay accounts for ionic leakage through the membrane of neurons, since they do not behave as perfect capacitors; we take it into account in a Quorum Percolation with Decay (DQP) where we model the decay by a discrete time disintegration process of the membrane potential of the neurons.

2. Phase diagram and critical behavior of the original quorum percolation model

The networks we deal with includes N neurons, where each of them is a two-level system which can be either active or at rest. A directed network is constructed by randomly choosing, for each neuron i, k incoming links among the $N-1$ other neurons according to an in-degree probability distribution p_k. Experimental results and their interpretation through the original

Quorum Percolation model suggest that the connectivity of mature *in vitro* cultures can be approximated by a random oriented graph with Gaussian distribution of incoming links [10]; hence, in the following, we will restrict ourselves to such networks, where $\overline{k}$ denotes the mean value of p_k and σ_k its standard deviation. The construction of such networks does not require information on the geometrical location of neurons in the physical space. Thus, we deal with percolation on a random graph without taking in account any spatial metric. Starting at time $t = 0$ from an initial state of the network where a given fraction f of randomly chosen neurons is set active, information spreads through the network according to an excitability threshold rule. A neuron i becomes active if a given number m (called quorum) of its k incoming neighbors are active. The activation process of the network is described by a discrete-time dynamics with a step Δt during which each neuron integrates the signals sent by its incoming neighbors. A discrete variable $V_i(t)$ – accounting for the membrane potential – is assigned to each neuron i. The transition from one time step to the next obeys the following rules:

(1) Every neuron i activated between $t - \Delta t$ and t sends at time t one signal to each of its neighbors through its outgoing links; no further signals will be sent by such an activated neuron at later times. Each sent signal has the same weight and is associated to an integer increment equal to $+1$.

(2) The variable $V_i(t)$ of each target neuron at rest is incremented by the sum of the inputs received at time t.

(3) If $V_i(t)$ is greater than or equal to the activation threshold m, the neuron i fires, which means that it switches from the resting state (at time t) to the active state (at time $t + \Delta t$).

(4) Once a neuron has been activated, it remains in the active state until the end of the process.

The macroscopic activity state of a network at time t is the fraction of its active neurons. Once a random network, and a random initial state have been drawn, the discrete-time dynamics described above is deterministic, monotonically increasing, and leads to an equilibrium state of the network characterized by a final fraction Φ of active neurons. Explicit simulations aim at directly calculating the response Φ of a finite-size network of N neurons, from an initial excitation parametrized by the fraction f of initially activated nodes. Given a set of parameters $\{\overline{k}, \sigma_k\}$, and m, a Monte-Carlo run consists of the following steps:

(i) A random directed network $\mathcal{G}$ is constructed according to the incoming links probability distribution p_k.

(ii) A fraction f of neurons is randomly activated.

(iii) The discrete time process described above goes on until the number of active neurons stops increasing, i.e. when the stationary state has been reached.

The average value of Φ is then calculated over several runs. A typical phase diagram is shown on Fig. 1 where two regimes can be distinguished as m varies, for fixed values of $\overline{k}$ and σ_k. Such a phase diagram provides a good description of experiments carried out in the group of E. Moses [4, 13]. Below some critical value m_C, the final fraction of activated neurons presents a discontinuity when we vary the control parameter f (the initial fraction of activated neurons), whereas it remains continuous above m_C. The sudden jump occurring at f^* in the global activity Φ is associated with a percolation transition on the network $\mathcal{G}$, where a very small variation of f results in the appearance of a giant cluster, whose normalized size g is given by the difference between the lower and upper values of ϕ at the discontinuity. Once p_k is fixed, following the usual concepts of percolation on lattices [14], the normalized mean cluster size $\langle g \rangle$ can be considered as an order parameter whose behavior in the vicinity of m_C is given by a power law: $\langle g \rangle \sim \left(\frac{m_C - m}{m_C}\right)^{\beta}$.

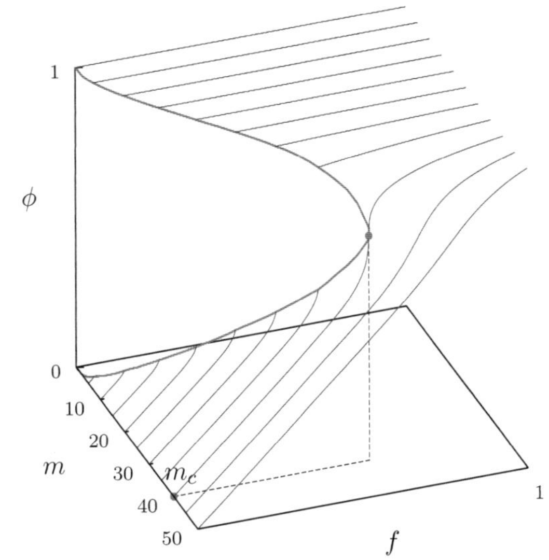

Figure 1. Phase diagram of the Quorum Percolation model, for a Gaussian "in-connectivity" with $\bar{k} = 50$ and $\sigma_k = 10$. When the quorum m is smaller than m_C, a jump in the fraction of active neurons Φ occurs when increasing the fraction f of initially activated neurons from zero. The height $\langle g \rangle$ of the jump at the discontinuity is the normalized size of the percolation giant cluster.

As a main result, it should be noticed that the critical behavior of $\langle g \rangle$ provides information on the connectivity of the network [10, 13]. It is tricky to calculate the critical exponent β, since the quorum percolation model is discrete; however we were able to derive an extension of the QP model to continuous values of m with the help of a mean-field approach in good agreement with Monte-Carlo simulations [15, 16]. A numerical resolution of the associated highly nonlinear self-consistent equation led to a value very close to the classical mean-field exponent $\beta = 1/2$ when p_k is Gaussian. This value, associated with a Gaussian distribution of incoming links, is in agreement with experimental results. Furthermore, the critical value has been shown to depend on $\bar{k}$ and σ_k as $m_C \approx \bar{k} - 1.3\sigma_k$. Hence the position of the critical point provides quantitative information on the mean value and width of the Gaussian connectivity distribution.

3. A quorum percolation model with quorum variability

Unlike the original QP model where the quorum takes the same value over the whole set of neurons, we introduce disorder on the excitability by randomly setting each node's quorum to an integer value according to some probability distribution $\mathcal{P}_m$ – with no correlation to other network properties. Thus, the Monte-Carlo algorithm described in section 2 includes an additional stage just after the first one, where such a disorder on the quorums is implemented. ϕ must be averaged over the three sets of associated configurations, that is a set of initial configurations associated to f, a set of quorums related to $\mathcal{P}_m$ and a set of networks realizations based on p_k.

3.1. Mean-field theory

An alternative approach for calculating ϕ can be deduced from a mean-field treatment: the probability for a neuron to be active at equilibrium corresponds to the probability to be either active through initial stimulation or to be activated during the QP discrete time process; the activation probability of a neuron – given m and k – can be approximated by a binomial process depending upon ϕ. Since the activation of a neuron can occur only if at least m incoming links are linked to active neighbors, such an activation probability reads $\sum_{l=m}^{\infty} \binom{k}{l} \Phi^l (1-\Phi)^{k-l}$. In the end, we obtain the following self-consistent equation:

$$\frac{\Phi - f}{1-f} = \sum_{m=1}^{\infty} \mathcal{P}_m \sum_{k=m}^{\infty} p_k \sum_{l=m}^{k} \binom{k}{l} \Phi^l (1-\Phi)^{k-l} = \sum_{k,m}^{\infty} \Pi(m,k) \sum_{l=m}^{\infty} \binom{k}{l} \Phi^l (1-\Phi)^{k-l} \quad (1)$$

where the right-hand term accounts for the total activation probability $\mathcal{P}^{act}(\Phi)$ of a neuron of the network. The solutions of the self-consistent equation (1) are given by the intersection points of $\mathcal{P}^{act}(\Phi)$ with the line of slope $1/(1-f)$ passing through the point of coordinates $(1,1)$ in the $\{\Phi, \mathcal{P}^{act}_m(\Phi)\}$ plane. Assuming that $\mathcal{P}_m$ and p_k are Gaussian probability distributions with respective average values $\overline{m}$ and $\overline{k}$ and variances σ_m and σ_k, the self-consistent equation involves a truncated bidimensional Gaussian probability distribution. As a first result, it turns out that the qualitative behavior of the solutions for Φ is close to the one observed in the absence of disorder. When numerically solving equation (1) in the physically meaningful range [0,1] of f, two regimes can be distinguished: For m smaller than a critical value m_C now depending on the additional parameter σ_m, there is a range $f \in]f_0, f^*[$ where three different real values of Φ satisfy (1). For $m > m_C$, a single real value of f satisfies (1). Since the QP process requires Φ to be an increasing function of f, the physical behavior of Φ resolves the existence of an unstable branch below m_C in this range by a discontinuity at f^* associated with the appearance of the giant cluster; the normalized size of this cluster is equal to the difference between the lower Φ^- and the upper Φ^+ solutions of equation (1) at the border between the two regimes. Situations showing the evolution of the jump in Φ for two different values of σ_m are displayed on Fig. 2.

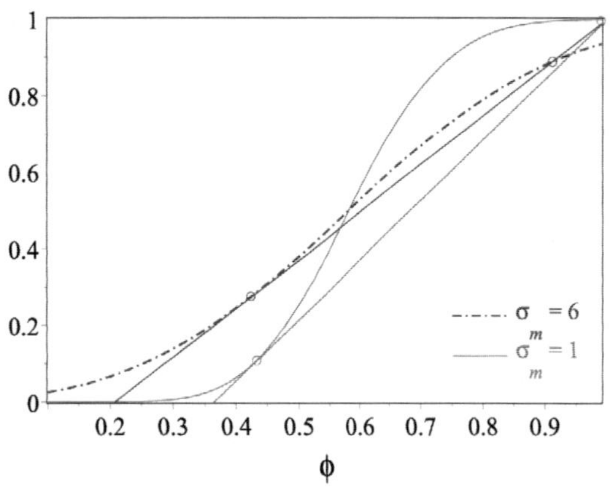

Figure 2. Evolution of $\overline{\mathcal{P}}^{act}(\Phi)$ for $\overline{k} = 25$, $\sigma_k = 3$ and $\overline{m} = 15$ and two different values of σ_m. The open circles represent the intersections of the curves $\overline{\mathcal{P}}^{act}(\Phi)$ with the lines $\mathcal{D}$ of slopes $1/(1-f)$ giving the two solutions ϕ^- and ϕ^+ of equation (1) associated with the percolation clusters.

3.2. Simulation results

Results reported on Fig. 3 provide a picture of the main conclusions that can be drawn out from a large set of Monte Carlo simulations and numerical resolutions of equation (1).

(i) There is a good agreement between the mean-field and Monte Carlo approaches at least within the range of physical parameters involved in the quantitative description of neuronal cultures with Gaussian in-degree.

(ii) For a fixed value of $\overline{m}$, increasing the variance σ_m shifts the position of the jump in Φ towards lower values of f and reduces the size of the giant cluster (unless $\overline{m}$ is "too small", in which case a slight bump can appear in the variation of g with σ_m).

(iii) $\overline{m}$ being fixed, a large enough amount of disorder (σ_m) on the excitability can destroy the percolation transition ($\sigma_m = 32$ on Fig. 3).

IOP Conf. Series: Journal of Physics: Conf. Series **905** (2017) 012008 doi:10.1088/1742-6596/905/1/012008

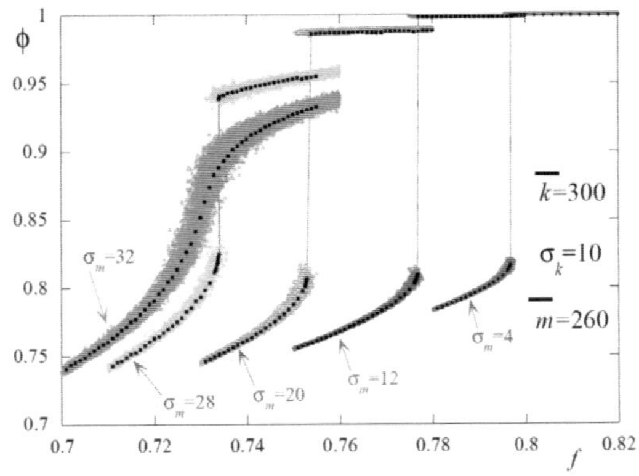

Figure 3. Evolution of the activity of the network when increasing σ_m at a fixed value of $\overline{m}$ obtained by explicit simulations with $N = 100\ 000$ neurons (clouds of points) and self-consistent equation (black dots). Notice the agreement between the two approaches and the vanishing of the percolation transition when the disorder width σ_m is large enough.

(iv) The position of the critical point m_C depends not only on $\overline{k}$ and σ_k, but also on σ_m; as long as the jump in ϕ survives, m_C decreases when σ_m increases.

An interpretation of some of the preceding conclusions can be done by looking at the truncated bidimensional Gaussian probability distribution $\Pi(m, k)$ involved in the activation probability. The initially excited nodes are uniformly drawn over the whole distribution, but the nodes liable to be involved in the network activation must satisfy $m \leq k$ in order to be excitable; hence, they lie under the bisecting line in the (m, k) plane. Two competitive effects arise when increasing σ_m from zero: a fraction of nodes, associated with the part of $\Pi(m, k)$ below $\overline{m}$ becomes more easily excitable, while the other fraction becomes less easy or even – when they cross the bisecting line – impossible to excite. The neurons below $\overline{m}$ are responsible for the shift of $\langle f^* \rangle$: the ignition mechanism of the giant cluster needs a smaller fraction of initially excited nodes, but a larger spread of activity throughout the network since $(\Phi^- - \langle f^* \rangle)$ increases with σ_m as shown in the example displayed Fig. 3. Moreover, as can be seen on Fig. 2, the sigmoids associated with the activation probability $\mathcal{P}^{act}(\Phi)$ become less steep when σ_m increases, leading to smaller values of the slope of the line tangent to this curve at the point Φ^-, hence to a decrease in f^*. Since Monte Carlo simulations showed that we can rely on the mean field theory to describe a Quorum Percolation with excitability disorder, a prolongation to non integer values of m with the help of Beta functions enables to investigate properly the critical region [15]. As a main result a normal form treatment of the prolonged self-consistent equation leads to the same power law behavior as in the case without disorder, that is $\langle g \rangle \propto \left(\frac{m_C(\sigma_m) - m}{m_C(\sigma_m)} \right)^{1/2}$. Nevertheless it is worth noticing that m_C depends on the additional parameter σ_m; hence the relation $m_C \approx \overline{k} - 1.3\sigma_k$ cannot be used to infer the values of the connectivity parameters.

3.3. Finite size analysis of the fluctuations

A detailed study of finite-size effects is of great interest from an experimental point of view, since measurements are always carried out on finite neuronal populations. A finite-size scaling in the vicinity of the critical point cannot be done from the standard point of view of percolation [14], since it relies on a comparison between the linear size of the network and some correlation length, quantities which do not make sense in the present case of percolation on a graph, where the dimensionality of the system and the metric are not defined. Recalling that, in the presence of quorum disorder, the physical quantities calculated from Monte Carlo simulations are averaged over three sets of configurations, special attention must be paid to the study of sample to sample fluctuations. Moreover, such fluctuations which are linked to self-averaging properties can exhibit very unusual properties in the vicinity of a critical point [17, 18]. A large set of parameters $\{\overline{k}, \sigma_k, \overline{m}, \sigma_m\}$ has been investigated by means of intensive Monte Carlo simulations for different sizes ranging from $N = 10^3$ to $N = 2.5 \times 10^6$. As a main result we found that the relative fluctuations decrease as power laws of the network sizes: $\langle \Delta g \rangle / \langle g \rangle \sim N^{-\gamma}$ and $\langle \Delta f^* \rangle / \langle f^* \rangle \sim N^{-\zeta}$ and that fluctuations of the order parameter f are always larger than the fluctuations of the jump positions; moreover, the fluctuations in g increase as $\overline{m}$ increases, approaching the critical value m_C as expected from a second order phase transition. On the other hand, the fluctuations in g and f^* increase as the threshold disorder σ_m is increased (for a given value of $\overline{m}$), the exponents associated with the power laws exhibit an universal character, since no significant difference in these exponents can be brought out from the set of simulations we carried out: $\zeta = 0.495(10)$ and $\gamma = 0.29(15)$. Therefore, a finite-size analysis of the fluctuations does not enable the direct detection of disorder on the quorum. Let us recall that, in the case of networks with a linear size L in a D dimensional space, a quantity $\mathcal{O}$ is said to be strongly self-averaging if $[\langle \Delta \mathcal{O} \rangle / \langle \mathcal{O} \rangle]^2 \sim 1/N = L^{-D}$ and weakly self-averaging if $[\langle \Delta \mathcal{O} \rangle / \langle \mathcal{O} \rangle]^2 \sim L^{-a}$ where $0 < a < D$ [17]. Hence, it turns out that f^* is practically strongly self-averaging since $[\langle \Delta f^* \rangle / \langle f^* \rangle]^2 \sim N^{-1}$ whereas the order parameter is weakly self-averaging independently of the physical parameters, in particular the threshold disorder.

3.4. Subcritical behavior: disorder-independent fixed points
We investigated the effects of threshold disorder on the behavior of the network activity for values of $\bar{m}$ and σ_m such that no percolation occurs anymore. An example of the results obtained by Monte Carlo simulations is shown on Fig. 4.

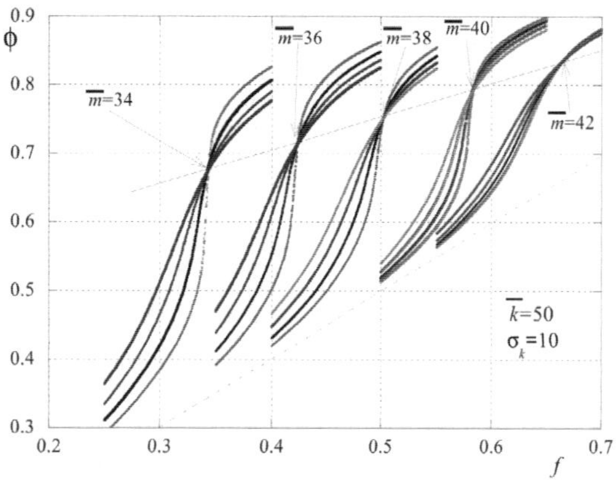

Figure 4. Disorder independent fixed points in the supercritical region with $\bar{k} = 50$, and $\sigma_k = 10$ for different values of $\overline{m}$ indicated inside the figure and each time four different values of σ_m (increasing from red to blue): $[11, 14]$ when $\overline{m} = 34$, $[9, 12]$ when $\overline{m} = 36$, $[7, 10]$ when $\overline{m} = 38$, $[3, 6]$, when $\overline{m} = 40$, $[2, 5]$, when $\overline{m} = 42$. The dotted line corresponds to $\Phi = f$.

The most striking result is the emergence of disorder independent fixed points: the mean activity of the network, for a given $\overline{m}$ and a given f_{FP} is independent of the width of the threshold distribution over a large range of σ_m. Furthermore, it turns out that the activity Φ_{FP} at the fixed point and f_{FP} follow a universal law, since they line up along the straight line $\Phi_{FP} = \frac{1}{2}(1 + f_{FP})$ independently of $\bar{k}$ and σ_k. Hence the fixed points occur right when the activation probability is equal to $\frac{1}{2}$. This can be interpreted if we remember the evolution of $\mathcal{P}^{act}(\Phi)$ with σ_m: the evolution observed on Fig. 2 has the same profile when Eq. (1) has a single solution. Below Φ_{FP}, an increase in σ_m enhances the activity propagation, while it has the opposite effect above. Therefore, the two competitive effects, arising between the more easily and less easily excitable populations when σ_m is varied, balance exactly at the fixed point.

3.5. Discussion: connectivity, excitability and disorder
With regard to experiments carried out on neuronal networks, m was tuned by drugs [13] and $\bar{k}$ and σ_k deduced from simulations fitting the experimental data in the critical region assuming an uniform excitability. Since the critical value of the quorum depends on σ_m, the hypothesis that no threshold disorder is present can lead to wrong estimations of the connectivity parameters

of the network. The results reported in section 3 show that the excitability and connectivity distribution widths quantified by σ_k and σ_k are intricately connected. There is some kind of equivalence between connectivity and excitability, which was noted by J. P. Eckmann et al. [19]. In fact, as already pointed out in the case of neuronal cultures, connectivity and excitability can be seen as two overlapping aspects of the same reality: the addition of synaptic blockers – used to increase the control parameter m – can also be interpreted as a weakening of the network functional connectivity [4, 13]. This is locally reflected in the model: a node whose threshold goes from m to $m + 1$ when introducing disorder needs a larger number of incoming links to fire. Roughly speaking, what matters in describing the qualitative behavior of the model is the ratio of excitability to connectivity $\overline{m}/\overline{k}$. However, the accessible physical quantities that can be brought out from experiments involve averaging processes from which the detailed respective roles of the connectivity driven by p_k and the excitability driven by $\mathcal{P}_m$ are very difficult to discriminate.

4. A quorum percolation model with decay

The membrane of biological neurons can be compared to a capacitor that supports electric potential difference through ionic charge separation. Active neighbors will inject ionic currents into this capacitor, changing the electric potential difference across the membrane until it eventually passes a threshold value (associated with m in the framework of the QP model) when the neuron fires. This membrane is not a perfect capacitor as it is continuously leaking ions; hence, without new input, the membrane potential decays exponentially to its resting value with a time constant τ. In the limit where the time interval between each received signals is much larger that τ, they won't add up at all. The state of a real neuron is thus not only determined by the number of received signals, but also by their arrival times. We take into account the decay by building an extended model, (called DQP for Decay Quantum Percolation) in which each discrete accumulated signal can disintegrate independently with a probability $d \in [0, 1]$ at each step of the percolation process. Thus, the evolution of the neuronal activity is described by a discrete time stochastic process with a step Δt involving two competitive mechanisms: the reception of new signals sent by activated neurons and the decay of the accumulated signals with a characteristic time τ. The quorum is here assumed to be the same for all neurons.

A scheme of The Monte Carlo DQP process is provided on Fig. 5, and goes as follow:

(1) Every neuron j activated between $t - \Delta t$ and t sends at time t one signal to each of its out-neighbors; no further signals will be sent by such an activated neuron at later times. Each sent signal has the same weight and is associated to an integer increment equal to 1.

(2) The variable $V_i(t)$ of each target neuron at rest is incremented by the sum of the number of signals it has received at time t.
   - If $V_i(t)$ is greater than or equal to the activation threshold m, the neuron i switches from the state at rest (at time t) to the state active (at time $t + \Delta t$).
   - If $V_i(t)$ is smaller than m, each integer element of this potential is submitted to a Bernoulli trial of parameter d (with $0 \leq d \leq 1$) and is disintegrated if the trial is positive; the potential of the neuron at rest has decayed from its value $V_i(t)$ to $V_i(t + \Delta t)$.

(3) Once a neuron has been activated it remains in such a state until the end of the process.

A DQP Monte Carlo run follows the same algorithm as the one described in section 2, where the discrete process of the item (3) is replaced by the process in item (2). Thus, it is as if the sum of accumulated signals decays on average exponentially with a time constant $\tau = -\Delta t / ln(1-d)$, within a time step Δt. If we normalize the time constant τ to the duration of an iteration *in vitro*, that is, the minimal time interval necessary to transmit activity from a neuron to another inside a culture, we can reckon the value of the decay parameter d. Taking into account the action

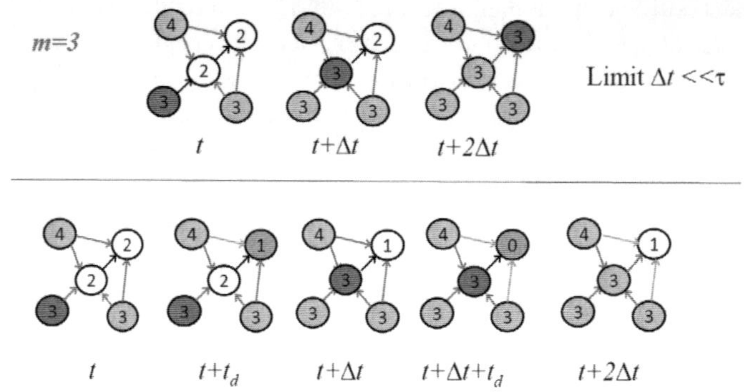

Figure 5. Mechanism of DQP model (below); the QP model is recalled above. At time t, the red neuron (left down) is activated; the orange ones (left up and down right have been activated before). At $t + t_d$, the green one (up right) encounters a decay; its potential is decreased from 2 to 1. The network is updated at $t + \Delta t$: the central neuron is activated. At $t + \Delta t + t_d$ the upper right neuron encounters once more a decay, and its potential is decreased from 1 to 0; in its updated state at $t + 2\Delta t$ it receives a signal from the centered neuron, but it is not in the same state as in the absence of decay (when $\Delta t << \tau$).

potential duration and propagation speed, the size of a typical culture and the synaptic delay, we can estimate that Δt lies between 1 and 10 ms and d between 0.1 and 0.01. Nevertheless, from a mean-field point of view, we were able to establish a recursive relation enabling us to fully describe the stochastic dynamics in the presence of decay [22]. The striking point is the idea that everything goes on as if the decay changes the connectivity of the network all along the process. At time t let us consider a neuron with k incoming neighbors, a potential $V_i(t) = s$ and x active neighbors ($x \geq s$ according the DQP rules); this neuron has undergone $(x - s)$ decrements due to the decay. Hence it is as if $(x - s)$ among the k incoming links had been erased. This remark enables to define a time dependent effective connectivity according to the Quorum Percolation without decay; this neuron will behave as a neuron experiencing an effective connectivity defined as $k_{eq} = k - (x - s)$. Details on the derivation of this recursive relation based on this equivalence can be found in [22].

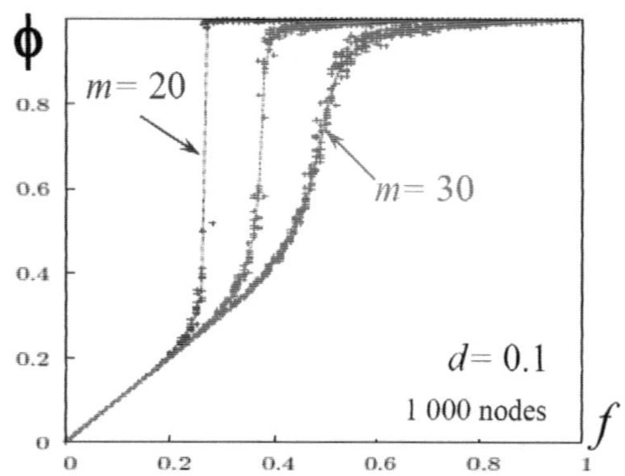

Figure 6. Comparison of a network calculated by Monte Carlo simulations (lines) and numerical resolution of the mean-field algorithm (dots); points of the same color correspond to 10 different equivalent Monte Carlo simulations. $\overline{k} = 50$, $\sigma = 10$, $d = 0.1$

Fig. 6 displays typical results where the agreement between Monte Carlo simulations and the mean-field approach can clearly be seen even when the size of the network is rather small. The effect of the decay is shown on Fig. 7 where Monte Carlo simulations are gathered together.

As a main result, the decay softens the transition, reduces the apparent size of the discontinuity, and shifts its position towards higher values of f.

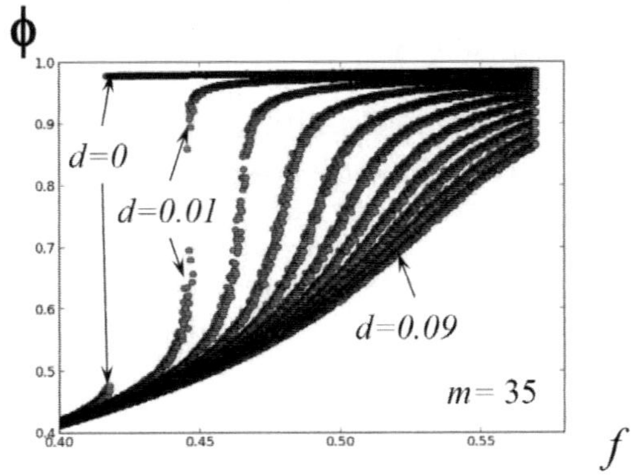

Figure 7. Evolution of the activity of a network when increasing the decay d at m constant calculated by Monte Carlo simulations. Note the decrease of the apparent size of the discontinuity with d and the vanishing of the transition if d is large enough; m_C should be equal to 0.58 without decay. $\overline{k} = 50$, $\sigma = 10$, m=35.

The size of the discontinuity g as a function of m was previously used [13] to infer connectivity in real neuronal networks; since decay is part of the physics of these networks, we wanted to evaluate to what extent the decay parameter d was changing those results. The value of the step ϵ between successive values of f is crucial in evaluating the size of the discontinuity; when dealing with Monte Carlo simulations on networks with N neurons, the increment in f associated with a single neuron imposes a lower bound $1/N$ for ϵ. For a given value of m, we define the apparent size of the discontinuity $g_\epsilon(m)$ as the maximum value of the difference $[\Phi(f+\epsilon) - \Phi(f)]$ with respect to f . In order to characterize properly the transition in the thermodynamical limit, when $N \to \infty$, numerical calculations of $g_\epsilon(m)$ are done in the framework of the mean-field approach; hence the evolution of $g_\epsilon(m)$ as a function of ϵ can be studied with ϵ as small as wished, approaching the real size of the discontinuity. Surprisingly, when d is non-zero, continuously decreasing ϵ leads to a steady gradual reduction of $g_\epsilon(m)$: $\forall m, \lim_{\epsilon \to 0} g_\epsilon(m) = 0$. The convergence is faster when the decay d is strong and the $m/\overline{k}$ ratio is high, but the phenomenon was observed in every computationally accessible case, as long as d was non-zero. This strongly suggests that $g_\epsilon(m)$ always converges towards 0 when $d > 0$ and $m > 1$.

Thus, the main result of our set of computations is the conjecture that the critical point expected from the classical QP model in the thermodynamical limit vanishes in the framework of the DQP model. It also reveals how the distinction between continuous and discontinuous transitions depends both on the size of the network and the accuracy in the control of external stimulation; consequently the decay, although being part of the networks dynamics, may remain unnoticed. Lastly, a consequence of the shift in the apparent size of the discontinuity is that a model which does not take decay into account leads to an underestimation of m_C, introducing a bias in the estimation of the network connectivity parameters, since the relation $m_C \approx \overline{k} - 1.3\sigma_k$ does not hold anymore when $d \neq 0$.

5. Percolation in dynamical situations of bursting cultures

The QP model was initially designed to study bursting activity in neuronal cultures; we will discuss here how it can be applied, beyond the study of "forced" systems, to analyze spontaneous activity in neuronal cultures. We simulated networks of oscillatory excitatory neurons using the adaptive exponential integrate-and-fire model [23]. Each neuron is connected to others from a Gaussian distribution of average 100 and standard deviation 5, and the spikes are transmitted between neurons with a constant delay of 1 ms. Figure 8 shows the simulated activity, which is composed of periodic bursts with a specific inner structure. A burst is indeed composed of a succession of *synchronous burst slices* (SBS) which are the base units that we will describe, using the introduced formalism, as several distinct percolation events. Let us first describe and explain how a burst is initiated, develops and terminates. Burst initiation comes from the intrinsic behavior of the neurons which are oscillators [8]: their membrane potential slowly depolarizes under the influence of a persistent sodium current $I_{Na,p}$ until a first spike is initiated. After an initial synchronization of the population due to phase rest and positive feedback [24], the first spikes of a burst occur concomitantly in a relatively short time period (e.g. 4510–4512 ms) and involve a large fraction of the population. This initial SBS acts as the intrinsic counterpart of the external excitation in the Quorum Percolation model since it is what activates the first neurons of the following SBS (the [4514–4516] ms slice on the inset). Thus we obtain a series of SBSs, each initiated by the input of the previous one. Moreover, the successive SBSs get wider and more sparse because of adaptation mechanisms and fatigue, which increase the quorum necessary for one neuron to activate. Since the effect of the SBS is spread over several milliseconds, taking decay into account, we eventually reach a point where the cumulative effect of the previous SBS in no longer sufficient for neurons to reach their increased quorum and the burst terminates.

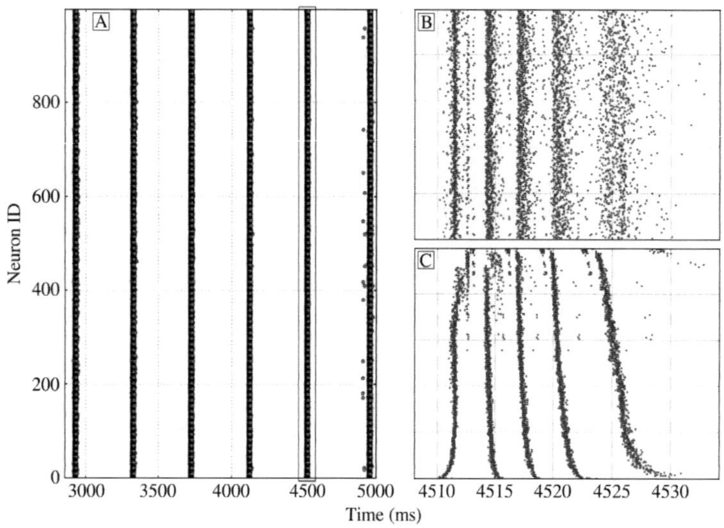

Figure 8. A: spike raster of a 1000-neuron network with Gaussian in-degree distribution $\mathcal{N}(100, 5)$, displaying periodic spiking behavior. **B**: detailed dynamics of 5th burst (boxed), with the successive SBSs. **C**: same time-window as **B** but with neurons ordered by increasing in-degree; we can clearly see the activity propagate between groups with different connectivity profiles.

The detailed structure of the burst becomes apparent if we sort the neurons based on their in-degree. Figure 8 C represents the same time window as **B** and shows how strongly the spiking times of the neurons correlate to their in-degree. Indeed, higher-degree nodes will reach their quorum more easily, thus firing earlier than the rest of the network. This effect becomes more significant as the average quorum increases; on the last SBS of **C**, we can clearly see the sigmoidal shape as the percolation front propagates from the higher to the lower in-degree nodes. Eventually, it should be stressed that, though the percolation formalism helps us understand the inner structure of the burst, only the last SBSs can be described as a "pure" percolation phenomenon. This can be understood from the 2nd SBS: because of the extension of the 1st SBS, the resulting "initially activated fraction" is not clearly defined and it looks like two percolation processes are interfering. We can see that the structure becomes clearer on the last SBSs where the activity of the whole population occurs on a unique and longer timescale, in a decreasing in-degree order, and follows the initial activation of the highest in-degree neurons.

6. Conclusion

We elaborated extensions to the original Quorum Percolation model by introducing two additional neurobiological properties; we studied their effects on the activity of the networks. In each case, we were able to construct a mean-field theory in good agreement with Monte Carlo explicit simulations within a wide parameter range, which is specified in this paper. The central idea that enabled us to derive these mean-field approaches is the mapping of the network onto equivalent ones related to the original QP model, but exhibiting a different connectivity. A main point is the close relation between excitability and connectivity, between decay and connectivity. These two effects impact the position of the critical point in a manner that can remain unnoticed. Hence deciphering the functional connectivity of the network using percolation methods is more difficult than expected. Yet, the study of bursts with the help of a two dimensional dynamical model gives evidence that the ideas of percolation are worth being kept in mind within such a framework. At last, although the Quorum Percolation models we set out in this paper have been designed to describe what triggers burst of activity in neuronal networks, they are also relevant in other scientific fields (propagation of rumors, ecology, sociology...) involving information propagation throughout networks with a threshold rule.

Acknowledgments
This work was granted access to HPC resources of IDRIS (Orsay) under the allocation 2015057043 made by GENCI (Grand équipement national de calcul intensif). We are grateful to Elisha Moses, Yaron Penn for helpful discussions.

References
[1] J P Eckmann, O Feinerman, L Gruendlinger, E Moses 2007 J Soriano, and T Tlusty, *Phys. Rep.* **449** 54
[2] S Stern, M Segal and E Moses 2015 *EBioMed.* **2** (9) 1048
[3] R Renault 2015 *Emergent design of neuronal devices* PhD thesis, Université Paris Diderot
[4] I Breskin, J Soriano, E Moses and T Tlusty 2006 *Phys. Rev. Lett.* **97**, 1
[5] G W Gross, B K Rhoades, D L Reust, F U Schwalm 1993 *J. Neurosci. Methods* **50** 131
[6] K R Gee, K A Brown, W N Chen, J Bishop-Stewart, D Gray, and I. Johnson 2000 *Cell Calcium* **27** 97
[7] R Renault, N Sukenik, S Descroix, L Malaquin, J L Viovy, J M Peyrin, S Bottani, P Monceau, E Moses and M Vignes 2015 *PLoS ONE* **10** 4: e0120680
[8] Y Penn, M Segal, E Moses 2016 *PNAS* **113** (12) 3341
[9] R Renault J B Durand J L Viovy and C Villard *Lab Chip* **16** 2188
[10] J Soriano, M Rodriguez Martinez,T Tlusty and E Moses 2008 *PNAS* **105** (37), 13758
[11] E M Izhikevich 2007 *Dynamical systems in neuroscience* (Cambridge Massachusetts: MIT Press)
[12] F A Rodrigues, T K D M Peron, Peng Ji, J Kurths 2016 *Phys. Rep.* **610** 1
[13] O Cohen, A Keselman, E Moses, M Rodriguez Martinez, J Soriano, and T Tlusty 2010 *Eur. Phys. Lett.* **89** 18008
[14] D Stauffer and A Aharony 1994 *Introduction to percolation theory* (London: Taylor and Francis)
[15] R Renault, P Monceau, S Bottani and S Métens 2014 *Physica* A **414** 352
[16] S Métens, P Monceau, R Renault and S Bottani 2016 *Phys. Rev.* E **93** 032112
[17] S Wiseman and E Domany 1998 *Phys. Rev. Lett.* **81** 22
[18] P Monceau 2011 *Phys. Rev.* E **84** 051132
[19] J P Eckmann, E Moses, O Stetter, T Tlusty and C Binden, *Front. Comput. Neurosci.* **4**, 132
[20] M Isokawa 1997 *Brain Res.* **1**(2) 114
[21] J R Geiger, J Lübke, A Roth, M Frotscher, and P Jonas 1997 *Neuron* 18(6) 1009
[22] R Renault, P Monceau, and S Bottani 2013 *Phys. Rev.* E **88** 062134
[23] R Brette and W Gerstner 2005 *J. Neurophysiol.* **94**(5) 3637
[24] S Bottani 1996 Phys Rev E . 1996 textbf 54(3) 2334-2350.
[25] F Lombardi, H J Herrmann, D Plenz and L De Arcangelis 2014 **8** Frontiers in Neuroscience doi: 10.3389/fnsys.2014.00204
[26] V S Sohal and J R Huguenard 2003, **2003** : 8978-8988;

Measurement of the main and critical parameters for optimal laser treatment of heart disease

FB Kabeya[1], H Abrahamse[2], AE Karsten[3]

[1]University of Johannesburg, Faculty of Health Sciences, Laser Research Centre, South Africa
[2]University of Johannesburg, Faculty of Health Sciences, Laser Research Centre, South Africa
[3]National Metrology Institute of South Africa, South Africa

Francoiskb177@gmail.com; habrahamse@uj.ac.za; akarsten@nmisa.org

Abstract. Laser light is frequently used in the diagnosis and treatment of patients. As in traditional treatments such as medication, bypass surgery, and minimally invasive ways, laser treatment can also fail and present serious side effects. The true reason for laser treatment failure or the side effects thereof, remains unknown. From the literature review conducted, and experimental results generated we conclude that an optimal laser treatment for coronary artery disease (named heart disease) can be obtained if certain critical parameters are correctly measured and understood. These parameters include the laser power, the laser beam profile, the fluence rate, the treatment time, as well as the absorption and scattering coefficients of the target treatment tissue. Therefore, this paper proposes different, accurate methods for the measurement of these critical parameters to determine the optimal laser treatment of heart disease with a minimal risk of side effects. The results from the measurement of absorption and scattering properties can be used in a computer simulation package to predict the fluence rate. The computing technique is a program based on the random number (Monte Carlo) process and probability statistics to track the propagation of photons through a biological tissue.

1. Introduction

Laser light treatment modalities in the health and beauty fields are increasing rapidly worldwide. Some of these modalities need to penetrate through some layers of the target treatment organ [1]. A large number of them fail to achieve the goal and sometimes bring serious side effects such as burning the no-atherosclerotic cells, scarring and permanent depigmentation [2]. Some succeed after a very long time of treatment [3]. Until now, the reasons for laser light treatment failure are not well understood.

The aim of this paper is to address these challenges, and to provide a method to address the optimal laser treatment of heart disease or any disease with reduced risk of side effects. For this, the authors estimate that the better measurement of the six key parameters (the initial laser power, the laser beam profile, the fluence rate, treatment time, and the absorption and scattering coefficients of the tissue) and better application of the results (from the measurements) to patients will bring the expected results. The main challenge in the measurement of these critical parameters remains on the methods used for achieving the correct results. Another challenge in the application can be attributed to the quality of the clinicians in charge of the patients' treatment.

2. The parameters involved in laser treatment of heart disease

A good understanding of the parameters involved in the process remains critical for the optimal use of laser as a treatment modality [4]. The optical properties of the coronary artery tissue (CAT) and the laser light propagation within the three layers of the arteries have been investigated by a number of researchers, [5]; [6]; [7]; [8]; [9]; [10]; [11]; [12; [13]; [14]; [15]; [16]; [17]. These properties are important as they determine the reduction in the laser fluence rate as it propagates through the CAT.

In discussing the optical properties of the coronary artery, it is important to distinguish between the fundamental optical properties and the transport properties used in the modelling work [18] as listed in Tables 2.1 and 2.2.

Table 2.1 Fundamental properties of tissue

Quantity	Symbol	Unit
Refractive index	η	Dimensionless
Scattering coefficient	μ_s	cm^{-1}
Absorption coefficient	μ_a	cm^{-1}
Anisotropy of scatter	g	dimensionless

Table 2.2 Transport properties of tissue

Quantity	Symbol	Unit
Reduced scattering coefficient	$\mu'_s = (1\text{-}g)\ \mu_s$	cm^{-1}
Transport Mean Free Path (MFP)	$MPF = 1/\ \mu_{a+}\mu_s$	cm

According to Vo-Dinh [19], the three photo-physical properties (refraction, scattering and absorption) are important to describe the propagation of light in biological tissue; and only the absorption coefficient, scattering coefficient, and reduced scattering coefficient are recommended for the measurement of the critical parameters for an optimal treatment of the coronary artery.

The following paragraphs will explain the critical parameters involved in laser treatment of heart disease.

According to Welch [20], the fluence rate is defined as the rate of energy delivered per unit area at a specific position. It is expressed in W/cm2 and is determined by the optical properties of the tissue. In order to provide the correct laser irradiance dose (J/cm2), it is advisable to determine the fluence rate reaching a required depth into the CA.

The beam profile is known as the effective area or diameter of the laser beam [21]. The initial laser power is the power measured as emitted by the laser.

The absorption of a tissue is defined as the ability of the tissue to absorb the photons. It is expressed in cm-1, which is not an SI unit but practical unit. Mathematically, it is defined as follows: $\mu a = \rho a \sigma a$ where ρa is the volume density of chromophores contained in the tissue; and σa is an effective cross-sectional area [22], [23]. Scattering in tissue is generally caused by morphological variations in the tissue density and the refractive index [23]. It is mathematically expressed as, where and σs are respectively the volume density and effective cross-sectional area of scatterers [22], [23]. The last critical parameter for optimal laser treatment named 'treatment duration' is defined as the time that laser is on (applied) during the treatment.

IOP Conf. Series: Journal of Physics: Conf. Series **905** (2017) 012009 doi:10.1088/1742-6596/905/1/012009

3. Measurement methods of the critical parameters for optimal laser treatment of heart disease

In this section, the main focus of this paper is briefly shown, that is the theory and results that are based on the measurement or calculation of the required six critical parameters for an optimal laser treatment. Currently, there are more than three hundred thousand types of laser on the market [21]. Not all of them can be unusable for CA treatment. Therefore, the initial power and beam profile are important criteria in the choice of those that can be used to achieve the goal

3.1. Initial laser power measurement

Here, the authors advise the use of a calibrated power meter to determine the initial power of the laser.

3.2. Laser beam profile measurement

According to Stratan [21], charge-coupled device camera-based beam profilers can be used as a method to measure the diameter of the laser beam. The challenge in this method is the presence of two factors: the background noise and the integration area selected in the software, which affect the accuracy of the measurement. Using MATLAB, these two challenges can be evaluated and reduced. The diameter of the atherosclerotic tissue can also be measured using the routine technology method named coronary angiography, which is a procedure performed to view the inside of the vessels after injecting them with a radiopaque dye that outlines them on x-rays [24]. This method is considered an 'indirect method' because the measured organ's diameter is not that of the atherosclerotic substance that blocks the artery's lumen but the interior diameter of the artery. From several experiments done at the Council for Scientific and Industrial Research (CSIR), the laser beam profile diameter must logically be lightly smaller or equal to the diameter of the atherosclerotic tissue.

Using the correct laser beam profile diameter is important. The non-atherosclerotic cells will be damaged when the laser beam is bigger than the atherosclerotic substance. The treatment will take more time when its diameter is smaller than that of the blockage.

3.3. Fluence rate determination

In vivo, measurements to determine the fluence rate are seldom practical [25]. A computer modelling has capability to predict the fluence rate at any level into the CA tissue once the optical properties of the target treatment tissue are introduced [26]. This prediction of the fluence rate with a computer modelling is done by tailoring the properties of laser to the properties of the tissue.

3.4. Treatment time measurement

Treatment time is calculated mathematically using the thermal relaxation of the tissue being treated [27]. It is determined through the thermal relaxation time (time after which the heated tissue has capability to drop half of its heat) of the tissue under consideration. This thermal relaxation time (TRT) depends on the diameter of the tissue to be treated. Using an integrating sphere experimental technique, the values of the optimal properties of the CAT can be determined. With these values, the TRT will be calculated by using the layered computer model.

3.5. Absorption and scattering coefficient measurement

One of the following methods can be used to measure the important recommended optical properties for the coronary artery treatment: optical coherence topography or integrating sphere. Figure 1 shows the integrating sphere method used by the authors in December 2015.

Figure 1. Measurement of the optical properties of a coronary artery using a computer package (December 2015).

4. Conclusion

As heart disease is the principal cause of death worldwide [28], and its treatment requires more attention, the authors have presented the critical parameters that must be considered when laser treatment is recommended for a patient. They have also presented different estimated methods to use for their measurements. Therefore, the authors believe that the correct measurement of the optical properties (absorption, scattering) of the coronary artery tissue and the six parameters estimated critical for the treatment are critical for optimal heart treatment. Another vital component to achieve optimal heart treatment is the application or administration of all the results from measurements to patients.

Due to a lack of data available on the absorption and scattering coefficients for CAT, different prototypes of CAT and of any atherosclerotic substance can be prepared in order to standardise and validate the technology. This means that the phantoms' optical properties must be predictable and adjustable to represent different layers of the CA and atherosclerotic substance. According to Dhanaliwala [29], the standardization requires phantoms that mimic actual tissue optical properties with a biological contrast agent, and are durable in order to serve as a reference to compare systems over time, and possess elastic properties similar to that of arteries.

For this reason, the authors of this article advise to fabricate the phantoms with the intravascular optical coherence tomography method (IV OCT) because this method is acknowledged for clinical use in different developed countries such as United States and Japan, as well as continents like Europe [6]. Different types of artery phantoms have been previously developed and presented by some researchers [8]. To present the optical and physical properties of the three layers (intima, media and adventitia) of the coronary artery, the authors advise to use the same support chemical material used previously by the researchers cited above for the responding phantoms. The chemical materials advised to be used include silicone (Sylgard 184, Dow Corning) and pure poly (polydimethyl siloxane) (PDMS) (Fluid 200, 50 CST, Dow Corning).

The computer model results must be compared favourably with laser measurements performed on the CA. These results from the computer model depend on both the accuracy of the model and the accuracy of the input data (optical properties). If the data on absorption coefficient for different prototypes is not readily available, it will necessitate the development of a diffuse reflectance probe system, which can be used to measure the absorption coefficient of the CA *in vivo*.

References

[1] Karsten AE, Singh A and Braun JS, 2012: 42. *The effect of skin phototype on laser propagation through skin*. South Africa.

[2] Tanzi EL, Alster TS 2003. *Laser surgery in dark skin* [Online]. Available from www.ncbi.nlm.nih.gov/pubmed/1467330. The Washington Institute of Dermatologic Laser. [Accessed: 15/11/2015].

[3] Tomi 2007:147 – 158. Intravascular lasers in the treatment of varicose veins [Online]. Available from *https://www.researchgate.net/.../6665123_Intravascular_**lasers**_in_the_**treatment**_of_vari*. Accessed: 10 /10 / 2015

[4] Manuskiatti W, Richard E, Fitzpatrick MD; Mitchel P and Goldman MD 2001. *Er:YAG Lasers in the Same Treatment Session* [Online]. Available from www.ncbi.nlm.nih.gov/pubmed/1467330. [Accessed: 02 / 04 / 2016]. USA.

[5] Fotakis C, Papazoglou TG, Kalpouzos C. 2012. *Optics and Lasers in Biomedicine* and *Artery Phantomes for intravascular optical coherence tomography [Online].* Available from https://www.osapublishing.org/boe/abstract.cfm?uri=boe-2-9-2599. [Accessed: 15/06, 2016] Beijing. China.

[6] Bisaillon CE, Dufour, ML, and Lamouche, G 2011. *Multilayered disease-mimicking bladder phantom with realistic surface* [Online]. Available from www.proceedings.spiedigitallibrary.org/proceeding.aspx?articleid=1840539. [Accessed: 01 /02 / 2016].

[7] Nordstrom, RJ, & SPIE (Society). (2010). Design and performance validation of phantoms used in conjunction with optical measurement of tissue II [Online]. Available from https://searchworks.stanford.edu/catalog/citation?id=8651424. [Accessed: 23/01/2016].

[8] Pogue BW and Patterson MS 2006 *Review of tissue simulating phantoms for optical spectroscopy, imaging and dosimetry*. Journal of Biomedical Optics 11, no 4 (2006): 041102 1-041102-16.

[9] Cheong WF, Prahl SA and Welch AJ 1990 *A Review of the Optical Properties of Biological Tissues* IEEE J. Quantum Electronics 26 (1990): 2166-2185 USA.

[10] Martelli F, Del Bianco S, Ismaelli A and Zaccanti G 2010 Light propagation through Biological Tissue and Other Diffusive Media ISBN 978-0-7658-6. SPIE Press.

[11] Bissaillon CE, Marc LD and Lamouche G 2011 *Artery phantoms for intravascular optical coherence tomography:healthy arteries* [Online] Available from www.ncbi.nlm.nih.gov/pm/a [Accessed: 11 / 01 / 2014].

[12] Metwally MK, El-Gohary SH, and Byun KM 2014 *influence of optical fluence distribution on photoacoustic imaging [Online]* Available from www.ncbi.nlm.nih.gov/pm/a [Accessed: 11 /11 / 2015].

[13] Suter MJ, Nadkami SK and Weisz G 2011 *Intravascular optical imaging technology for investigating the coronary artery* [Online] Available from https://scholar.google.co.za. [Accessed: 05 / 05 /2015].

[14] *Campbell EEB, Hansen K, Hoffmann K, Korn G 2000 The photoelectron spectra of C 60 ionized using a 790 nm laser with pulse durations varying from 25 fs to 5 ps* [Online]. *Available from http//:spie.orgx648.xml?product_id=21777832. Accessed: 16/06/2016.*

[15] Wang XC, Wang S, Karunasiri G, Xu SJ, Chua SJ, Mei T, Zhang XH, Fan WJ, Wang CH and Jiang J 2015. *Photonics, Nanophotonic Structures and Materials.* David LA China.

[16] Tuchin V 2007 *Tissue Optics: Light Scattering Mehodsand instruments for Medical Diagnostics. 2nd edition, p 165-175. SPIE Press*

[17] Jiang L, Xin Li, Xi-Peng Zhang and Qiang-Hua Chen 2003 *Laser Micro/Nano Fabrication*

Laboratory. Available from http://web.mst.edu/tsai/published%papera/jiang- Accessed: 20/03/2016.

[18] Jacques S, 2008(b) *Modelling tissue optics using Monte Carlo modelling: a tutorial.* "Edited by Roach WP, Thomas RJ

[19] Vo-Dinh T 2003 *SERODS Optical Data Storage with Parallel Signal Transfer.* US

[20] Welch A and van Gemert MJC 2011:29 *Optical-Thermal* Response of Laser-Irradiated Tissue. 2nd Springer

[21] Stratan A, Alexandru Zorila, Laurentiu Rusen, L, and Gearge N: 2014. "*Measuring effective are of spots from pulsed laser beams". [*Online]. Available from www.opticalengineering.spiedigitalli. Accessed: 26 / 05 / 2016

[22] Jacques SL and Prahl SA 1998 (a) *introduction to Biomedical Optics [*Online]. Available from http://omlc.ogi.edu/classroom/ece532/class3/index.html (accessed: November 2015).

[23] Dam JS, Dalgaard P, Fabricius PE and Anderson E 2000 (b) *Multiple polynomial regression method for determination of biomedical optical properties from integrating sphere measurements.* Applied Optics 39 (2000(a)): 1202-1209.

[24] Novelline, RA, Squire, LF 2004. *Bronchoscopy 594 Coronary Calcium Scoring 596 Coronary CT Angiography 598.* [Online]. Available from https://book.google.co.za/books?hl=en&id..... Accessed: 03/ 03/ 2016.

[25] Peng Q, Jezeniene A, Chen J, O'Svaasand L, Warloe T, Giercksky K and Moan J 2008 *Lasers in Medicine. Rep Prog Physics 71 2008: 1 – 28 (doi: 10.1088/0034-4885/71/5/056701).*

[26] Karsten 2012: 2-4 *Experimental verification and validation of a computer model for light-tissue interaction.* Lasers Med Sci 27 (2012a): 79-86.

[27] Alexandru Z, Stratan A and Laurentiu R 2014. *Measuring effective area of spots from pulsed laser beams* (Journal Paper) [Online]. Available from http//:spie.orgx648.xml?product_id=21777832. Accessed: 16/06/2016.

[28] David R 2009 *Laser therapy in cardiovascular disease* [Online] Available from http://dx.doi.org/10.1117/12.808050 Accessed: 10 / 10 / 2015.

[29] Dhanaliwala AH 2013 *liquid Flooded Flow-Focusing Microfluidic device for in Stu...NCBI.*[Online]. Available from www.ncbi.nih.gov.NCBI.Literature.

Accurate analytical expression of the electrostatic potential close to a grid placed between two plates

G Orjubin

F´SATI, Cape Peninsula University of Technology, Bellville 7535, South Africa

Email: orjubing@cput.ac.za

Abstract. The potential distribution between a grid and two plates is an electrostatic problem already solved for MultiWire detectors used in Nuclear Physics, where theoretical expressions are obtained assuming linear charge distributions. In this paper, the accuracy of the line model is investigated close to the wires, using both analytical and numerical approaches. In the symmetric case where the grid is at equal distances between two grounded plates, it is shown that the error can be modeled using a quadrupole charge. After solving the electrostatic problem for the asymmetric case, a larger discrepancy is found, and is interpreted as a consequence of a dipole charge on the electrode. Two different models of this dipole are then implemented, leading to an accurate theoretical expression of the potential.

1. Introduction

The potential distribution created by a conductive grid is a classical electrostatic problem solved several decades ago [1]. In engineering, its solution is used to model systems containing a grid, *e.g.* electrostatic precipitators [2]. Another application is the MultiWire Proportional Counter (MWPC) invented in the '70s for Nuclear Physics [3]. This case has been intensively investigated, the analytical expression of the potential distribution being derived from a line model of the electrodes [4], thus considering wires as lines instead of cylinders.

The validity of this assumption is discussed in this paper. As a reference for precise determination, Laplace's equation is solved numerically using Comsol Multiphysics®, a commercial Finite Element Model (FEM) package.

Recent literature on this topic puts a focus on the numerical approach [5], particularly useful when space charge effects are considered [6]. However, accurate analytical expressions are always necessary to design a prototype containing conductive grids, justifying the investigation presented herein. Moreover, while expressions given in [4] only apply when the grid is placed between two grounded plates, our results are readily usable for an arbitrary configuration.

The paper is organized as follows: in section 3, a classical formula of the symmetric case (*i.e.* grid at equal distance between grounded plates) is recalled, and a first study of the potential on the conductor surface is done using both theoretical and numerical approaches. Then the general case, or asymmetric, is considered in section 4: after solving the electrostatic problem, it is seen that a dipole behavior of higher strength characterizes the line expression used for MWPC. In order to cope with this small but not negligible error, two models of this dipole are then presented and discussed.

2. Geometry

The geometry under investigation is described in figure 1.

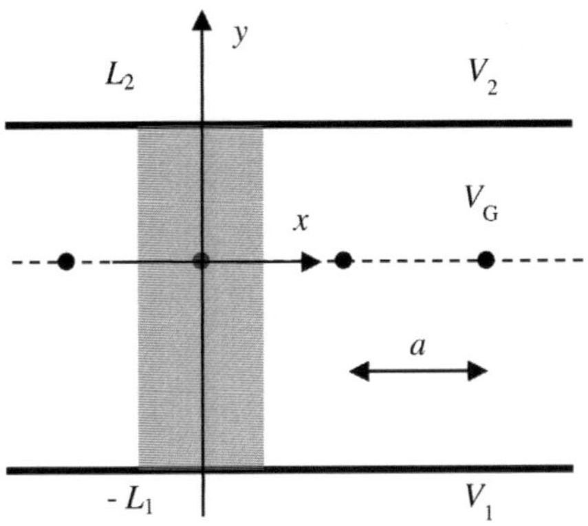

Figure 1. Cross-section of the grid with the two parallel plates.

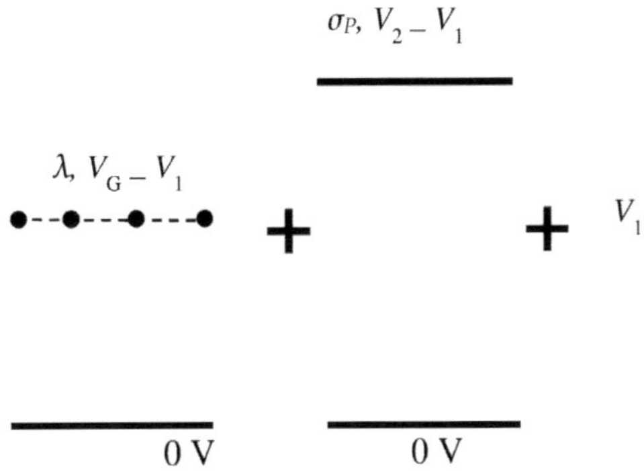

Figure 2. Linear superposition used in the appendix.

Electrodes are situated on the line $y = 0$ at $x = ka$ where k is an integer. These wires are cylinders out-of-the plane x-y with radius r_0. The distance between two parallel wires is a. Their charge per unit length, λ, depends on the potentials V_1, V_2 and V_G applied to the conductors.

To solve this 2D electrostatic problem it is convenient to use the complex-valued variable

$$\xi = \pi \frac{z}{a} \tag{1}$$

with $z = x + jy$, as well as the complex electrostatic potential $\Phi(z)$ such as $V(x, y) = \mathrm{Re}[\Phi(z)]$.

Throughout this paper, $a = 1\,\mathrm{mm}$ and $r_0 = 12.5\,\mu\mathrm{m}$.

3. Symmetric case

3.1. General results

In the symmetric case, the grid is located midway between the two grounded plates, thus $L_1 = L_2 = L$ and $V_1 = V_2 = 0$. The expression of the complex potential can be found in [4], where unit linear charge and CGS units were adopted. With a linear charge λ and MKS units, this potential is

$$\Phi_{line} = \frac{L\lambda}{2\varepsilon_0 a} - \frac{\lambda}{2\pi\varepsilon_0} \log\left(2 \frac{\sin\xi - q^2 \sin 3\xi + q^6 \sin 5\xi - ...}{1 - 2(q\cos 2\xi - q^4 \cos 4\xi + ...)} \right) \tag{2}$$

with $q = \exp\left(\frac{-4\pi L}{a} \right)$.

In our case, $L = 10\,\mathrm{mm}$, so $q = 2.7\times10^{-55}$, a negligible value that leads to the following approximation

$$\Phi_{line} \cong \frac{L\lambda}{2\varepsilon_0 a} - \frac{\lambda}{2\pi\varepsilon_0} \log(2\sin\xi). \tag{3}$$

The potential on the wire surface is $V_{line}(z = r_0\, e^{j\theta})$ and can be expressed using the fact that $r_0 << a$; the first order approximation $\sin\xi \cong \xi$ yields a uniform distribution, *i.e.* V_{line} does not depend on θ. This constant is the potential V_G applied to the conductor:

$$V_G = \frac{L\lambda}{2\varepsilon_0 a} - \frac{\lambda}{2\pi\varepsilon_0}\log\rho \tag{4}$$

with $\rho = \frac{2\pi r_0}{a}$.

3.2. High order approximation on the wire

The third order approximation of $\sin\xi$ in equation (3) actually leads to

$$\Phi_{line} \cong \frac{L\lambda}{2\varepsilon_0 a} - \frac{\lambda}{2\pi\varepsilon_0}\log\rho + \frac{j\theta\lambda}{2\pi\varepsilon_0} + \frac{\lambda}{12\pi\varepsilon_0}\left(\frac{\pi r_0}{a}\right)^2 e^{2j\theta}. \tag{5}$$

The corresponding physical potential on the wire is the real part of this, and can be written as

$$V_{line}(z = r_0 e^{j\theta}) = V_G + V_{0Quad}\cos 2\theta, \tag{6}$$

with $V_{0Quad} = \frac{\lambda}{12\pi\varepsilon_0}\left(\frac{\pi r_0}{a}\right)^2$.

The second term of equation (6) can be interpreted as the contribution of an electric quadrupole [7], and finds its origin in the fact that equipotential lines derived from equation (3) are not exactly circles.

3.3. Numerical validation using FEM

The commercial FEM-based package Comsol Multiphysics ® is used to solve the Laplace´s equation in the shaded domain represented in figure 1. Boundary conditions are V_G = 10 V on the grid, 0 V on the plates, and the zero charge condition $\vec{E}.\vec{n} = 0$ is implemented on the lateral faces. In order to improve the resolution, the finer mesh is selected, and 4th order shape functions are used during discretization. Finally, the numerical problem with 81×10^3 degrees of freedom is solved in 3 s CPU time (dual core I7, 32 Go RAM) using a direct solver. The accuracy of the numerical analysis is 10^{-9}.

The difference between the almost exact numerical FEM solution and the approximate line model is plotted in figure 3.

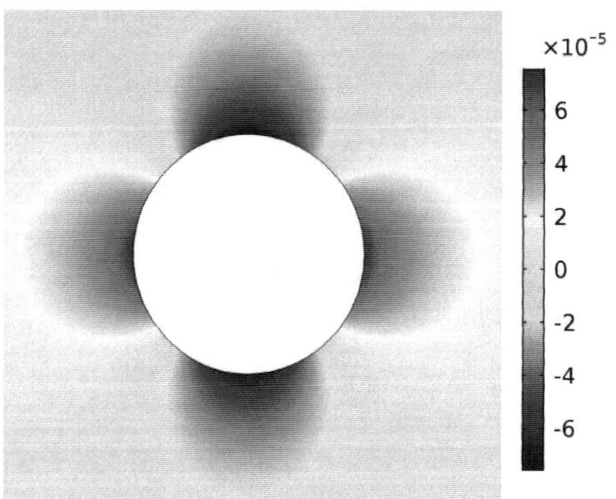

Figure 3. Surface plot of $V_{FEM} - V_{line}$. Close view around the wire for a symmetric case.

This plot illustrates the quadrupole behaviour previously explained in section 3.2. The amplitude $V_{0Quad} = 6.5 \times 10^{-5}$ V of this quadrupole is well predicted by equation (6).

4. Asymmetric case

The asymmetric case, where the grid at potential V_G is not situated midway between two parallel plates at potentials V_1 and V_2, is now presented.

4.1. General equations

As mentioned in [4], equation (3) is the expression for an isolated grid and the constant term enables a matching on the wire. Because the field created by an isolated grid is symmetric, this kind of expansion is not relevant for the asymmetric case. To solve this electrostatic problem, generic methods such as superposition, images and matrix of capacitance can be used [3]. Developments are done in the appendix, leading to the following expression of the complex potential created by this line distribution (placed at $y = 0$) under the influence of the plates

$$\Phi_{line} = V_1 - \frac{\lambda}{2\pi\varepsilon_0} \log \frac{\sin\xi}{\sin\left(\xi + \dfrac{2j\pi L_1}{a}\right)} - j\frac{\sigma_P}{\varepsilon_0}(z + jL_1), \tag{7}$$

where the induced charges λ and σ_P are given in equations (A4-5) as functions of V_G, V_1, and V_2.

4.2. Approximation on the wire

When approximating equation (7) on the wire, one finds that

$$\Phi_{line}(r_0\, e^{j\theta}) \cong V_1 + \frac{\lambda L_1}{\varepsilon_0 a} - \frac{\lambda \log\rho}{2\pi\,\varepsilon_0} + \frac{\sigma_P L_1}{\varepsilon_0} - \frac{\lambda}{2\pi\varepsilon_0}\left(-j\frac{\pi}{2} + j\theta + j\xi - \frac{1}{6}\xi^2\right) - j\frac{\sigma_P r_0\, e^{j\theta}}{\varepsilon_0}. \tag{8}$$

Using the expression of the potential V_G on the grid given in equation (A3), the real potential becomes

$$V_{line}(r_0 e^{j\theta}) \cong V_G + \frac{r_0}{\varepsilon_0}\left(\frac{\lambda}{2a} + \sigma_P\right)\sin\theta + \frac{\lambda}{12\pi\varepsilon_0}\left(\frac{\pi\, r_0}{a}\right)^2 \cos 2\theta. \tag{9}$$

Similar to the symmetric case, the line model presents a quadrupole contribution $V_{quad} = V_{0Quad}\cos 2\theta$ with the same amplitude seen in equation (6) for the symmetric case. Moreover, it is interesting to notice that a dipole component

$$V_{dip} = V_{0Dip}\sin\theta \tag{10}$$

also exists, with the amplitude $V_{0Dip} = \dfrac{r_0}{\varepsilon_0}\left(\dfrac{\lambda}{2a} + \sigma_P\right)$.

It can be noted that for the symmetric case, equations (A4-5) lead to $\sigma_P = -\dfrac{\lambda}{2a}$, meaning that equation (9) reduces to equation (6).

4.3. FEM results

The studied asymmetric configuration is characterized by $L_1 = 10\,\text{mm}$, $L_2 = 5\,\text{mm}$, $V_1 = -5\,\text{V}$, $V_2 = 10\,\text{V}$, and $V_G = 2\,\text{V}$. Figure 4 shows the difference between the field V obtained by FEM and the theoretical one associated to a line distribution, *i.e.* equation (7).

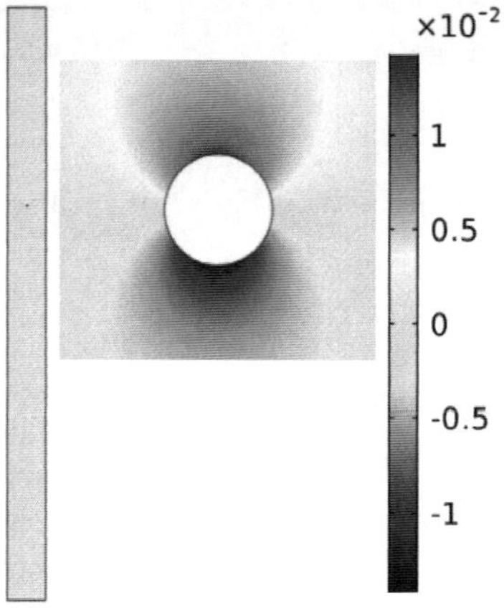

Figure 4. General and close view around the wire of $V_{FEM} - V_{line}$ for an asymmetric case.

Figure 4 exhibits a very small difference between V_{FEM} and V_{line} over the computational domain, except close to the wire where a dipole behaviour is noted. The maximum error is actually equal to $V_{0Dip} = 1.4 \times 10^{-2}$ V, which is in agreement with equation (10).

It is noticeable that this error is much larger than the one obtained for the symmetric case. It is therefore of interest to investigate this case and model this dipole behaviour.

4.4. Dipole modelling

It is well-known [1] that a grounded cylindrical conductor placed in an external electric field $\vec{E}_{ext}$ behaves like an electric dipole. In our case $\vec{E}_{ext} = -E_{ext}\,\vec{u}_y$ for $y < 0$. Solving the Laplace's equation for this classical problem gives $V = E_{ext}\left(r - \frac{r_0^2}{r}\right)\sin\theta$: We recognize the external potential $V_{ext} = E_{ext}\, r\sin\theta = E_{ext}\, y$ and an additive term $V_{dip} = -E_{ext}\frac{r_0^2}{r}\sin\theta$ that can be written as $V_{dip} = V_{0Dip}\frac{r_0}{r}\sin\theta$, with $V_{0Dip} = -E_{ext} r_0$. In terms of complex potential, this classical model of the dipole is described by

$$V_{dipA} = \mathrm{Re}\left(jV_{0Dip}\frac{r_0}{z}\right). \tag{11}$$

With this expression of the distribution created by the dipole, it becomes possible to eliminate the artefact of the variation of the potential on the wire. This artefact is due to the use of equation (7) that does not describe the wire geometry perfectly (it should be cylinder, not line). To check the validity of this model's dipole referred to as Model A, or *dipA*, the difference between $(V_{line} - V_{dipA})$ and V obtained by FEM is plotted in figure 5.

Compared to figure 4, figure 5 illustrates the improvement of the analytical expression but an error of 2×10^{-4} V still remains. It is also noticed that the error on the wire features a $V_{0Quad}\cos 2\theta$ variation, in agreement with our findings.

4.5. Improved dipole modelling

In spite of the good results found in the previous section, a discrepancy can be noted in figure 5 at large distances from the wire. In order to improve the modelling, the technique presented in [8] is adopted. First, the expression of the complex potential of a line distribution (with distributed charge λ placed at z_0) is recalled:

$$\Phi_l = \frac{-\lambda}{2\pi\varepsilon_0}\log(z - z_0). \tag{12}$$

A line dipole constituted of two opposite distributions placed at $z_0 \pm jy_0$ is then represented by

$$\Phi_d = \frac{-\lambda}{2\pi\varepsilon_0}\log\frac{z - jy_0}{z + jy_0}. \tag{13}$$

For $|z| >> |y_0|$, the following approximation holds

$$\Phi_d \cong \frac{p_0}{2\pi\varepsilon_0}\frac{1}{z}, \tag{14}$$

where $p_0 = 2jy_0\lambda$ is purely imaginary as the dipole is on the *y*-axis.

It has been noted in [8] that the distribution Φ_d of the line dipole is proportional to $-\frac{d\Phi_l}{dz}$, a property also reported in [9]. Generalizing to our case, one first considers the line model of the grid described by expression (A1); then the line dipole of the grid is obtained through differentiation:

$$\Phi_{dip} = \frac{p_0}{2\pi\varepsilon_0}\frac{\pi}{a}\frac{\cos\xi}{\sin\xi}. \tag{15}$$

Adopting this expression, the approximation of the complex potential on the wire is

$$\Phi_{dip}(r_0 e^{j\theta}) \cong \frac{p_0}{2\pi\varepsilon_0}\frac{\pi}{a}\frac{1}{\xi} = \frac{2jy_0}{2\pi\varepsilon_0}\frac{1}{z}, \tag{16}$$

leading to

$$V_{dip} = V(r_0 e^{j\theta}) \cong \frac{y_0}{\pi\varepsilon_0}\frac{1}{r_0}\sin\theta. \tag{17}$$

Identifying equation (17) with equation (10) gives $\frac{y_0}{\pi\varepsilon_0 r_0} = V_{0Dip}$. Then expression (15) becomes $\Phi_{dip} = jV_{0Dip}\frac{\pi r_0}{a}\cot\xi$, and finally

$$V_{dipB} = V_{0Dip}\frac{\pi r_0}{a}\mathrm{Re}(j\cot\xi). \tag{18}$$

Equation (18) expresses the dipole contribution when the conducting cylinder is placed in the non-homogeneous external field created by the grid between the plates.

In order to check the validity of this model, the surface plot of the difference between V_{FEM} and $(V_{line} - V_{dipB})$ is given in figure 6.

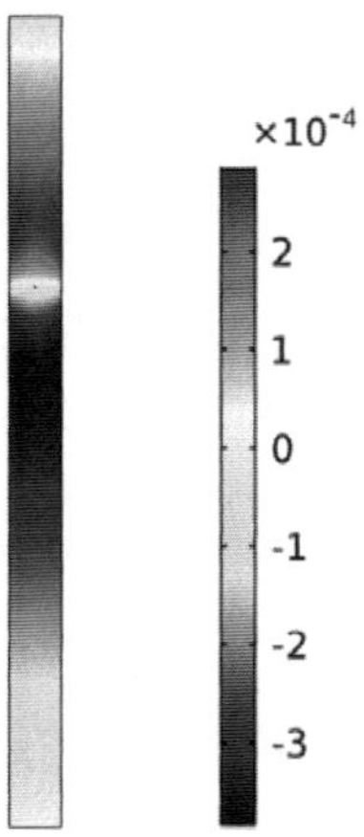

Figure 5. Model A of the dipole for an asymmetric case: surface plot of $V_{FEM} - (V_{line} - V_{dipA})$.

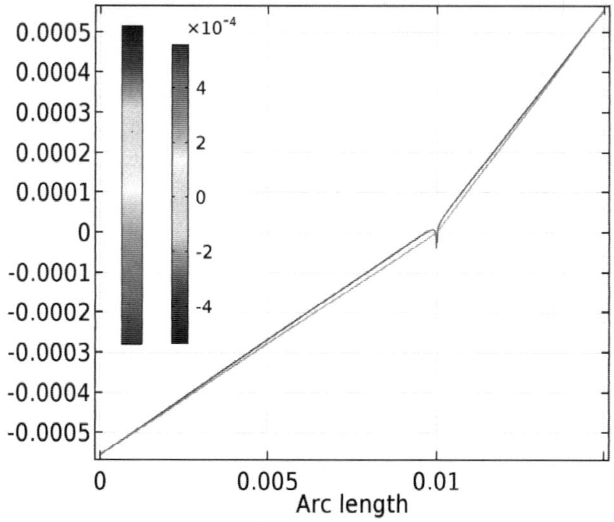

Figure 6. Model B of the dipole for an asymmetric case. $V_{FEM} - (V_{line} - V_{dipB})$ along the line $x = 0$. Inset: General view over the computational domain.

The comparison between figures 5 and 6 shows that both models A and B yield an error with same order of magnitude. However, the error of model B features a linear variation due to the fact that $\mathrm{Re}(j\cot\xi)\xrightarrow[|y|\to\infty]{}\pm 1$. Model B can then be improved by subtracting a linear term V_{dip_corr} of amplitude $V_{0Dip}\dfrac{\pi r_0}{a}$. The residual error is shown in figure 7.

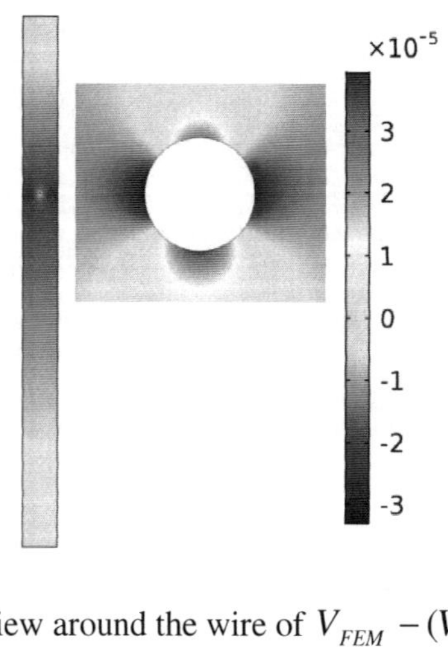

Figure 7. General and close view around the wire of $V_{FEM} - (V_{line} - V_{dipB} - V_{dip\_corr})$ for an asymmetric case

This new model has improved the theoretical determination that now leads to an error of 3×10^{-5} V.

IOP Conf. Series: Journal of Physics: Conf. Series **905** (2017) 012010 doi:10.1088/1742-6596/905/1/012010

5. Conclusion

The potential distribution around conductors of a grid has been studied from a well-known formula given by Erskine for MultiWire detectors. In the symmetric case where the grid is placed midway between two grounded plates, it has been shown that the error due to the almost but not exactly circular equipotential has a quadrupole-like contribution. This point has been checked through Finite Element Model analysis. After solving the electrostatic problem for the general geometry, it has been proved that a dipole contribution also exists in this case with a rather large amplitude. This dipole was then studied using a classical model and an advanced technique suitable for this geometry has been proposed. Finally this paper provides corrective terms of Erskine´s equation that yield at least a 10^{-6} accuracy, as reported in table 1.

Table 1. Order of magnitude of the relative error between FEM (with 10^{-9} error) and different models.

Symmetric	Equation (3)	10^{-6}
Symmetric with model of the quadrupole	Equations (3), (6)	10^{-8}
Asymmetric	Equation (7)	10^{-3}
Asymmetric with model A of the dipole	Equations (7), (11)	10^{-5}
Asymmetric with model B of the dipole	Equations (7), (18)	10^{-6}

Table 1 also shows that the error for the symmetric case is negligible, justifying why in the past, little attention was paid to the modelling issue. This does not apply to the general (asymmetric) case anymore.

6. Appendix

The general case where the grid is not centralized is somewhat more complicated than the symmetric one. Special functions can be invoked [1,3,7], but some of them are not embedded in Comsol Multiphysics®. Another method consists in assuming that the total potential is due to the one created by an isolated grid and superposing an external field [8]. This method gives good insights into the potential but it does not solve the electrostatic problem entirely, since the external electric field also depends on the potential. We can also find in [10] an expansion of the potential using separation of variables and eigenfunctions of the Laplace problem.

In their book [3], the authors present a technique based on the easy-to-obtain configuration of a grid above an infinite plane. As sketched in figure 2, the general case is decomposed into a grid (with linear charge λ) over a grounded plate, plus a two-plate capacitor with surface charge σ_P.

First is recalled the expression of complex potential for an isolated grid placed in the plane $y = 0$:

$$-\frac{\lambda}{2\pi\varepsilon_0}\log\left(2\sin\xi\right). \tag{A1}$$

For a grid placed above a grounded plate situated at $z_0 = -jL_1$, the application of image theory gives

$$-\frac{\lambda}{2\pi\varepsilon_0}\log\frac{\sin\xi}{\sin\left(\xi+\xi_0\right)}, \tag{A2}$$

with $\xi_0 = 2j\pi\frac{L_1}{a}$. The different approximations of expression (A2) are easy to derive and are indicated in [3].

Finally, adding to (A2) the constant offset V_1 and the linear expression of the potential for a two-plate capacitor, the superposition represented in figure 2 is expressed by equation (7).

IOP Conf. Series: Journal of Physics: Conf. Series **905** (2017) 012010 doi:10.1088/1742-6596/905/1/012010

The boundary conditions $V_{line}(0) = V_G - V_1$ and $V_{line}(z = jL_2) = V_2 - V_1$ are two equations depending on the unknowns λ and σ_P. This 2×2 system is written in a matrix form, using a convenient equivalent charge density of the grid $\sigma = \frac{\lambda}{a}$:

$$\begin{pmatrix} V_G - V_1 \\ V_2 - V_1 \end{pmatrix} = \frac{1}{\varepsilon_0} \begin{pmatrix} L_1\left(1 - \frac{a}{2\pi L_1}\log\rho\right) & L_1 \\ L_1 & L_1 + L_2 \end{pmatrix} \begin{pmatrix} \sigma \\ \sigma_P \end{pmatrix}. \tag{A3}$$

Inversion of this system gives the expression of the charges λ and σ_P as functions of V_G, V_1, and V_2:

$$\lambda = CV_0, \tag{A4}$$

$$\sigma_P = -\frac{L_1}{L_1 + L_2}\frac{C}{a}\left[V_G - V_1 + (V_2 - V_1)\left(1 - \frac{a}{2\pi L_1}\log\rho\right)\right], \tag{A5}$$

with $C = \frac{2\pi\varepsilon_0}{2\pi\frac{L_e}{a} - \log\rho}$, $V_0 = V_G - \frac{L_2V_1 + L_1V_2}{L_1 + L_2}$ and $L_e = \frac{L_1L_2}{L_1 + L_2}$.

It can be checked that, in the symmetric case, the surface charge on each plate is half the one distributed on the grid, *i.e.* $\sigma_P = -\frac{\lambda}{2a}$.

References

[1] Morse P M and Feshbach H 1953 *Methods of Theoretical Physics* McGraw Hill p 1241

[2] Sekar S and Stomberg H 1981 On the prediction of current-voltage characteristics for wireplate precipitators *J. Electrostat.* **10** 35-43

[3] Blum W and Rolandi L 1994 *Particle Detection With Drift Chambers* Springer-Verlag

[4] Erskine G A 1972 Electrostatic problems in multiwire proportional chambers *Nucl. Instr. And Meth.* **105** 565-72

[5] Read F H *et al.* 1999 Short and long-range penetration of fields and potentials through meshes, grids or gauzes, *Nucl. Instr. and Meth.* **427** 363-7

[6] Katagiri *et al.* 2010 Numerical analysis of the space charge effect in a MWPC *J. Plasma Fusion Res.* **9** 614-9

[7] Tomitani T 1972 Analysis of potential distribution in a gaseous counter of rectangular cross-section *Nucl. Instr. and Meth.* **100** 179-91

[8] Bunemann O *et al.* 1949 Design of Grid Ionization Chambers *Can. J. Res. A* **27** 191

[9] Durand E 1966 *Electrostatique, Tome II* Masson p 236

[10] McDonald K T 2003 Notes on Electrostatic Wire Grids http://hep.princeton.edu/~mcdonald/examples/grids.pdf

IOP Conf. Series: Journal of Physics: Conf. Series **905** (2017) 012011 doi:10.1088/1742-6596/905/1/012011

Device Simulation using Symmetric Smoothed Particle Hydrodynamics

K. Kitayama[1], **M. Toogoshi**[2], **and Y. Zempo**[2]

[1]Simulatio Corporation, Yokohama 222-0033 Japan

[2]Computer and Information Sciences, Hosei University, Tokyo 184-8584 Japan

E-mail: zempo@hosei.ac.jp

Abstract. We have applied symmetric smoothed particle hydrodynamics (SSPH) to electronic structure calculations for high electron mobility transistors (HEMTs). In layered structures such as field effect transistors (FETs), and especially HEMTs, the current density is mainly dependent on the electron mobility and the electronic field near the gate, where both can be taken to be constant. The relation between the channel current and the applied gate voltage can be obtained by a one-dimensional calculation. Then, it is easy to apply SSPH to evaluate the simple quantum properties of a device. We mainly focus on the *I-V* characteristics, which are typical device features. The electronic structure of a HEMT was calculated using both SSPH and finite-difference (FD) methods. The results from SSPH calculations are in good agreement with those from the FD method, and the accuracy of SSPH is similar to that of FD. In a simple example, where three particles are employed in the SSPH domain, we show there is an equivalence to the three-point method in FD.

1. Introduction

Smoothed Particle Hydrodynamics (SPH) is a typical mesh-free particle method, in which the system is represented by a finite set of arbitrarily distributed particles without using any mesh. SPH is widely applied to hydrodynamic problems that deal with complex shapes, large deformations and free surfaces. Recently, SPH-based methods have been applied to solve non-hydrodynamic partial differential equations such as the wave equation, the diffusion equation, Maxwell's equations and Poisson's equation [1, 2, 3, 4]. However, there are few studies for electronic structure calculations [5, 6].

An SPH-based method may lead to an efficient real-space technique because of arbitrarily distributed computation points. However, it is generally known that the standard SPH method has low accuracy [7]. In considering applications of mesh-free particle methods to practical electronic structure calculations, obtaining comparable accuracy to the results of finite-difference (FD) methods is one of the problems. Symmetric Smoothed Particle Hydrodynamics (SSPH) [8, 9] have been proposed as one of the improved techniques. This method successfully increases the accuracy of standard SPH by using a Taylor-series expansion.

We applied SSPH to a practical electronic device calculation for a high-electron-mobility transistor (HEMT). As field-effect transistors (FETs) incorporating heterojunctions, HEMTs have attracted attention due to features such as high gain, high switching speeds and extremely low noise values. These features are suitable for applications in industry such as high-frequency devices and low noise amplifiers [10]. In such a device, a very high current can be provided by the accumulated electrons, which can be considered as a two-dimensional electron gas (2DEG) [11]. For such a case, calculations of the electronic structure can be expressed using one-dimensional equations. This provides quite a nice example to use in checking the accuracy of SSPH calculations. Non-uniform distributions of computational points can be treated in the frame-work of SSPH, which we expect to reduce computational costs significantly. However, the SSPH calculation described in this paper was performed for a uniform particle distribution for simplicity.

The purpose of this paper is to evaluate SSPH for the simple quantum device of a HEMT. We mainly focus on the I-V characteristics, which is one of the typical device features. The paper is organized as follows. First, the HEMT structure we adopted is explained, and the formulation of SSPH follows in the next section. The numerical results are shown in Sec. 3. Finally, we summarize the conclusions in the last section.

2. Method

A typical, single-heterojunction HEMT is shown in Fig. 1. Usually such a device is fabricated as an epitaxial multi-layer structure. Using the band offset at the heterojunction, electronic devices such as FETs, and especially HEMTs, work quite efficiently. Such devices have been intensively studied both experimentally and theoretically [11, 12]. The electronic charge distribution in this system can be considered as a quasi-2D system. The electric current from the source to the drain is controlled by changing the voltage applied to the gate, which changes the electrostatic potentials around the gate. The current density of the heterostructure FET is determined by the ionized impurities, the gate electric field, and the density and the electron mobility. The electric field is approximately constant in the narrow region beneath the gate. Also, the electron mobility can be taken as constant in the long gate device. Then the current is proportional to the electron density. Thus, in a one-dimensional analysis, in a direction perpendicular to the surface that corresponds to the gate, it is possible to determine the device characteristics, e.g., the relation between the channel current and the applied gate voltage for a practical FET device [10]. The channel is narrow enough to be expressed in terms of quantized states, which are simply described in a one-dimensional (1D) equation.

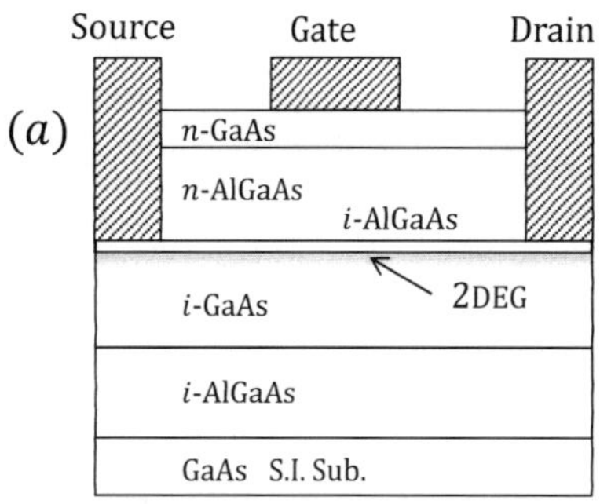

(b)

n^+- GaAs	2.6×10^{18} cm^{-3}	100Å
n^+ - AlGaAs	2.0×10^{18} cm^{-3}	300Å
i - AlGaAs		20Å
i - GaAs		500Å
i - AlGaAs	Buffer layers	3000Å
i - GaAs		1000Å
GaAs S.I. Substrate		

Figure 1. The structure of a HEMT. (a) The schematic device model, and (b) the layer structure we adopted in the calculation.

In our general analysis of the characteristic behavior of the device, we solved the effective one-dimensional, one-electron Schrödinger equation and Poisson's equation [15]. In the vertical direction beneath the gate, these equations can be written in 1D as follows;

$$-\frac{\hbar^2}{2}\frac{d}{dz}\left(\frac{1}{m^*(z)}\frac{d\psi_i(z)}{dz}\right)+V(z)\psi_i(z)=E_i\psi_i(z) \quad (1)$$

$$\frac{d}{dz}\left(\varepsilon(z)\frac{d\phi(z)}{dz}\right)=-(\rho_I+\rho_q), \quad (2)$$

where E_i is the i-th energy eigenvalue, and $\psi_i(z)$ is the corresponding wave function. The position-dependent effective mass and permittivity are expressed as $m^*(z)$, and $\varepsilon(z)$, respectively. The effective potential $V(z)$ can be obtained from Poisson's equation. The electrostatic potential V_H can be expressed by $V_H=-e\phi+\Delta v$, where Δv is the band offset of the two heterojunction interfaces in the HEMT structure. The potential profile ϕ is determined by the ionized charge ρ_I.

$$\rho_I=e\left[p(\phi)+N_D^+(\phi)-N_A^-(\phi)-n(\phi)\right], \quad (3)$$

and the carrier concentration $\rho_q(z)$, which exists mainly in the channel region, is given by

$$\rho_q(z)=\frac{e\,m^*(z)kT}{\pi\hbar^2}\sum_i g_i\left|\psi_i(z)\right|^2\ln\left[1+\exp\left(\frac{E_F-E_i}{kT}\right)\right]. \quad (4)$$

Here g_i is the degeneracy of the state, E_F is the Fermi level, T is the temperature, and k is Boltzmann's constant. The electrostatic potential distribution is obtained from a self-consistent solution of the Schrödinger and Poisson equations. Usually, Eqs. (1) and (2) are expressed in finite-difference (FD) form, and are solved subject to appropriate boundary conditions. In addition, matching conditions at the heterojunction interfaces are required for the derivatives of wave functions and electrostatic potentials. The solution can be obtained by solving each equation step-by-step until convergence is achieved [13, 14, 15, 16]. The electronic charge density determines the potential V, which can also include the exchange-correlation potential.

In SPH, on the other hand, the wave function is approximated using an integral form, and the system is represented by a finite number of particles. The concept of the integral representation of a wave function $\psi(r)$ starts from the following identity. The delta function is replaced by a kernel function $W(r,h)$.

$$\psi(z)=\int_\Omega\psi(z')\,\delta(z-z')\,dz'\simeq\int_\Omega\psi(z')\,W(z-z',h)\,dz', \quad (5)$$

where h is the smoothing length that defines the domain Ω. We adopt the Wendland function as the kernel function in the one-dimensional system we consider [17]. It is finite and is convenient for numerical calculations:

$$W(r,h)=\begin{cases}\frac{3r}{4h}\left(1-\frac{1}{2}\frac{r}{h}\right)^5\left\{2\left(\frac{r}{h}\right)^2+\frac{5r}{2h}+1\right\} & \left(0\leq\frac{r}{h}\leq 2\right)\\ 0 & \left(2\leq\frac{r}{h}\right)\end{cases} \quad (6)$$

where $r=|z-z'|$.

SSPH corrects the approximation of SPH using a Taylor-series expansion of $\psi(\mathbf{r})$. The right-hand side of Eq. (5) can then be rewritten as follows:

$$\int_\Omega(z-z_i)^k\psi(z)W(z-z_i,h)\,dz\simeq\sum_{l=0}^{m}\left(\int_\Omega(z-z_i)^{k+l}\,W(z-z_i,h)dz\right)\left(\frac{1}{l!}\frac{\partial^l\psi}{\partial z^l}\right)\Bigg|_{z_i}, \quad (7)$$

where $k = 0, 1, \ldots, m$. The integral equation is thus approximated by the summation over the particles in the domain Ω corresponding to the smoothing length h. In this version of SSPH, there is no derivative of the kernel function, which sometimes causes errors in SPH calculations. The accuracy of SSPH mainly depends on the order of the Taylor-series expansion [5]. Equation (7) has the form of simultaneous linear equation, $P = KD$, as given below:

$$\begin{aligned} P = (\ P_k\ ), \qquad P_k &= \sum_j^N (z_j - z_i)^k\, \psi(z_j) W_{ij} \Delta V_j \\ K = (\ K_{kl}\ ), \qquad K_{kl} &= \int_\Omega (z - z_i)^k\, (z - z_i)^l\, W(z - z_i, h) dz \\ D = (\ D_l\ ), \qquad D_l &= \left(\frac{1}{l!} \frac{\partial^l \psi}{\partial z^l} \right) \Bigg|_{z_i}, \end{aligned} \tag{8}$$

where $W_{ij} = W(z_j - z_i, h)$. Naturally, the Schrödinger equation in Eq. (1) is a generalized eigenvalue problem. The small volumes occupied by the j-th particles surrounding particle i are defined as ΔV_j, which is calculated the number density $n(z_j)$:

$$\Delta V_j = 1/n(z_j), \quad n(z_j) = \sum_k^m \delta(z_j - z_k) \simeq \sum_k^m W(z_j - z_k; h). \tag{9}$$

Defining $D_0 = \psi\big|_{z_i}$, $D_1 = \frac{\partial \psi}{\partial z}\big|_{z_i}$, and $D_2 = \frac{1}{2}\frac{\partial^2 \psi}{\partial z^2}\big|_{z_i}$, we can write the Schrödinger and Poisson equations Eqs. (1) and (2) at the position of the i-th particle in the form of the SSPH discretization. For example, in the simplest case of a constant effective mass, the Schrödinger equation, Eq. (1), for the i-th particle in SSPH can be written as

$$\left(V(z_i) \quad 0 \quad -\frac{\hbar^2}{m^*} \quad \ldots \quad 0 \right) D = (E \quad 0 \quad 0 \; \ldots \; 0)\, D. \tag{10}$$

Poisson's equation in Eq. (2), with constant permittivity, also can be simply described in a manner similar to the derivation of Eq. (10). From the electronic charge density, the electrostatic potential and the exchange-correlation potentials are also determined. In SSPH, the particle position can be arbitrarily distributed so as to obtain a specified level of accuracy. This is a particular feature of this method, which we expect to reduce the computational cost significantly. However, we have to calculate K^{-1} in each particle position, keeping the matrix K regular.

3. Results and discussion

We have calculated device characteristics both in SSPH and FD, and have compared the results. In the SSPH calculation, we adopted the smoothing length $h = 0.2$ Å and assumed constant particle distributions over the distance $\Delta x = 0.2$ Å. For the FD mesh size, we used the same value $\Delta x = 0.2$ Å as for the particle distribution in SSPH. Figure 2 shows that band structure and the charge profile of the results from both the SSPH and FD calculations, plotted in the vertical direction. Both results from SSPH are in good agreement with those from the FD calculation.

As mentioned above, the current is determined by the carrier density. The sheet carrier density N_s and dN_s/dV_g are plotted in Fig. 3 as functions of the gate voltage V_g. These provide the I-V characteristics and the transconductance g_m, respectively. Changing the applied gate voltage corresponds to calculating the characteristics using different boundary conditions at the

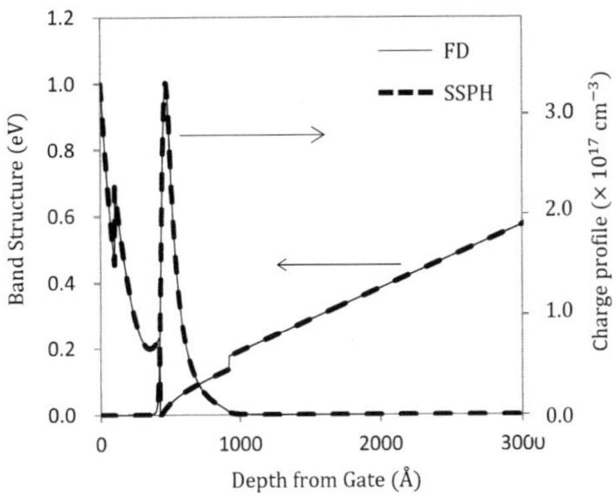

Figure 2. Conduction band structure and charge profile, calculated using SSPH (dashed line) and FD (solid line).

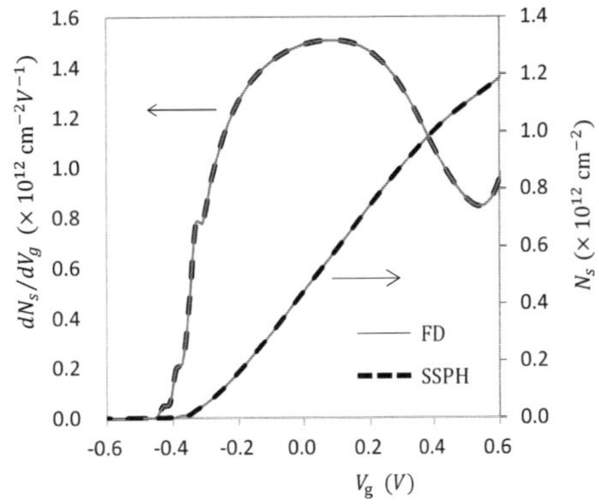

Figure 3. Characteristic properties and sheet carrier density N_s and dN_s/dV_g, calculated using SSPH (dashed line) and FD (solid line).

surface $z = 0$. We have calculated the charge profile for different gate voltages in the steps of $\Delta V_g = 0.005$. The sheet carrier density N_s integrated over the whole region and is plotted in Fig. 3. At each point, we also calculate dN_s/dV_g using ΔV_g in a difference scheme; this is also plotted in Fig. 3. We can see that the results of SSPH are in good agreement with those of FD. The accuracy is sufficient for practical device simulations. In the SSPH calculation, we have also tried the different smoothing length $h = 0.3$ Å, which includes five particles in the calculation domain, rather than three. The results were the same as for the calculation performed with the smoothing length $h = 0.2$ Å. Within this difference, the accuracy of SSPH does not change.

There is a significant difference in the form of the SSPH and FD equations. The former is based on integral equations, and the latter on differential equations. However, there is an equivalence in the simplest case, in which SSPH uses three particles in the computational domain, and FD uses a three-point method. The reason is that, from the discretization point of view, there is no difference between SSPH and FD in this case. Let us consider a typical case, as

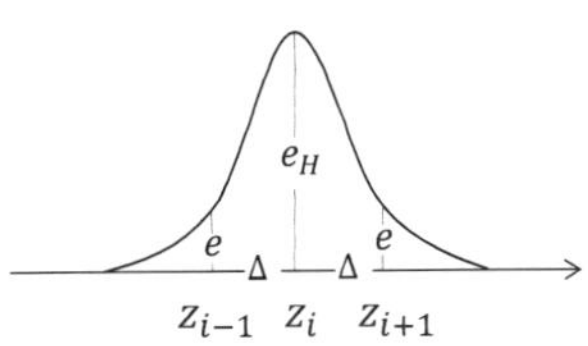

Figure 4. Example of the kernel function W in the case of a constant distribution Δ of three particles over the smoothing length. The values at each point are $W(z_i) = e_H$, and $W(z_{i-1}) = W(z_{i+1}) = e$, respectively.

shown in Fig. 4. Three particles are distributed over a constant spacing Δ in the computational domain defined by the smoothing length h. In this case, the matrix size of K reduces to 3×3. The inverse matrix K^{-1} can be calculated manually, and the equation $D = K^{-1}P$ can then be expressed as

$$\begin{pmatrix} \psi\big|_i \\ \frac{\partial}{\partial x}\psi\big|_i \\ \frac{1}{2}\frac{\partial^2}{\partial x^2}\psi\big|_i \end{pmatrix} = \begin{pmatrix} \frac{1}{1-2e} & 0 & \frac{1}{(1-2e)\Delta^2} \\ 0 & \frac{1}{2e\Delta^2} & 0 \\ \frac{1}{(1-2e)\Delta^2} & 0 & \frac{1}{2e(1-2e)\Delta^4} \end{pmatrix} \begin{pmatrix} e\psi_{i-1} + e_H\psi_i + e\psi_{i+1} \\ e\Delta\left(-\psi_{i-1} + \psi_{i+1}\right) \\ e\Delta^2\left(\psi_{i-1} + \psi_{i+1}\right) \end{pmatrix}. \quad (11)$$

We can immediately see the equivalence by substituting each element – that is, $D_0 = \psi_i$, $D_1 = (-\psi_{i-1} + \psi_{i+1})/(2\Delta)$, and $D_2 = (\psi_{i-1} - 2\psi_i + \psi_{i+1})/(2\Delta^2)$ – into Eq. (10). The practical formula obtained from three-point FD can thus be reproduced through the SSPH procedure.

4. Summary

We introduced SSPH as a discretization technique for device simulations. SSPH was applied to an electronic structure calculations for a HEMT. In comparing the characteristics of HEMTs calculated using SSPH and FD, we find that the results from SSPH are in good agreement with those obtained using FD. The accuracy of SSPH is similar to that of FD. From the technical point of view, the computational effort increases as the order of the Taylor-series expansion becomes higher. However, SSPH and FD were also shown to be equivalent in a simple case corresponding to the three particles in the SSPH computational domain and to a three-point FD method. If the number of SSPH particles is large enough to describe steep changes in the one-dimensional Schrödinger equation, reasonable accuracy can be obtained. Our results indicate that SSPH can be applied to other electronic-structure calculations that require high accuracy.

Acknowledgments

The authors would like to express their gratitude to Professor S. S. Kano for useful discussions and encouragement. This work was partially supported by JSPS Grants-in-Aid for Scientific Research (C) Grant number 16K05047, Sumitomo Chemical Co., Ltd. and Simulatio Corporation. Computations were partially performed at the Research Institute for Information Technology, Kyushu University and the Supercomputer Center.

References

[1] Laguana P 1995 *Astrophys. J.* **439**, 89
[2] Stranex T and Wheaton S 2011 *Comput. Method. Appl. Math.* B**200**, 392
[3] Tascano E, Balasi G D and Tortorici A 2012 *Appl. Math. Comput.* **218**, 8906
[4] Ala G, Francomano E, Tortorici A, Tascano E and Viola F 2006 *J. Comput. Appl. Math.* **191**, 194
[5] Sugimoto S and Zempo Y 2014 *J. Phys.: Conf. Ser.* **510**, 012037
[6] Zempo Y and Sugimoto S 2015 *J. Phys.: Conf. Ser.* **640**, 012031
[7] Lucy L B 1977 *Astron. J.* **439**, 1013
[8] Batra R C and Zhang G M 2008 *Compt. Mech.* **41**, 527
[9] Zhang G M and Batra R C 2009 *Compt. Mech.* **43**, 321
[10] Hata M, Fukuhara N, Sasajima Y, and Zempo Y, 2000 *Sumitomo Kagaku* **2000**-I, 10
[11] e.g. Sze M, 1981 *Physics of Semiconductor Devices, 2nd ed.* (Wiley, New York)
[12] Adachi S, et. al. 1982 *Rev. Mod. Phys.***54**, 437 (Wiley, New York,).
[13] Stern F, 1972 *Phys. Rev.* B**5**, 4891
[14] Venter B, 1978 *Sol. Stat. Comm.* **28**, 861
[15] Stern F and Sarma S, 1984 *Phys. Rev.* B**30**, 840
[16] Tan I.-H, Snider G L Chang L D Hu E L, 1990 *Appl. Phys.* **68**, 4071
[17] Wendland H 1995 *Adv. Comput. Math.* **4**, 389

IOP Conf. Series: Journal of Physics: Conf. Series **905** (2017) 012012 doi:10.1088/1742-6596/905/1/012012

Computational study of TiO_2 Brookite (100), (010) and (210) surface doped with Ruthenium for application in Dye Sensitised Solar Cells

R.S Dima[1,2], N.E Maluta[1,2], R.R Maphanga[2,3], V Sankaran[1]

[1]Department of physics, University of Venda, P/Bag X 5050, Thohoyandou, 0950, South Africa
[2]National Institute for Theoretical Physics (NITheP), Gauteng, South Africa
[3]Materials Modelling Centre, University of Limpopo, P/Bag X 1106, Sovenga, 0727

E-mail: dimaratshilumelasteve@gmail.com

Abstract. Titanium dioxide (TiO_2) polymorphs are widely used in many energy-related applications due to their peculiar electronic and physicochemical properties. The electronic structures of brookite TiO_2 surfaces doped with transition metal ruthenium have been investigated by ab initio band calculations based on the density functional theory with the planewave ultrasoft pseudopotential method. The generalized gradient approximation (GGA) was used in the scheme of Perdew-Burke-Ernzerhof (PBE) to describe the exchange-correlation functional. All calculations were carried out with CASTEP (Cambridge Sequential Total Energy Package) code in Materials Studio of Accelrys Inc. The surface structures of Ru doped TiO_2 were constructed by cleaving the $1 \times 1 \times 1$ optimized bulk structure of brookite TiO_2. The results indicate that Ru doping can narrow the band gap of TiO_2, leading to the improvement in the photoreactivity of TiO_2, and simultaneously maintain strong redox potential. The theoretical calculations could provide meaningful guide to develop more active photocatalysts with visible light response.

1. Introduction

In the last decades, properties for titanium dioxide (TiO_2) polymorphs have been the subject of many experimental and computational studies, the most common polymorphs being the minerals rutile, anatase, and brookite [1]. A large number of applications of TiO_2 in materials science is, almost without exception, ultimately a result of the facile electron-transfer processes that occur at the interface between the semiconductor and adsorbed molecules [1–7]. When photons are excited, TiO_2 is a good electron and hole donor and can, therefore promote photocatalytic processes at its interface [3]. Its photocatalytic properties, in addition to its abundance, low cost, stability, and low toxicity, are the basis for its use in solar cells [4]. However, brookite is the rarest of the natural occurring TiO_2 polymorphs, and it is the most difficult phase to prepare in the laboratory [1]. As a result, the properties of pure brookite are poorly known. TiO_2 can only show photocatalytic activity under ultraviolet (UV) light irradiation ($\lambda < 387.5$ nm) that accounts for only a small portion of solar energy (approximately 5%), in contrast to visible light for a major part of solar energy (approximately 45%), but can be photosensitized by the adsorption of chromophores that, when excited, inject electrons into the TiO_2 conduction band [8]. In spite of a large number of publications on pure and doped TiO_2 many aspects of its optical and electronic properties and their modifications after doping still remain unclear. In particular, a key problem of selecting proper impurities to meet particular requirements for future applications is related to the location of the impurity's energy levels with respect to the band structure of a particular host [9, 10]. Consistent analysis of the changes induced by impurities implies a detailed comparison of the calculated properties of the undoped and doped materials [8,10,11]. Such an approach is used in the present work, whereby first principle method is employed to calculate structural and electronic properties of brookite TiO_2 surfaces for both pure and Ru-doped systems.

2. Structures

Brookite, which is the third natural form together with rutile and anatase polymorphs of TiO_2, has a complex structure. It has eight formula units in the orthorhombic cell with space group Pbca. Brookite formation may be pictured as the joining of distorted TiO_6 octahedral sharing three edges. The surface structures of brookite have been modelled by cleaving the optimized TiO_2 brookite bulk structure and constructing slab models. The relaxation of these surfaces has been taken into account by optimizing all of the slabs. This study explores the low index (100) and (010) surfaces for the brookite phase and (210) surface. The (100) and (010) surfaces are represented by 12 atoms while the (210) surface is composed of 24 atoms A schematic representation of the three low-index surfaces is depicted in Figure 1. The truncation of the octahedral renders different coordination combinations for the outermost titanium cations. Depending on the slice of the brookite surface, the (010) and (010) surfaces may be either oxygen- or titanium terminated, the oxygen-terminated surfaces are found to be the most stable surfaces. Hence, only the oxygen-terminated surfaces are considered in this study. The band structures of the studied surface slab models (100), (010), and (210) surfaces, in addition to the first Brillouin zone with the high symmetry points present a direct gap, as in the bulk.

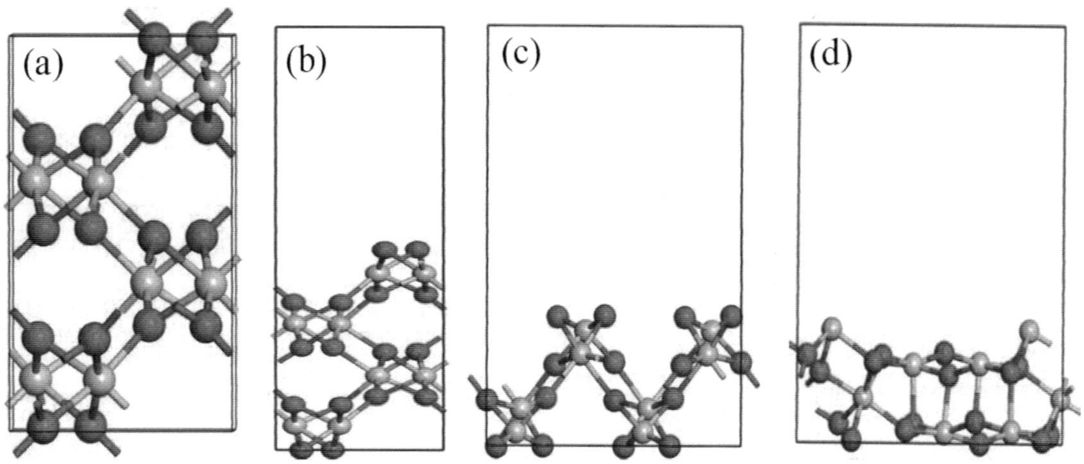

Figure 1. Theoretical models. (a) bulk TiO_2 brookite, (b) (100) TiO_2, (c) (010) TiO_2 and (d) (210) TiO_2 surfaces.

3. Computational Method

The structural optimizations, band structure, and electronic density of states of doped and undoped brookite TiO_2 (100), (010) and (210) surfaces were performed using the first-principles

IOP Conf. Series: Journal of Physics: Conf. Series **905** (2017) 012012 doi:10.1088/1742-6596/905/1/012012

calculations based on density functional theory (DFT) using a plane-wave pseudo-potential method. The surfaces structures of transition metal ruthenium-doped TiO_2 were constructed by cleaving the 1 × 1 × 1 optimized bulk structure of brookite TiO_2 with one Ti atom being replaced by Ru atom. We used the generalized gradient approximation (GGA) in the scheme of Perdew-Burke-Ernzerhof (PBE) to describe the exchange-correlation functional. All calculations were carried out using CASTEP (Cambridge Sequential Total Energy Package) code in Materials Studio of Accelrys Inc [12]. The Monkhorst-Pack scheme k-point grid sampling was set as 4 × 4 × 2 for the irreducible Brillouin zone and the cut-off potential energy of plane waves was set to 650 eV. The convergence criteria for structural optimization were set to be medium quality with the tolerance for the self-consistent field (SCF), energy, maximum force, and maximum displacement of $2x10^{-6}$ eV/atom, $2x10^{-5}$ eV/atom, 0.05 eV/Å and $2x10^{-3}$ Å, respectively. The doping criteria considered was to replace one titanium atom at the top of the surface with Ru.

4. Results and discussions

4.1. Structural optimization

The optimized structures of doped and un-doped brookite TiO_2 were calculated to find the lattice parameters with the lowest energy. Table 1, shows the lattice parameters a, b and c of the current work, which are compared with the reported experimental values and previous theoretical results. It can be observed that our calculated results are in agreement with both the experimental and reported theoretical results [13]. The lattice parameters obtained in this work are comparable to the experimental data, with deviation of 0.1967% along the a-axis, 0.203% along the b-axis and 0.196% along the c-axis. In comparison with the previous theoretical results reported in [14], the calculated results are more accurate, and thus validating the starting models and method used.

Table 1. Optimized structural parameters for bulk brookite TiO_2 compared with experimental and previous theoretical results

	Experimental [13]	This work		Literature [14]	
		Results	Deviation % %	Results	Deviation % %
a(Å)	9.184	9.166	0.196	9.157	0.294
b(Å)	5.447	5.436	0.202	5.430	0.202
c(Å)	5.145	5.135	0.195	5.122	0.448

4.2. Electronic properties

In order to conveniently investigate the electronic structures of Ru-doped brookite TiO_2 surfaces, we set the same k-points mesh to sample the first Brillouin zone for pure and doped systems. An energy band gap of 2.386 eV was obtained for the bulk brookite structure and is shown in Figure 2(a). Both the conduction band minimum (CBM) and the valence band maximum (VBM) are located at G. Thus, the brookite TiO_2 can be considered as a direct band gap semiconductor. But is underestimated compared with the experimental value of 3.23 eV, due to the limitation of DFT: the discontinuity in the exchange correlation potential is not taken into account within the framework of DFT. However, our discussions about energy gap will not be affected because only the relative energy changes are of concern. The upper valence band, VB, lying in the range of 2.1-4.2 eV consists of O 2p orbitals hybridized with Ti 3d orbitals as illustrated in Figure 2. The band structures of the studied surface slab models (100), (010), and (210) surfaces, in addition to the first Brillouin zone with the high symmetry points present a direct band gap, as in the bulk. The corresponding values are 2.317, 2.080 and 3.405 eV for (100), (010), and (210) surfaces, respectively. The lower value of band gap energy for the (010)

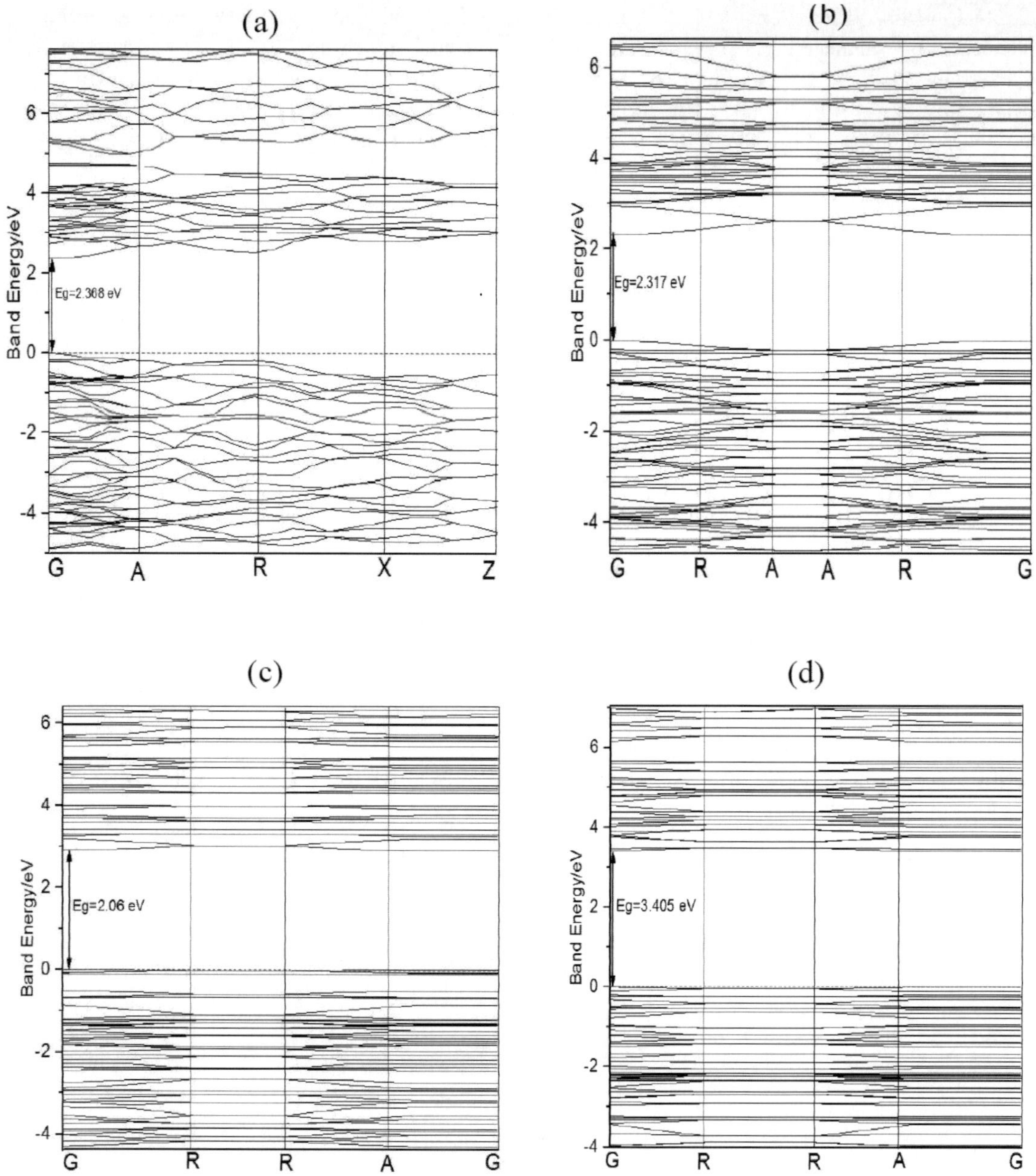

Figure 2. The energy band structures of (a) bulk brookite TiO_2; (b) (100) TiO_2; (c) (010) TiO_2 and (d) (210) TiO_2 surfaces. The dashed line indicates the Fermi level.

surface can be attributed to the minor stabilization of the Fermi energy showing the most flat VB topology. After doping with Ru all the surfaces present a direct gap, as in the bulk (the bottom of the CB and the top of the VB are located at G). The band gaps for the TiO_2 brookite surfaces (100), (010) and (210) after doping with Ru are determined as 0.256 eV, 0.275 eV and 0.177 eV, respectively. The lower value of band gap energy for the (210) surface can be attributed to the minor stabilization of the Fermi energy showing the most flat VB topology. The total density

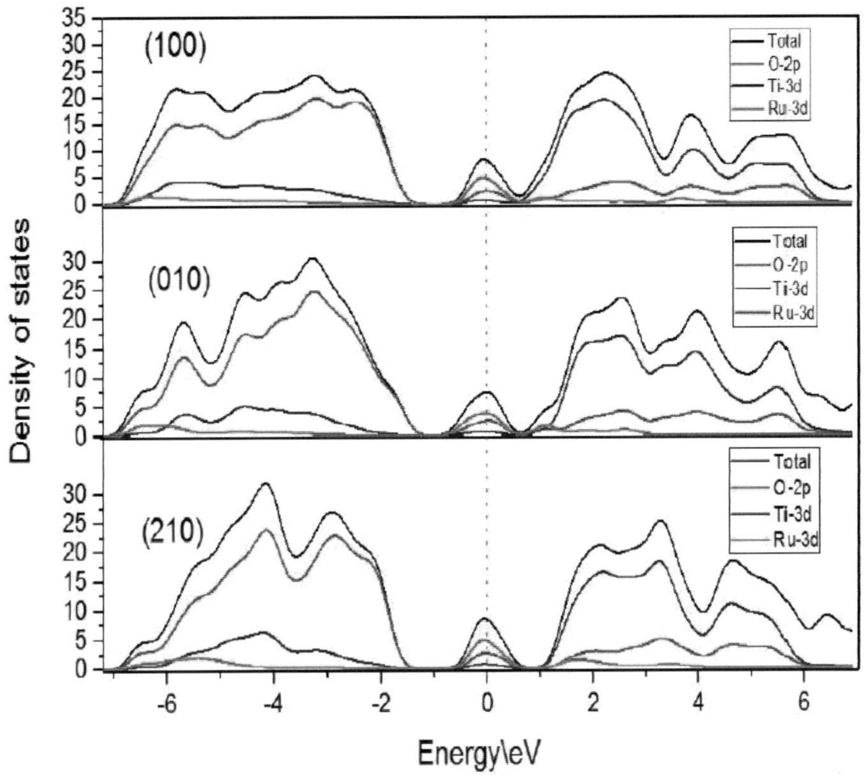

Figure 3. The TDOS and PDOS of the doped TiO_2. The red dashed line represents the position of the Fermi level.

of states (TDOS) and partial density of states (PDOS) of Ru-doped brookite TiO_2 are shown in Figure 3, which were drawn using Origin Lab software [15]. The TDOS shape of doped TiO_2 becomes broader than that of pure TiO_2, which indicates that the electronic nonlocality is more obvious, owing to the reduction of crystal symmetry. The 3d or 4d states of Ru are somewhat delocalized, which contributes to the formation of impurity energy levels (IELs) by hybridizing with O 2p states or Ti 3d states. Such hybrid effect may form energy levels in the band gap or hybrid with CBM/VBM, providing trapping potential well for electrons and holes. It gives a contribution to separation of photogenerated electronhole pairs, as well as in favour of the migration of photoexcited carriers and the process of photocatalysis. The decrease in the band gap attribute to the fact that TiO_2 band gap can be engineered to absorb light in visible region of the spectrum. The reduction in energy band gap could possibly increase the efficiency of the solar cell when TiO2 is used.

5. Conclusions

First principle calculations based on the density functional theory were performed successfully to understand the effect of doping brookite (100), (010) and (210) TiO_2 surfaces. The structural and electronic properties of Ru doped on different surfaces were fully investigated. The energy band gaps for various surfaces doped with ruthenium are predicted and Ru doped (210) surface showed a good redshift found to have lowest energy band gap as compared to other surfaces. The analysis of the electronic structure for the studied surfaces of brookite revealed that all the

IOP Conf. Series: Journal of Physics: Conf. Series **905** (2017) 012012 doi:10.1088/1742-6596/905/1/012012

surfaces have a direct band gap. The findings of this comprehensive study on brookite TiO_2 surfaces shed insights and probe further investigation on defect properties, impurities, thin films, surfaces, and interfaces of TiO_2 system, particularly brookite polymorph.

Acknowledgements
This work is financially supported through the NRF Thuthuka program and also by National Institute for Theoretical Physics (NiTheP). We should also like to thank the University of Venda for the support to carry out this research and Centre for High performance Computing (CHPC) for using their facilities.

References

[1] Zallen R, and Moret M.P. The optical absorption edge of brookite TiO2.Solid State Commun, 2006, 137 154-157
[2] Yang K, Dai Y, Huang B, Han S: Theoretical study of N-doped TiO2 rutile crystals. J Phys Chem B 2006, 110:2401124014.
[3] Grätzel M, and ORegan B. A Low-Cost, High-Efficiency Solar Cell Based on dye sensitized Colloidal TiO2 Films Nature, 353, 737-740, 1991.
[4] Grätzel M, Bach U, Lupo D, Comte P, Moser J. E, Weissrtel F, Salbeck J, Spreitzer, H. Solidstate Dye-sensitized Mesoporous TiO2 Solar Cells with High Photon-to-electron Conversion Efficiencies. Nature, 395, 583-585, 1998.
[5] Grätzel M. Photo electrochemical Cells. Nature 414, 338-344, 2001.
[6] Grätzel M. Recent Advances in Mesoscopic Solar Cells. Acc. Chem. Res. 42, 1781-1798, 2009.
[7] Tesfamichael T, Will G, Bell J, Prince K, and Dytlewski N. Characterization of a commercial dye-sensitised titania solar cell electrode. Solar Energy Materials and Solar Cells, 76(1):25 35, 2003.
[8] Li SP, Lin SW, Liao JJ, Pan NQ, Li DH, Li JB: Nitrogen-doped TiO2 nanotube arrays with enhanced photoelectrochemical property. Int J Photoenergy 2012, 2012-794207.
[9] Luo W, Yu T, Wang Y, Li Z, Ye J, Zou Z. Enhanced photocurrent-voltage characteristics of WO3/Fe2O3 nano-electrodes. J Phys D Appl Phys 2007, 40-1091.
[10] Beltran A, Gracia L and J. Andres J. Density functional theory study of the brookite surfaces and phase transitions between natural titania polymorphs. J. Phys. Chem. B 2006, 2341723423
[11] Ma J.G, Zhang C.R, Gong J.J, Wu Y.Z, Kou S.Z, Yang H, Chen Y.H, Liu Z.J and Chen H.S 2015 Mater 8(8) 5508-5525
[12] Clark S.J, Segall M.D, Pickard C.J, Hasnip P.J, Probert M.I.J, Refson K and Payne M.C 2005 Z Kristallogr 220 567570
[13] Meagher E. P and Lager G.A 1979 Mineral 17 77-85
[14] Beltran A, Gracia L and Andres J 2006. J. Phys. Chem. B, 110(46) 23417-23423
[15] Origin (OriginLab, Northampton, MA)

DFT calculations of Anatase TiO_2 (1 0 1) Surface Doped with Ruthenium for Application in Dye Sensitised Solar Cell

Nemudzivhadi[1] H., Maluta[1, 2] N.E., Maphanga[3] R.R., Sankaran[1]V.

[1]University of Venda, Department of physics, P/Bag X 5050, Thohoyandou, 0950

[2]National Institute for Theoretical Physics (NITheP), Gauteng, South Africa

[3]Material Modelling Centre, University of Limpopo P/Bag X 1106, Sovenga, 0727.

Email: hulisaninemudzivhadi@gmail.com

Abstract. Titanium dioxide (TiO_2) is considered to be an ideal semiconductor for photocatalysis because of its high stability, low cost and safety towards both humans and the environment. Doping TiO_2 with different elements has attracted much attention as the most important way of enhancing the energy band gap and the adsorption on different wavelength region, in order to improve the efficiency of the dye sensitized solar cells (DSSCs). In this study, first principle density function theory was used to investigate the reaction mechanisms between anatase TiO_2 (1 0 1) surface doped with ruthenium (Ru) atom as a dopant. Ru-doped anatase TiO_2 (1 0 1) surface was geometrically optimized using the generalized gradient approximation implemented within the CASTEP CODE. The calculated density of states and band structures for Ru-doped systems shows that the synergistic effects of Ru impurities at lattice titanium sites lead to a sharp band gap narrowed to 0.345 eV.

1. Introduction

The Energy is the basic resource of today's world. Along with the increase of the population, the demand of the amount of energy resource gets higher rapidly [1]. The solution to the growing demand and environmentally sustainable energy resources in the world might lie in the ability to capture and utilize renewable solar energy. To capture and utilize solar energy for a sustainable development on a large scale, development of cheaper technologies will be an added advantage [2]. In photovoltaic technologies, the DSSCs is a promising approach for direct conversion of sunlight into electrical energy at low cost and high-energy conversion efficiency [2]. The dye molecules as a sensitizer absorbs solar radiation and transfers the photo-excited electron to the wide band gap of TiO_2 semiconductor [3, 4]. The molecule dye is normally adsorbed on the surface of various TiO_2 polymorphs, which is usually on a nanometer size particle [3, 4]. Kamat reported that the inertness to chemical environment and long-term photostability has made TiO_2 an important material in many practical applications and in commercial products ranging from drugs to foods, cosmetics to catalysts, paints to pharmaceuticals and sunscreens to solar cells in which TiO_2 is used as a desiccant, brightener, or reactive mediator [5]. TiO_2 has an added advantage to be widely applied in the DSSCs because of its wide band gap [6]. But for the applications in paint material, the narrowing of the TiO_2 band gap is not desired. The problem encountered with the wide band gap of TiO_2 is that it crucially limits the practical application involving the photons absorption and electrons transport in DSSCs [7, 8]. To enhance the efficiency of DSSCs, it is essential to reduce band gap for TiO_2 material, such that the absorption properties will match well with visible solar spectrum [7- 10].

TiO_2 has three commonly known different structural types, which are rutile, anatase and brookite as shown in figure 1 [11]. All of these crystalline forms of TiO_2 occur in nature as mineral, but only rutile and anatase have been synthesized in pure form at low temperature [11]. Therefore, anatase and rutile phases, which are tetragonal crystal system, are widely studied, for their syntheses, characterizations and photocatalytic properties [9, 11]. So far, anatase phase is known to exhibit higher photocatalytic efficiency compared to rutile phase [12, 13]. Anatase is one of the most popular and promising materials in photocatalytic applications due to the strong oxidizing power of its holes, high photo stability and redox selectivity, with its band gap of (3.2 eV) [6, 11, 12, 13].

The light absorption is limited in the ultraviolet (UV) region because of the wide band gap, but Tian et al. reported that doping is a good approach to get a visible light driven photocatalyst [14]. The surfaces of TiO_2 play an important role in many technical applications, since Fujishima and Honda showed that photocatalytic water splitting is possible [15]. Recently Navrotsky showed that the anatase surfaces play an important role for the phase stability of TiO_2-nanoparticles [16]. Esch et al. studied the surfaces structures and thermodynamics of low-index of rutile, anatase and brookite, using density function theory (DFT) method [17]. The study investigated low-index surfaces of rutile, anatase and brookite by determining surfaces stability. It was reported that atomic relaxations of rutile (1 1 0), (1 0 0), anatase (1 0 1), (1 0 0) and brookite (2 1 0) surfaces, are the most common and most stable surfaces. Chen et al. investigated anatase TiO_2 (1 0 1) surface doped with O vacancies and nitrogen (N) atoms using DFT. The findings revealed that the mixing of O vacancies and N dopants induced state with original Ti 3d and O 2p valence band attributes to the band gap narrowing [18]. Using DFT method. Chen et al. reported that doping anatase (1 0 1) surface with sulfur lead to band gap narrowing as compared un-doped anatase TiO_2 systems [19]. Labt et al. and Mino et al. both they reported that anatase TiO_2 (1 0 1) surface is the only surface that can be modelled with even-n-layer slabs and the most stable surface of anatase [20, 21].

In this study, adsorption mechanisms for anatase TiO_2 (1 0 1) surface doped with Ru was investigated using DFT to calculate electronic properties and analyses.

2. Structures

Anatase TiO_2 has a tetragonal structure with (-space group: I 41 141/AMD. i.e., local symmetry: D19 4h/. The system is composed of four Ti atoms and eight O atoms in the primitive cell and its lattice parameters are a = b = 3.776 Å and c = 9.486Å) [11].

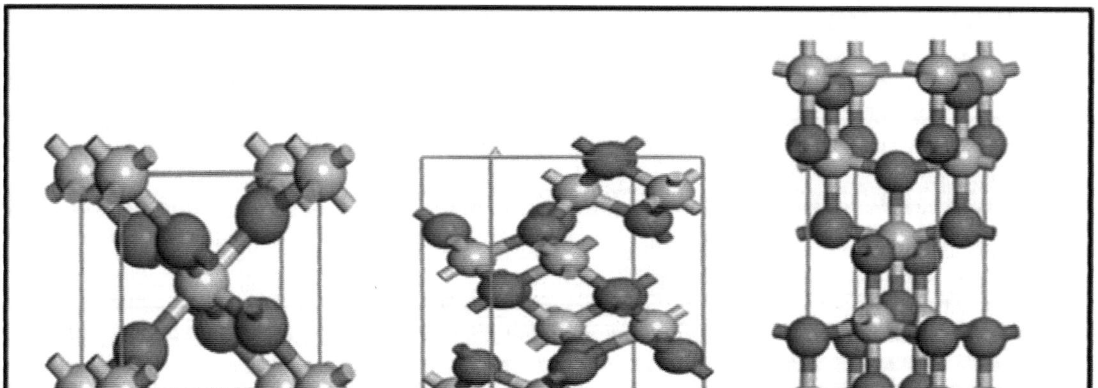

Figure 1. The primitive cells of rutile, brookite and anatase (from left to right), where the red balls and grey balls represent oxygen (O) and titanium (Ti) respectively.

3. Methodology and computational details

The first-principles calculations were performed using the Cambridge Serial Total Energy Package (CASTEP) code in materials studio developed by Accelrys Software Inc [22]. The energy plane-wave pseudopotential total calculation method based on DFT was used to carry out all the calculations. For structural optimization, the generalized-gradient approximation (GGA) with the Perdew-Burke Ernzerhof (PBE) exchange-correlation functional was employrd for all the calculations to obtain the electronic structures and accurate formations energies [22]. In these calculations, the energy cut-off for the plane-wave basis was set to be 600 eV and the Brillouin zone for the surface model was represented by a Monkhorst-Pack (MP) scheme for the generation of k-points of (4 x 6 x 1).

The convergence threshold for self-consistent tolerance was set to $2.0e^{-6}$ eV/atom and geometric relaxation was obtained with the Broyden-Fletcher-Goldfarb-Shanno (BFGS) algorithm until the atomic relaxations were below 0.05 eV/Å. The above mentioned parameters were suitable for getting well-converged total energy and geometry optimization results. In the geometry optimization, the convergence criteria for structural optimization were set to be medium quality with the energy change, maximum stress and maximum displacement of $2.0e^{-5}$ eV/atom, 0.1 GPqa and 0.002 Å, respectively. We cleaved surface with the fractional and angstrom thickness of 1.0 Å and 3.508 Å respectively and employ a slab unit cell with the same reconstruction on the top and bottom of the unit cell surface. In order to reduce the systematic error in comparing energies of different distances, we use the same computational setup as much as possible in all calculations. The vacuum orientation was c, thicknesses of the vacuum layers were 10.00 Å and crystal thickness was 13.3609 Å.

The doping effects were modelled by varying the distances between the ruthenium atom and atom on the anatase TiO_2 (1 0 1) surface. The Ru position was varied at an interval from 0.5 Å to 2.5 Å.

4. Results and discussion

Ru-doped anatase TiO_2 (1 0 1) surface were geometrically optimized until all the parameters were below the suitable tolerance for converged systems. The total energies as a function of Ru displacement are presented. Tables 1 present the initial (before geometry optimization) and final (after geometry optimization) atomic distance between the anatase TiO_2 (1 0 1) surface and the adsorbed Ru atom together with total energies. It is noted that Ru-doped atom is displaced away from the surface after optimizing with average bond distance between Ti and Ru atoms being 3.151 Å. The system that gave the most stable energy was then considered for further calculation on electronic properties. Thus, the optimized system whereby Ru atom was placed as 1.000 Å from the surface is considered and its surface structure is shown in Figure 2.

Total 1. Energies of the anatase TiO_2 (1 0 1) surface after optimization with Ru atom placed at various positions.

Initial distance (Å)	Final distance (Å)	Total energies (eV)
0.500	2.712	-7571.643
1.000	3.623	-7573.722
1.500	3.263	-7573.169
2.000	3.098	-7572.414
2.500	3.059	-7572.410

IOP Conf. Series: Journal of Physics: Conf. Series **905** (2017) 012013 doi:10.1088/1742-6596/905/1/012013

Figure 2. Cleaved structures of undoped and Ru-doped anatase TiO_2 (1 0 1) surface, where Ru adsorbed at a distance of 1.000 Å away from one Ti on the surface (from left to right).

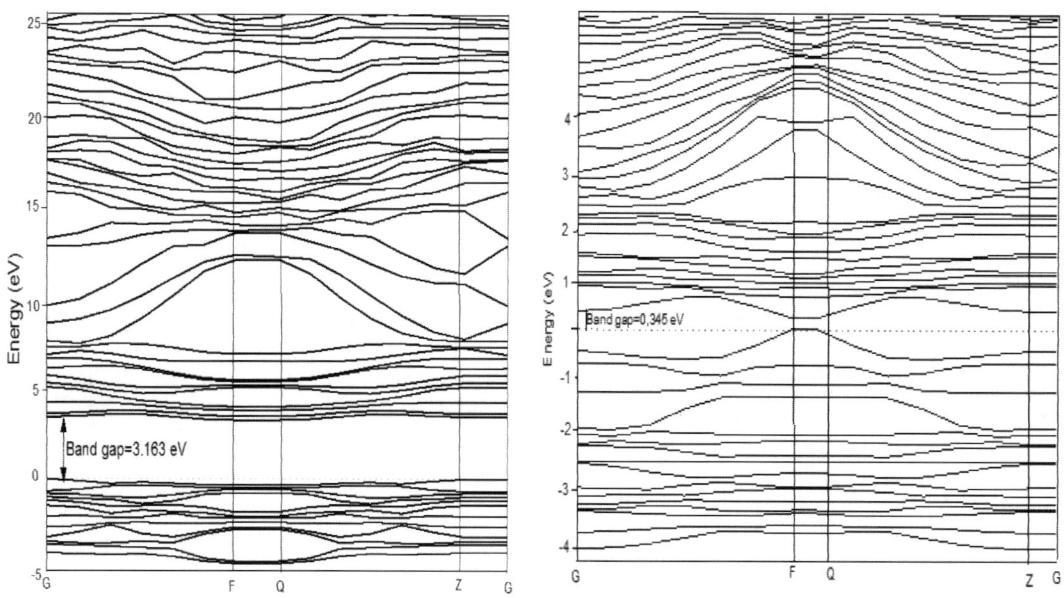

Figure 3.The band gaps of undoped and Ru-doped anatase TiO_2 (1 0 1) surface (from left to right).

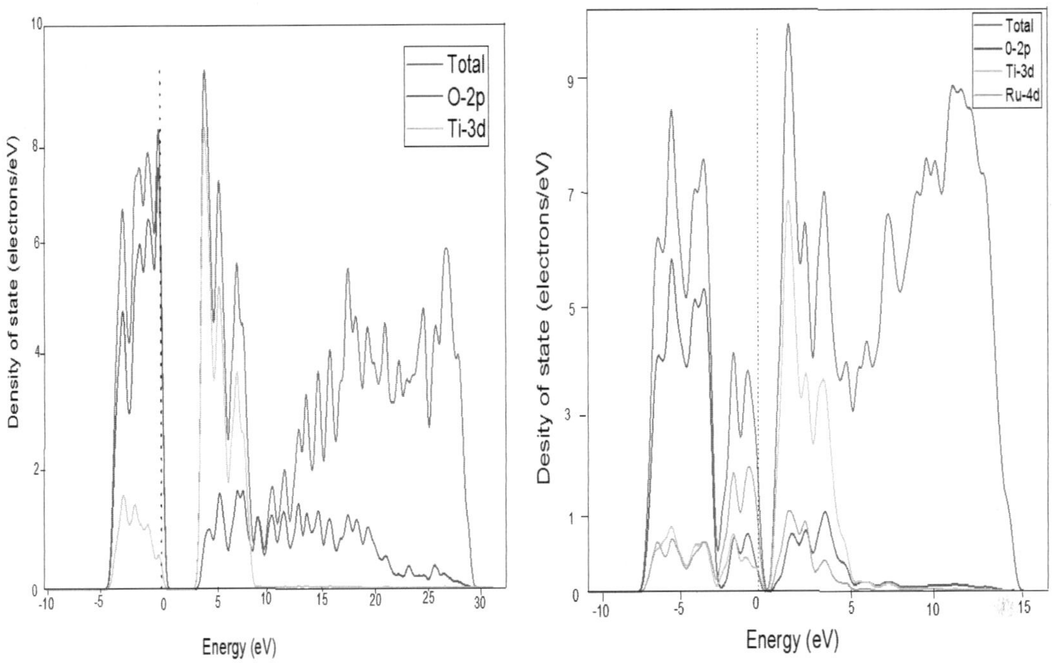

Figure 4. The partial and total density of state of undoped and Ru-doped anatase TiO_2 (1 0 1) surface (from left to right). Fermi level is at 0 eV.

To study the influence of doping on the electronic structures of TiO_2 (1 0 1) surface, the energy band structures of Ru-doped and undoping were calculated and shown in Figure 3. For undoped anatase (1 0 1) surface, the band gap was found to be 3.163 eV which is in agreement with results reported by [23], but with both lower than the experimental result of 3.230 eV [6, 11]. This difference of the calculated band gap from the experimental value can be due to the selection of exchange-correlation energy in DFT. For undoped system, the valance band (VB) near the Fermi level is mainly dominated by O 2p states and the conduction band (CB) is contributed by Ti 3d states, exhibiting that anatase TiO_2 has an indirect band gap. Moreover, Ru-doped systems show a pronounced narrowing of band gap to 0.345 eV as compared to band gap 3.163 eV of undoped (1 0 1) surface of anatase. Furthermore, the study suggests that the rate of the electron transferring from the VB to the CB is enhanced, due to the reduction in band gap doped system as compared to undoped system. The total density of states (TDOS) and partial density of states (PDOS) graph of undoped and Ru-doped TiO_2 is plotted in Figure 4. The valence band is dominantly composed of O 2p states and Ru 4d states while the conduction band is dominantly composed of Ti-3d. After doping the Fermi level is shifts near to the valence band, this shifts of the valence band indicate that Ru-doped anatase TiO_2 (1 0 1) surface is p-type semiconductor.

5. Conclusion

The energy plane-wave pseudopotential method implemented in DFT has been employed successfully to investigate the electronic properties of TiO_2 anatase (1 0 1) surface doped with Ru. When Ru atom is placed at various distance relative to Ti atom it prefer to move away from the surface suggesting a weaker bond between Ru and Ti. The calculations showed that doping TiO_2 with Ru-doping reduces the energy band gap. Thus calculated total and partial density of states for Ru-doped anatase TiO_2 (1 0 1) surface show a sharp decrease in the energy band gap from 3.163 eV to 0.345 eV, this is due to the synergistic effect of Ru.

6. Future work

Calculating optical properties (i.e. reflectivity, absorption, refractive index, dielectric function, conductivity and loss function) of undoped and Ru-doped anatase (1 0 1) surface. Investigating anatase TiO_2 low index surfaces [i.e. (1 0 0), (1 1 0), (1 1 1) and (1 0 1)] doped with other elements, such as alkaline-earth metallic (AEM) using substitution and adsorption methods and compare their electronic and optical properties results.

Acknowledgements

The financially supporter, from National Research Foundation (NRF). The University of Venda and University of Limpopo for the support to carry out this research. Centre for High performance Computing (CHPC) for using their facilities.

References

[1] Torres-Martínez J, Turner C, Uchiyama Y, Vuori S.J.V, Wamukonya N and Zhang X, (2007): In Climate Change, 4, 252-322.
[2] O'Regan B and Grätzel M (1991): Nature, 353, 737-740.
[3] Khalil E.J (2012): Sains malaysiana 41, 1011–1016.
[4] Bisquert J and Vikhrenko V. S (2004): Physics chemistry, B108, 2313-2322.
[5] Chen L.C (2015): Taipei, Taiwan, 1, 110-126.
[6] Kamat P (2012): Journal of Physical Chemistry Letters, 116, 11849–11851.
[7] Tang X.H, Zhang Y, Yin G.F, Zhou D.L, Liu H and Zheng C.Q (2004): Rare metal engineering, 33, 864–868.
[8] Tanaka K, Lapule M.F.V and Hisanaga T (1991): Chemical Physics Letters, 187, 73–76.
[9] Park J.Y, Lee C, Jung K.W and Jung D (2009): Bulletin of the Korean Chemical Society, 2 (30), 402-404.
[10] Linsebigler A.L, Lu G and Yates J.T (1995): Chemical Reviews, 95, 735–758.
[11] Addamo M, Bellardita M, Paola A.D and Palmisano L: (2006), Chemical Communications, 4943–4945.
[12] Qili C, Min L, Kaihua H and Bo L (2014), International Journal of Photoenergy, 2014, 1-7.
[13] Ibhadon A.O and Fitzpatrick P: (2013), Catalysts, 3 (1) 189-218.
[14] Tian F, Liu C, Zhao, Wang X, Wang Z and Yu C.J (2011): Journal of Computational Science and Engineering, 1, 33-41.
[15] Fujishima A and Honda K (1972): Nature, 238 (5358), 37–38.
[16] Navrotsky A (2011): Chem PhysChem, 12 (12), 2207–2215.
[17] Esch T.R Gadaczek I and Bredow T (2014): Applied Surface Science 288, 275– 287.
[18] Chen Q, Liu M, He K and Li B: (2004), Photoenergy, 2014, 1-7.
[19] Chen Q, Tang C and Zheng G (2009): Physical B, Condensed Matter, 404, 1074–1078.
[20] Labat F, Baranek P and Adamo C (2008): Journal of Chemical Theory and Computation, 4 (2), 341–352.
[21] Mino L, Ferrari A.M, Lacivita V, Spoto G, Bordiga S and Zecchina A (2011):, Journal of Physical Chemistry. C, 115 (15), 7694–7700.
[22] Clark S.J, Segall M.D, Pickard C.J, Hasnip P.J, M. J. Probert, Refson,M. C, Payne K, (2005): Zeitschrift fuer Kristallographie, 220(5-6). 567-570.
[23] Lin Y.M, Jiang Z.Y, Hu X.Y, Zhang X.D, Fan J, Miao H and Shang Y.B (2012): Chinese Physics. B, 21(3), 033103 -1-033103-6.

IOP Conf. Series: Journal of Physics: Conf. Series **905** (2017) 012014 doi:10.1088/1742-6596/905/1/012014

Anisotropic solutions in modified gravity

Amare Abebe[1], Davood Momeni[2] and Ratbay Myrzakulov[2]

[1] Department of Physics, North-West University, Mmabatho 2735, Mafikeng, South Africa

[2] Eurasian International Center for Theoretical Physics and Department of General & Theoretical Physics, Eurasian National University, Astana 010008, Kazakhstan

E-mail: amare.abbebe@gmail.com

Abstract. Anisotropic but homogeneous, shear-free cosmological models with imperfect matter sources in $f(R)$ gravity are investigated. The relationship between the anisotropic stresses and the electric part of the Weyl tensor, as well as their evolutions in orthogonal $f(R)$ models, is explored. The late-time behaviour of the de Sitter universe (as an example of a locally rotationally symmetric spacetimes in orthonormal frames) in $f(R)$ gravity is examined. By taking initial conditions for the expansion, acceleration and jerk parameters from observational data, numerical integrations for the evolutionary behavior of the Universe in the Starobinsky model of $f(R)$ have been carried out.

1. Introduction

Despite General Relativity's (GR) great successes in explaining many cosmological and astrophysical scenarios, it miserably fails to provide:

- an elegant solution to the early and current accelerated expansion phases of the Universe,
- the mechanism for dark matter production (if one is convinced that dark matter exists),
- a consistent framework to combine gravity with the other three forces of nature.

As a result, recently a large number of alternative, modified or generalized propositions to GR have emerged, one of which involves the inclusion of higher-order curvature invariants in the Einstein-Hilbert action [1, 2, 3, 4, 5]:

$$A = \frac{1}{2}\int d^4x\sqrt{-g}\left[f(R) + 2\mathcal{L}_m\right] , \qquad (1)$$

where $f(R)$ is some differentiable function in the Ricci curvature scalar R, g is the determinant of the spacetime metric g_{ab}, $\mathcal{L}_m$ is the standard matter Lagrangian, and the natural units ($\hbar = c = k_B = 8\pi G = 1$) have been used. The generalised Einstein Field Equations (EFEs), obtained from variational principles, is given by

$$G_{ab} = \tilde{T}^m_{ab} + T^R_{ab} \equiv T_{ab} , \qquad (2)$$

where the modified matter energy-momentum tensor is given by

$$\tilde{T}^m_{ab} \equiv \frac{T^m_{ab}}{f'} , \quad T^m_{ab} \equiv \frac{2}{\sqrt{-g}}\frac{\delta(\sqrt{-g}\mathcal{L}_m)}{\delta g_{ab}} , \qquad (3)$$

and the energy-momentum tensor of the *curvature fluid* can be defined as

$$T^R_{ab} \equiv \frac{1}{f'}\left[\tfrac{1}{2}(f - Rf')g_{ab} + \nabla_b\nabla_a f' - g_{ab}\nabla_c\nabla^c f'\right] \text{ ; with } f' \equiv df/dR, \text{etc .} \tag{4}$$

The energy-momentum tensor of standard matter is given by

$$T^m_{ab} = \frac{2}{\sqrt{-g}}\frac{\delta(\sqrt{-g}\mathcal{L}_m)}{\delta g^{ab}} = \mu_m u_a u_b + p_m h_{ab} + q^m_a u_b + q^m_b u_a + \pi^m_{ab} \,, \tag{5}$$

where we have defined

- $\mu_m \equiv T^m_{ab} u^a u^b$ as the energy density
- $p_m \equiv \frac{1}{3}(T^m_{ab} h^{ab})$ as the isotropic pressure
- $q^m_a \equiv -T^m_{bc} u^b h^c_a$ as the heat flux
- $\pi^m_{ab} \equiv T^m_{cd} h^c{}_{\langle a} h^d{}_{b\rangle}$ as the anisotropic pressure

of matter. The four-vector $u^a \equiv \frac{dx^a}{dt}$ is the normalized 4-velocity of fundamental observers comoving with the fluid. The covariant time derivative OF any tensor $S^{a..b}_{c..d}$ along an observer's worldlines is defined as

$$\dot{S}^{a..b}_{c..d} = u^e \nabla_e S^{a..b}_{c..d} \,, \tag{6}$$

whereas the fully orthogonally projected covariant derivative for any tensor $S^{a..b}_{c..d}$ is given by

$$\tilde{\nabla}_e S^{a..b}_{c..d} = h^a_f h^p_c ... h^b_g h^q_d h^r_e \nabla_r S^{f..g}_{p..q} \,, \tag{7}$$

with total projection on all the free indices. Here r$h_{ab} \equiv g_{ab} + u_a u_b$ is known as the projection tensor. We extract the orthogonally projected symmetric trace-free part of vectors and rank-2 tensors using

$$V^{\langle a\rangle} = h^a_b V^b \,, \quad S^{\langle ab\rangle} = \left[h^{(a}_c h^{b)}_d - \tfrac{1}{3}h^{ab}h_{cd}\right] S^{cd} \,, \tag{8}$$

and the volume element for the restspaces orthogonal to u^a is given by [6]

$$\varepsilon_{abc} = u^d \eta_{dabc} = -\sqrt{|g|}\delta^0_{[a}\delta^1_b\delta^2_c\delta^3_{d]}u^d \Rightarrow \varepsilon_{abc} = \varepsilon_{[abc]}, \; \varepsilon_{abc}u^c = 0, \tag{9}$$

where η_{abcd} is the 4-dimensional volume element satisfying the conditions

$$\eta_{abcd} = \eta_{[abcd]} = 2\varepsilon_{ab[c}u_{d]} - 2u_{[a}\varepsilon_{b]cd}. \tag{10}$$

The covariant spatial divergence and curl of vectors and rank-2 tensors are given as [7]

$$\text{div}V = \tilde{\nabla}^a V_a \,, \quad (\text{div}S)_a = \tilde{\nabla}^b S_{ab} \,, \quad curlV_a = \varepsilon_{abc}\tilde{\nabla}^b V^c \,, \quad curlS_{ab} = \varepsilon_{cd(a}\tilde{\nabla}^c S_{b)}{}^d \,. \tag{11}$$

In this formalism, u^a can be split into its irreducible parts as

$$\nabla_a u_b = -A_a u_b + \tfrac{1}{3}h_{ab}\Theta + \sigma_{ab} + \varepsilon_{abc}\omega^c, \tag{12}$$

where $A_a \equiv \dot{u}_a$, $\Theta \equiv \tilde{\nabla}_a u^a$, $\sigma_{ab} \equiv \tilde{\nabla}_{\langle a} u_{b\rangle}$ and $\omega^a \equiv \varepsilon^{abc}\tilde{\nabla}_b u_c$ are the 4-acceleration, (volume) expansion, shear and vorticity of the fluid. The thermodynamic quantities for the curvature

IOP Conf. Series: Journal of Physics: Conf. Series **905** (2017) 012014 doi:10.1088/1742-6596/905/1/012014

fluid can be defined similarly to the standard matter ones:

$$\mu_R = \frac{1}{f'}\left[\frac{1}{2}(Rf' - f) - \Theta f''\dot{R} + f''\tilde{\nabla}^2 R\right] , \tag{13}$$

$$p_R = \frac{1}{f'}\left[\frac{1}{2}(f - Rf') + f''\ddot{R} + f'''\dot{R}^2 + \frac{2}{3}\left(\Theta f''\dot{R} - f''\tilde{\nabla}^2 R - f'''\tilde{\nabla}^a R\tilde{\nabla}_a R\right)\right] , \tag{14}$$

$$q_a^R = -\frac{1}{f'}\left[f'''\dot{R}\tilde{\nabla}_a R + f''\tilde{\nabla}_a\dot{R} - \frac{1}{3}f''\Theta\tilde{\nabla}_a R\right] , \tag{15}$$

$$\pi_{ab}^R = \frac{1}{f'}\left[f''\tilde{\nabla}_{\langle a}\tilde{\nabla}_{b\rangle}R + f'''\tilde{\nabla}_{\langle a}R\tilde{\nabla}_{b\rangle}R - \sigma_{ab}\dot{R}f''\right] , \tag{16}$$

whereas the total thermodynamics of the matter-curvature fluid composition is described by

$$\mu \equiv \frac{\mu_m}{f'} + \mu_R , \quad p \equiv \frac{p_m}{f'} + p_R , \quad q_a \equiv \frac{q_a^m}{f'} + q_a^R , \quad \pi_{ab} \equiv \frac{\pi_{ab}^m}{f'} + \pi_{ab}^R . \tag{17}$$

The Weyl conformal curvature tensor

$$C^{ab}{}_{cd} \equiv R^{ab}{}_{cd} - 2g^{[a}{}_{[c}R^{b]}{}_{d]} + \frac{R}{3}g^{[a}{}_{[c}g^{b]}{}_{d]} \tag{18}$$

can be split into its "gravito-electric" (GE) and "gravito-magnetic" (GM) parts, respectively:

$$E_{ab} \equiv C_{agbh}u^g u^h, \qquad H_{ab} = \tfrac{1}{2}\eta_{ae}{}^{gh}C_{ghbd}u^e u^d. \tag{19}$$

The GE and GM components influence the motion of matter and radiation through the geodesic deviation for timelike and null-vector fields, respectively [6]. The GM has no Newtonian analogue, and is responsible for gravitational radiation.

By covariantly $1+3$-splitting the Bianchi and Ricci identities

$$\nabla_{[a}R_{bc]d}{}^e = 0 , \quad (\nabla_a\nabla_b - \nabla_b\nabla_a)u_c = R_{abc}{}^d u_d \tag{20}$$

for the total fluid 4-velocity u^a, we obtain the following field (propagation and constraint) equations. The propagation equations uniquely determine the covariant variables on some initial hypersurface S_0 at $t = t_0$:

$$\dot{\mu}_m = -(\mu_m + p_m)\Theta - \tilde{\nabla}^a q_a^m - 2A_a q_m^a - \sigma_b^a\pi_{a,m}^b , \tag{21}$$

$$\dot{\mu}_R = -(\mu_R + p_R)\Theta + \frac{\mu_m f''}{f'^2}\dot{R} - \tilde{\nabla}^a q_a^R - 2A_a q_R^a - \sigma_b^a\pi_{a,R}^b , \tag{22}$$

$$\dot{\Theta} = -\tfrac{1}{3}\Theta^2 - \tfrac{1}{2}(\mu + 3p) + \tilde{\nabla}_a A^a - A_a A^a - \sigma_{ab}\sigma^{ab} + 2\omega_a\omega^a , \tag{23}$$

$$\dot{q}_a^m = -\tfrac{4}{3}\Theta q_a^m - (\mu_m + p_m)A_a - \tilde{\nabla}_a p_m - \tilde{\nabla}^b\pi_{ab}^m - \sigma_a^b q_b^m - A^b\pi_{ab}^m - \varepsilon_{abc}\omega^b q_m^c , \tag{24}$$

$$\dot{q}_a^R = -\tfrac{4}{3}\Theta q_a^R + \frac{\mu_m f''}{f'^2}\tilde{\nabla}_a R - \tilde{\nabla}_a p_R - \tilde{\nabla}^b\pi_{ab}^R - \sigma_a^b q_b^R - (\mu_R + p_R)A_a - A^b\pi_{ab}^R - \varepsilon_{abc}\omega^b q_R^c , \tag{25}$$

$$\dot{\omega}_a = -\tfrac{2}{3}\Theta\omega_a - \tfrac{1}{2}\varepsilon_{abc}\tilde{\nabla}^b A^c + \sigma_a^b\omega_b , \tag{26}$$

$$\dot{\sigma}_{ab} = -\tfrac{2}{3}\Theta\sigma_{ab} - E_{ab} + \tfrac{1}{2}\pi_{ab} + \tilde{\nabla}_{\langle a}A_{b\rangle} + A_{\langle a}A_{b\rangle} - \sigma_{\langle a}^c\sigma_{b\rangle c} - \omega_{\langle a}\omega_{b\rangle} , \tag{27}$$

$$\dot{E}_{ab} + \tfrac{1}{2}\dot{\pi}_{ab} = \varepsilon_{cd\langle a}\tilde{\nabla}^c H_{b\rangle}^d - \Theta\left(E_{ab} + \tfrac{1}{6}\pi_{ab}\right) - \tfrac{1}{2}(\mu + p)\sigma_{ab} - \tfrac{1}{2}\tilde{\nabla}_{\langle a}q_{b\rangle}$$

$$+ 3\sigma_a^{\langle c}\left(E_{b\rangle c} - \tfrac{1}{6}\pi_{b\rangle c}\right) - A_{\langle a}q_{b\rangle} + \varepsilon_{cd\langle a}\left[2A^c H^d_{b\rangle} + \omega^c(E^d_{b\rangle} + \tfrac{1}{2}\pi^d_{b\rangle})\right] , \tag{28}$$

$$\dot{H}_{ab} = -\Theta H_{ab} - \varepsilon_{cd\langle a}\tilde{\nabla}^c E^d_{b\rangle} + \tfrac{1}{2}\varepsilon_{cd\langle a}\tilde{\nabla}^c \pi^d_{b\rangle}$$
$$+ 3\sigma_a^{\langle c}H_{b\rangle c} + \tfrac{3}{2}\omega_{\langle a}q_{b\rangle} - \varepsilon_{cd\langle a}\left[2A^c E^d_{b\rangle} - \tfrac{1}{2}\sigma^c_{b\rangle}q^d - \omega^c H^d_{b\rangle}\right] . \tag{29}$$

Restrictions on the initial data to be specified are provided by the constraint equations:

$$(C^1)_a := \tilde{\nabla}^b\sigma_{ab} - \tfrac{2}{3}\tilde{\nabla}_a\Theta + \varepsilon_{abc}\left(\tilde{\nabla}^b\omega^c + 2A^b\omega^c\right) + q_a = 0 , \tag{30}$$

$$(C^2)_{ab} := \varepsilon_{cd\langle a}\tilde{\nabla}^c\sigma_{b\rangle}{}^d + \tilde{\nabla}_{\langle a}\omega_{b\rangle} - H_{ab} - 2A_{\langle a}\omega_{b\rangle} = 0 , \tag{31}$$

$$(C^3)_a := \tilde{\nabla}^b H_{ab} + (\mu + p)\omega_a + \varepsilon_{abc}\left[\tfrac{1}{2}\tilde{\nabla}^b q^c + \sigma_{bd}\left(E^d{}_c + \tfrac{1}{2}\pi^d{}_c\right)\right]$$
$$+ 3\omega_b\left(E^{ab} - \tfrac{1}{6}\pi^{ab}\right) = 0 , \tag{32}$$

$$(C^4)_a := \tilde{\nabla}^b E_{ab} + \tfrac{1}{2}\tilde{\nabla}^b\pi_{ab} - \tfrac{1}{3}\tilde{\nabla}_a\mu + \tfrac{1}{3}\Theta q_a$$
$$- \tfrac{1}{2}\sigma^b_a q_b - 3\omega^b H_{ab} - \varepsilon_{abc}[\sigma^{bd}H^c_d - \tfrac{3}{2}\omega^b q^c] = 0 , \tag{33}$$

$$(C^5) := \tilde{\nabla}^a\omega_a - A_a\omega^a = 0 , \tag{34}$$

and the Gauß-Codazzi equations, given by

$$\tilde{R}_{ab} + \dot{\sigma}_{\langle ab\rangle} + \Theta\sigma_{ab} - \tilde{\nabla}_{\langle a}A_{b\rangle} - A_{\langle a}A_{b\rangle} - \pi_{ab} - \frac{1}{3}\left(2\mu - \frac{2}{3}\Theta^2\right)h_{ab} = 0 . \tag{35}$$

Here $\tilde{R}_{ab}$ is the Ricci tensor on 3-D spatial hypersurfaces, the trace of which gives the corresponding (3-curvature) Ricci scalar: $\tilde{R} = 2\mu - \frac{2}{3}\Theta^2 + 2\sigma^2$. The constraint equations must remain satisfied on any hypersurface S_t for all comoving time t.

In orthogonal cosmological models, the matter energy density μ_m and isotropic pressure p_m are measured by an observer moving with the velocity u^a. These models are characterised by the matter energy-momentum tensor representing an anisotropic fluid without heat fluxes [8]

$$T^m_{ab} = \mu_m u_a u_b + p_m h_{ab} + \pi^m_{ab} , \tag{36}$$

and by an irrotational and non-accelerated flow of the vector field u^a, $\omega_a = 0 = A_a$. The revised evolution and constraint equations for orthogonal models are now given by

$$\dot{\mu}_m = -(\mu_m + p_m)\Theta - \sigma^a_b\pi^b_{a,m} , \tag{37}$$

$$\dot{\mu}_R = -(\mu_R + p_R)\Theta + \frac{\mu_m f''}{f'^2}\dot{R} - \tilde{\nabla}^a q^R_a - \sigma^a_b\pi^b_{a,R} , \tag{38}$$

$$\dot{\Theta} = -\tfrac{1}{3}\Theta^2 - \tfrac{1}{2}(\mu + 3p) - \sigma_{ab}\sigma^{ab} , \tag{39}$$

$$\dot{q}^R_a = -\tfrac{4}{3}\Theta q^R_a + \frac{\mu_m f''}{f'^2}\tilde{\nabla}_a R - \tilde{\nabla}_a p_R - \tilde{\nabla}^b\pi^R_{ab} - \sigma^b_a q^R_b , \tag{40}$$

$$\dot{\sigma}_{ab} = -\tfrac{2}{3}\Theta\sigma_{ab} - E_{ab} + \tfrac{1}{2}\pi_{ab} - \sigma^c_{\langle a}\sigma_{b\rangle c} , \tag{41}$$

$$\dot{E}_{ab} + \tfrac{1}{2}\dot{\pi}_{ab} = \varepsilon_{cd\langle a}\tilde{\nabla}^c H^d_{b\rangle} - \Theta\left(E_{ab} + \tfrac{1}{6}\pi_{ab}\right) - \tfrac{1}{2}(\mu + p)\sigma_{ab} - \tfrac{1}{2}\tilde{\nabla}_{\langle a}q^R_{b\rangle}$$
$$+ 3\sigma_a^{\langle c}\left(E_{b\rangle c} - \tfrac{1}{6}\pi_{b\rangle c}\right) , \tag{42}$$

$$\dot{H}_{ab} = -\Theta H_{ab} - \varepsilon_{cd\langle a}\tilde{\nabla}^c E^d_{b\rangle} + \tfrac{1}{2}\varepsilon_{cd\langle a}\tilde{\nabla}^c\pi^d_{b\rangle} + 3\sigma_a^{\langle c}H_{b\rangle c}$$
$$+ \tfrac{1}{2}\varepsilon_{cd\langle a}\sigma^c_{b\rangle}q^d_R , \tag{43}$$

$$(C^{*1})_a := \tilde{\nabla}^b\sigma_{ab} - \tfrac{2}{3}\tilde{\nabla}_a\Theta + q_a^R = 0\ , \tag{44}$$

$$(C^{*2})_{ab} := \varepsilon_{cd(a}\tilde{\nabla}^c\sigma_{b)}{}^d - H_{ab} = 0\ , \tag{45}$$

$$(C^{*3})_a := \tilde{\nabla}^b H_{ab} + \varepsilon_{abc}\left[\tfrac{1}{2}\tilde{\nabla}^b q_R^c + \sigma_{bd}\left(E^d{}_c + \tfrac{1}{2}\pi^d{}_c\right)\right] = 0\ , \tag{46}$$

$$(C^{*4})_a := \tilde{\nabla}^b E_{ab} + \tfrac{1}{2}\tilde{\nabla}^b\pi_{ab} - \tfrac{1}{3}\tilde{\nabla}_a\mu + \tfrac{1}{3}\Theta q_a^R - \tfrac{1}{2}\sigma_a^b q_b - \varepsilon_{abc}\sigma^{bd}H_d^c = 0\ . \tag{47}$$

We notice that a new constraint

$$(C^{*5})_a := \tilde{\nabla}_a p_m + \tilde{\nabla}^b\pi_{ab}^m = 0 \tag{48}$$

comes out of equation (24) as a result of the orthogonality assumption.

2. Shear-free anisotropic models with an imperfect fluid

For imperfect fluids, the the thermodynamic evolution equation for the anisotropic pressure is given by [9, 10]

$$\tau\dot{\pi}_{ab} + \pi_{ab} = -\lambda\sigma_{ab}\ . \tag{49}$$

Here τ and λ are, respectively, relaxation and viscosity parameters. For negligible τ and a positive constant λ, the equation of state between the shear and anisotropic pressure is given by [8]

$$\pi_{ab} = -\lambda\sigma_{ab}\ . \tag{50}$$

Making use of equations (17) and (16), equation (50) can now be rewritten:

$$\pi_{ab}^m + f''\tilde{\nabla}_{\langle a}\tilde{\nabla}_{b\rangle}R + f'''\tilde{\nabla}_{\langle a}R\tilde{\nabla}_{b\rangle}R = \sigma_{ab}\left(\dot{R}f'' - \lambda f'\right)\ . \tag{51}$$

This implies that shear-free in the case of shear-free fluid spacetimes, the above equation and the Gauß-Codazzi equations (35) simplify, respectively, to

$$\pi_{ab}^m = -f''\tilde{\nabla}_{\langle a}\tilde{\nabla}_{b\rangle}R - f'''\tilde{\nabla}_{\langle a}R\tilde{\nabla}_{b\rangle}R\ , \tag{52}$$

$$\tilde{R}_{ab} - \tfrac{1}{3}\tilde{R}h_{ab} = \pi_{ab} = \frac{1}{f'}\left(\pi_{ab}^m + f''\tilde{\nabla}_{\langle a}\tilde{\nabla}_{b\rangle}R + f'''\tilde{\nabla}_{\langle a}R\tilde{\nabla}_{b\rangle}R\right)\ . \tag{53}$$

These results show that even in the case of vanishing anisotropic pressure from matter, spacetime geometries are not necessarily of constant curvature and hence not necessarily FLRW universes. If we allow the matter anisotropic pressure to be nonzero despite a vanishing shear, constant-curvature models are allowed, unlike in GR, provided

$$f''\tilde{\nabla}_{\langle a}\tilde{\nabla}_{b\rangle}R + f'''\tilde{\nabla}_{\langle a}R\tilde{\nabla}_{b\rangle}R = 0\ . \tag{54}$$

In the case of shear-free fluid spacetimes, we notice from equation (41) that the tidal effect (represented by the EM component of the Wely tensor) on the anisotropic stress is given by

$$\pi_{ab} = 2E_{ab}\ . \tag{55}$$

Thus the anisotropic stresses are related to the electric part of the Weyl tensor in such a way that they balance each other, a necessary and sufficient condition for the shear to remain zero if initially vanishing [8, 11].

For nonzero, second-order shear contributions, equation (41) can be approximated by

$$\dot{\sigma}_{ab} \approx -\tfrac{2}{3}\Theta\sigma_{ab}\,, \quad \left(\sigma^2\right)^{\cdot} \approx -\tfrac{4}{3}\Theta\sigma^2. \tag{56}$$

This clearly shows that small perturbations of shear are damped in the class of orthogonal $f(R)$ models in $f(R)$. In agreement with GR results [8], these models are stable if expanding.

Shear-free orthogonal models satisfying equation (55) are purely EM, i.e., $H_{ab} = 0$. Thus, equation (43) reduces to an identity

$$\varepsilon_{cd\langle a}\tilde{\nabla}^c E^d_{b\rangle} = \tfrac{1}{2}\varepsilon_{cd\langle a}\tilde{\nabla}^c \pi^d_{b\rangle}\,, \tag{57}$$

whereas using equations (42) and (47), it can be shown that the evolution and divergence of the EM Weyl tensor are given by

$$\dot{E}_{ab} = -\tfrac{2}{3}\Theta E_{ab} - \tfrac{1}{4}\tilde{\nabla}_{\langle a}q^R_{b\rangle}\,, \quad \tilde{\nabla}^b E_{ab} = \tfrac{1}{6}\left(\tilde{\nabla}_a\mu - \tfrac{1}{3}\Theta q^R_a\right). \tag{58}$$

The decaying of the EM Weyl tensor, and hence of the anisotropic stress tensor, with the expansion is demonstrated by the relation

$$\left(E^2\right)^{\cdot} = -\tfrac{4}{3}\Theta E^2 - \frac{1}{8}\left(\tilde{\nabla}_{\langle a}q^R_{b\rangle}E^{ab} + \tilde{\nabla}^{\langle a}q^{b\rangle}_R E_{ab}\right), \quad E^2 \equiv E_{ab}E^{ab}. \tag{59}$$

3. Illustration using the Starobinsky $f(R)$ model

As a simple illustration, we will try to integrate the Friedmann equation in locally rotationally symmetric spacetimes

$$3\frac{\dot{a}^2}{a^2} + \frac{k}{a^2} = \mu, \tag{60}$$

provided the barotropic EoS , $p_m = (\gamma_m - 1)\mu_m$. If we rewrite (60) using equations (17) and (37) (for shear-free cases, of course), we obtain the model-dependent equation

$$3\frac{\dot{a}^2}{a^2} + \frac{k}{a^2} = \mu^0_m a^{-3\gamma_m} + \frac{1}{f'}\left[\frac{1}{2}(Rf' - f) - \frac{3\dot{a}}{a}f''\dot{R}\right] \tag{61}$$

where μ^0_m is the matter density at the time $t = t_0$ and γ_m is the EoS parameter for the matter content. A qualitative analysis of the late-time behavior of the solutions for (61) in flat $k = 0$ spacetimes without matter gives a de Sitter (dS) solution, with $R = 6H_0^2$ and equation (61) solves to

$$H_0^2 = \frac{1}{6f'}(Rf' - f). \tag{62}$$

The Friedmann equation (60) for generic $f(R)$ model is, in general, a fourth-order ordinary differential equation (ODE). There are no known exact solutions for the full evolution history, but the equation can be solved numerically (such as in terms of quadratures) given appropriate initial conditions. For the purpose of our illustration, if we choose the Starobinsky model,

$$f(R) = R + \alpha R^2\,, \tag{63}$$

equation (61) reduces to the following differential equation:

$$3\frac{\dot{a}^2}{a^2} = \mu^0_m a^{-3\gamma_m} + \frac{\alpha}{2}\frac{R^2 - 12H\dot{R}}{1 + 2\alpha R}\,, \qquad R = 6\left(\frac{\dot{a}^2}{a^2} + \frac{\ddot{a}}{a}\right). \tag{64}$$

This is a third-oder ODE in $a(t)$. To solve it, let us use the cosmological initial conditions (ICs) for the Hubble H, deceleration q, jerk j, and snap s parameters:

$$q \equiv -\frac{\ddot{a}a}{\dot{a}^2}, \qquad j \equiv \frac{a^2}{\dot{a}^3}\frac{d^3a}{dt^3}, \qquad s \equiv \frac{a^3}{\dot{a}^4}\frac{d^4a}{dt^4} \tag{65}$$

evaluated at the present time $t = t_0$, such that

$$a(0) = a_0 = 1\,, \qquad \dot{a}(0) = H_0 a_0\,, \qquad \ddot{a}(0) = -H_0^2 a_0 q_0\,, \qquad \frac{d^3a}{dt^3}(0) = H_0^3 j_0 a_0^{-1}\,. \tag{66}$$

A series solution using these cosmographic parameters in equation (64), evaluated at $t = t_0$ can be given by

$$\begin{aligned} a(t) &= 1 + H_0\,(t-t_0) - 1/2\,{H_0}^2 q_0\,(t-t_0)^2 \\ &- \frac{1}{216}\,\frac{\left(-3\,{H_0}^2 + 54\,{H_0}^4\alpha + \mu_m + 12\,\alpha\,\mu_m {H_0}^2 - 12\,\alpha\,\mu_m^0 {H_0}^2 q_0 + 18\,\alpha\,{H_0}^4 {q_0}^2 + 36\,{H_0}^4\alpha\,q_0\right)}{\alpha\,H_0}\,(t-t_0)^3 \\ &+ \frac{1}{2592}\,\frac{(t-t_0)^4}{\alpha\,{H_0}^2} \times \Big(9\,{H_0}^4 + 162\,{H_0}^6\alpha - 12\,\mu_m^0 {H_0}^2 + 18\,{H_0}^4\alpha\,\mu_m^0 + 108\,{H_0}^4\alpha\,\mu_m^0 q_0 - 54\,{H_0}^6\alpha\,{q_0}^2 \\ &+ 324\,{H_0}^6\alpha\,q_0 - 108\,\alpha\,{H_0}^6 {q_0}^3 - 6\,\mu_m^0 {H_0}^2 q_0 + 9\,\mu_m^0\gamma_m {H_0}^2 + {\mu_m}^2 + 90\,\alpha\,\mu_m^0 {H_0}^4 {q_0}^2 - 12\,\alpha\,{\mu_m^0}^2 {H_0}^2 q_0 \\ &+ 12\,\alpha\,{\mu_m}^2 {H_0}^2 + 108\,\mu_m^0\gamma_m {H_0}^4\alpha - 108\,\mu_m^0\gamma_m {H_0}^4\alpha\,q_0\Big) + O\left[(t-t_0)^5\right] \end{aligned} \tag{67}$$

and can be used to check observational constraints. If we solve equation (64) numerically and plot the solutions versus time, we notice from figure 1 that H is an oscillatory function which can be identified in the late-time as the ΛCDM era.

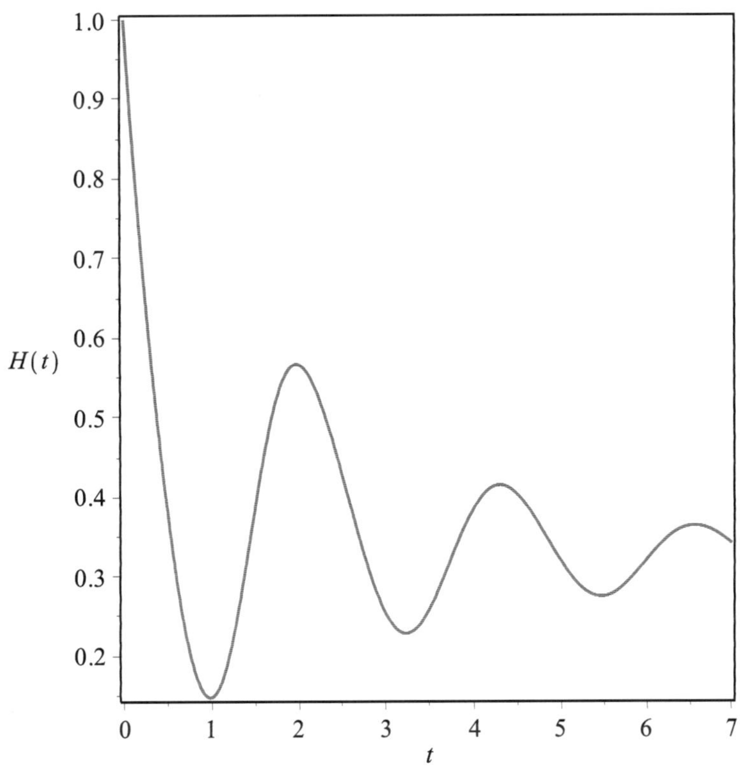

Figure 1. Numerical solution for $H(t)$. Model: $\alpha = 0.02$, $a_0 = H_0 = 1$, $q_0 = -0.7$. The solution, which is oscillatory in nature, can be identified in the late-time as the ΛCDM era.

The Hubble parameter and its first, second and third derivatives of H are plotted in figure 2.

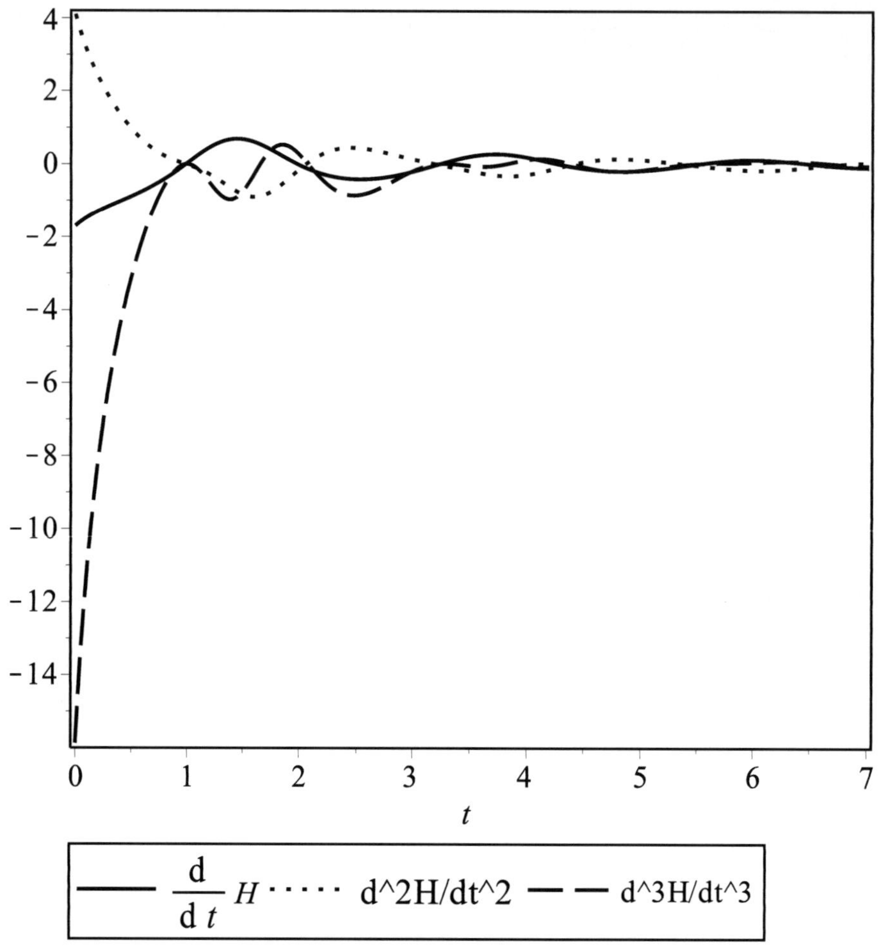

Figure 2. Numerical solution for the first three derivatives of the Hubble parameter. Note the singularity-free nature of the solution, as none of the higher derivatives of H diverges.

No higher derivatives of H diverges and, therefore, our solution is singularity free. Specializing to dust models, i.e., $\Omega_m^0 \equiv \frac{\mu_m^0}{3H_0^2} = 0.3$, $\gamma_m = 1$, we plot the phase portrait for the Starobinsky model in figure 3.

The model is well established as an attractor. Figure 4 shows that the model is a late-time or asymptotic attractor, the solutions to the equations of motion have a generic form independent of the initial conditions.

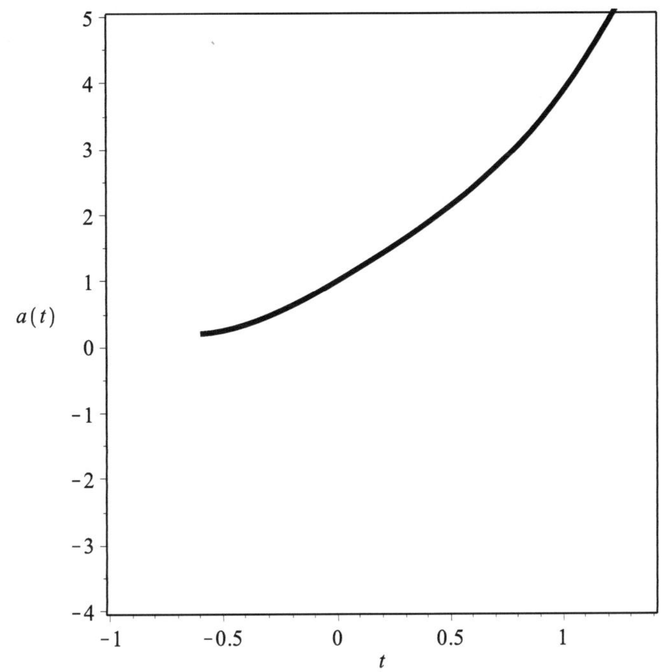

Figure 3. The phase portrait for Starobinsky's dust model. The scale factor $a(t)$ is a monotonically increasing function of cosmic time t.

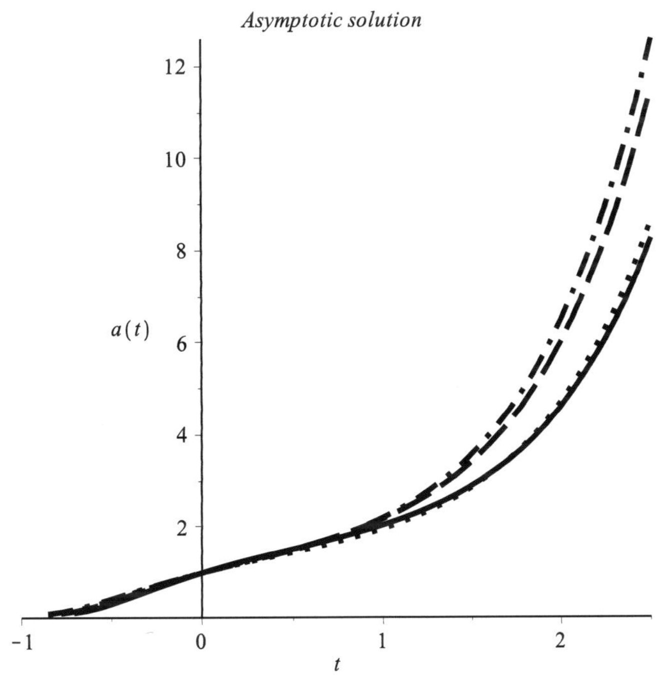

Figure 4. Late-time asymptotic attractors for Starobinsky's model of gravitation. The solutions to the equations of motion have a generic form independent of the initial conditions.

4. Conclusion

In this work we have looked at classes of shear-free anisotropic cosmological spacetimes in $f(R)$ gravity. Specializing to orthogonal models with irrotational and non-accelerated fluid flows without heat fluxes, we have derived the relationship between the anisotropic stresses and electric part of the Weyl tensor, which is the necessary and sufficient condition for the shear to be vanishing forever if vanishing initially.

Moreover, we have shown that within the class of orthogonal $f(R)$ models, small perturbations of shear are damped. Considering a subclass of locally rotationally symmetric spacetimes with barotropic equations of state, we have shown that the late-time behaviour of the dS universe in $f(R)$ gravity should satisfy equation (62).

Finally we have provided a power-series solution for $a(t)$ and studied the behavior of the expansion parameter $H(t)$ by numerically integrating the Friedmann equation (64), where the initial conditions for H_0 , q_0 and j_0 are taken from observational data.

A full computational implementation of the field equations under realistic initial conditions is left for a subsequent work.

Acknowledgments

AA acknowledges the Faculty Research Committee of the Faculty of Agriculture, Science and Technology of North-West University for financial support to attend the *28th IUPAP Conference on Computational Physics.*

References

[1] Buchdahl H A 1970 *Mon. Not. Roy. Astron. Soc.* **150** 1
[2] Starobinsky A A 1980 *Phys. Lett. B* **91** 99–102
[3] Capozziello S and De Laurentis M 2011 *Phys. Rep.* **509** 167–321
[4] Clifton T, Ferreira P G, Padilla A and Skordis C 2012 *Phys. Rep.* **513** 1–189
[5] De Felice A and Tsujikawa S 2010 *Living Rev. Rel.* **13** 1002–4928
[6] Ellis G and van Elst H 1999 Cosmological models *Theoretical and Observational Cosmology* (Dordrecht: Kluver) pp 1–116
[7] Maartens R and Triginer J 1997 *Phys. Rev. D* **56** 4640
[8] Mimoso J P and Crawford P 1993 *Class. Quantum Grav.* **10** 315
[9] Israel W 1976 *Ann. Phys.* **100**(1-2)
[10] Maartens R 1996 *arXiv preprint astro-ph/9609119*
[11] Abebe A, Momeni D and Myrzakulov R 2016 *Gen. Relativ. Gravit.* **48** 1–17

IOP Conf. Series: Journal of Physics: Conf. Series **905** (2017) 012015 doi:10.1088/1742-6596/905/1/012015

Cosmological Chaplygin gas as modified gravity

Maye Elmardi[1] and Amare Abebe[2]

[1] Astrophysics, Cosmology and Gravity Center, Department of Mathematics and Applied Mathematics, University of Cape Town, Rondebosch 7701, Cape Town, South Africa
[2] Department of Physics, North-West University, Mmabatho 2735, Mafikeng, South Africa

E-mail: maye.elmardi@acgc.uct.ac.za, amare.abbebe@gmail.com

Abstract. In the scramble for the understanding of the nature of dark matter and dark energy, it has recently been suggested that the change of behavior of the missing energy density might be regulated by the change in the equation of state of the background fluid. The Chaplygin Gas (CG) model in cosmology is one of the most profound candidates for this suggestion. This work aims to bring to light a geometric interpretation of the model by re-writing the different toy models in terms of exact $f(R)$ gravity solutions that are generally quadratic in the Ricci scalar with appropriate ΛCDM limiting solutions.

1. Introduction

The Chaplygin gas model in FLRW background provides a cosmic expansion history with a universe filled with an exotic background fluid. The model consists of a universe that transits from a decelerating matter-dominated phase to a late-time accelerated one. However, in its intermediate stages, it behaves as a mixture of a cosmological constant and a perfect fluid obeying the $p = w\rho$ equation of state. One can, therefore, mimic the background evolution history of the Universe, including the current observation of cosmic acceleration, using such models [1, 2, 3, 4, 5, 6].

The CG model was introduced first by Chaplygin [7] as a model for aerodynamical studies. In its original form, the mode has an equation of state in the form

$$p = -\frac{A}{\rho^{\alpha}}\,, \tag{1}$$

where p and ρ are respectively pressure and energy density in a comoving reference frame with $\rho > 0$, and A and α are positive constants. Chaplygin considered the case with $\alpha = 1$, but a more *generalised Chaplygin gas* (GCG) equation of state is obtained when $0 \le \alpha \le 1$ [8, 9, 10]. Under homogeneity considerations, the relativistic energy conservation equation

$$\dot{\rho} + 3\frac{\dot{a}}{a}(\rho + p) = 0 \tag{2}$$

in the context of FRW cosmology substituting for the equation of state yields an expression for the density in terms of the scale factor $a(t)$ given as

$$\rho = \left(A + \frac{B}{a^{3(1+\alpha)}}\right)^{\frac{1}{1+\alpha}}\,, \tag{3}$$

where B here is a positive integration constant. For small values of a ($a^6 \ll B/A$) the energy density is approximated by

$$\rho \sim \frac{\sqrt{B}}{a^3}\,, \tag{4}$$

which clearly corresponds to a dust-like dominated phase. For large values of the cosmological radius a, it follows that

$$p \simeq -\rho \Rightarrow \quad \rho \sim \sqrt{A} \Rightarrow \quad p \sim -\sqrt{A}\,, \tag{5}$$

which corresponds to an empty universe with a cosmological constant $\sqrt{A}$ and that is a de Sitter universe, through an intermediate regime described by the equation of state for stiff matter $p = \rho$. There is a possibility of interpreting the model as a "quintessential" model [6, 2, 11] with a critical density

$$\rho_c = (A + B)^{\frac{1}{1+\alpha}}\,, \tag{6}$$

whereas the Hubble parameter is given through the Friedmann equation by

$$H^2 = \left(\frac{\dot{a}}{a}\right)^2 = \frac{1}{3}\rho = \frac{1}{3}\left(A + \frac{B}{a^{3(1+\alpha)}}\right)^{\frac{1}{1+\alpha}}\,. \tag{7}$$

This model automatically leads to an asymptotic phase where the equation of state is dominated by a cosmological constant $\sqrt{A}$. Subsequently, it has been shown that this model admits, under appropriate conditions, an inhomogeneous generalisation, which can be regarded as a unification of dark matter and dark energy models [1]. We can express the perturbed pressure as

$$\delta p = \frac{A}{\rho^2}\delta\rho\,, \tag{8}$$

and for the density contrast we have

$$\delta = \frac{\delta\rho}{\rho} \propto t^{2/3}\,. \tag{9}$$

A very important point here is that the evolution of density perturbations in a universe dominated by the Chaplygin gas admits an initial phase of growing perturbations, with the same rate as in the dust case of the cosmological standard mode [12]. Furthermore, the model predicts an increasing value for the effective cosmological constant [2], *i.e*, in the context of a Chaplygin cosmology, once an expanding universe starts accelerating it cannot decelerate any more, a fact that we seem to be observing today.

For open or flat Chaplygin cosmologies ($k = -1, 0$), the Universe always evolves from a decelerating to an accelerating epoch. For the closed Chaplygin cosmological models ($k = 1$), Einstein static universes have

$$B = \frac{2}{3\sqrt{3A}}\,. \tag{10}$$

The GCG cosmological model, with no additional fluid components, is compatible with structure formation and large-scale structure only for α sufficiently small ($\alpha < 10^{-5}$), in which case it is indistinguishable from the ΛCDM model.

The GCG of Eq. (1) is important to cosmology and it is one of the promising candidates to explain the present accelerated expansion of the universe with simple model unifying dark matter and dark energy [1, 12, 13] as manifestations of a single cosmic fluid [14, 15, 16, 17].

The *modified Chaplygin gas* (MCG) model is often used to describe the acceleration phase of the Universe from the radiation era to the ΛCDM model. It includes a matter term [8, 18]

$$p = \gamma\rho - \frac{A}{\rho^{\alpha}}, \quad 0 \leq \alpha \leq 1, \tag{11}$$

where $0 < \gamma < 1/3$, $0 < \alpha < 1$ and B is a positive constant. It has been shown that $A = 1/3$ is the best fitted value to describe the evolution of the Universe from radiation regime to the Λ-cold dark matter regime [19]. The energy density for such a model is given by

$$\rho = \left[\frac{A}{1+\gamma} + \frac{C}{a^{3(1+\gamma)(1+\alpha)}}\right]^{\frac{1}{1+\alpha}}, \tag{12}$$

where C is an arbitrary integration constant. This model is a more appropriate choice to have constant negative pressure at low energy density and high pressure at high energy density. We can use this equation of state to describe low-surface brightness galaxies which are supposed to be dominated by dark matter [20]. The MCG equation of state is a suitable description for an ordinary linear barotropic fluid, but there are other barotropic fluids with equation of state being quadratic and higher orders. For example, recently the model has been extended [21, 22, 23, 24] so that the resulting equation of state can also recover a barotropic fluids with higher orders

$$p_c = \sum_{i=1}^{n} \gamma_i \rho_c^i - \frac{A}{\rho^{\alpha}}, \tag{13}$$

where p_c and ρ_c are the pressure and energy density of the extended Chaplygin gas which is the unification of the dark matter and dark energy. There is no general solution to this approach. But if we reduce $n = 1$ the above expression recovers the standard MCG. Barotropic fluids with quadratic equation of state can be recovered by setting $n = 2$, reducing Eq. (13) to

$$p_c = \gamma_1\rho_c + \gamma_2\rho_c^2 - \frac{A}{\rho_c^{\alpha}}, \tag{14}$$

where γ_1, γ_2, A and α are positive constants. These models give us the second-order solution

$$\rho_c = \left[\frac{A}{1+\gamma_1} + \frac{C}{a^{3(1+\gamma_1)(1+\alpha)}} e^{-(1+\alpha)(1+\gamma_1)f(\rho_c)}\right]^{\frac{1}{1+\alpha}}, \tag{15}$$

where $C = (1+\gamma_1)^{-1}$ and

$$\begin{aligned} f(\rho_c) = & \frac{\gamma_2\rho_c}{(1+\gamma_1)^2} - \frac{A\gamma_2\rho_c}{(1+\alpha)(1+\gamma_1)^2((1+\gamma_1)\rho_c^{1+\alpha} - A)} \\ & + \gamma_2 \int \frac{A(2+\alpha)}{(1+\alpha)(1+\gamma_1)^2((1+\gamma_1)\rho_c^{1+\alpha} - A)} d\rho_c . \end{aligned} \tag{16}$$

Note that in case of $\gamma_2 = 0$, we have a vanishing $f(\rho_c)$. With higher order n, we will recover a higher-order barotropic fluid.

The *generalized cosmic Chaplygin gas* (GCCG) models are those Chaplygin gas models that admit the equation of state given by [25]

$$p = -\rho^{-\alpha}\left[C + (\rho^{1+\alpha} - C)^{-\omega}\right], \tag{17}$$

where

$$C = \frac{\gamma}{1+\omega} - 1\,, \tag{18}$$

with γ a constant which now can take on both positive and negative values, and $0 > \omega > -l$, l being a positive definite constant which can take on values larger than unity. In the special case when $\omega = 0$ one can write that $C = \gamma - 1$. The speciality of this model is its stability so the theory is free from unphysical behaviours even when the vacuum fluid satisfies the phantom energy condition [19]. The above equation of state satisfies the following conditions:

- it becomes a de Sitter fluid at late time and when $\omega = -1$,
- it reduces to $p = \omega\rho$ in the limit that the Chaplygin parameter $\gamma \to 0$,
- it reduces to the equation of state of current Chaplygin unified dark matter models at high energy density,
- the evolution of density perturbations becomes free from any pathological behaviour of the matter power spectrum for physically reasonable values of the involved parameters at late times.

By integrating the continuity equation (2) we get for the energy density

$$\rho(a) = \left[C + \left(1 + \frac{A}{a^{3(1+\alpha)(1+\omega)}}\right)^{\frac{1}{1+\omega}}\right]^{\frac{1}{1+\alpha}}\,, \tag{19}$$

where B is a positive integration constant. B shows the effect of Chaplygin gas, and the cosmic effect represented by ω. A further extension of the CG model is called *modified cosmic Chaplygin gas* (MCCG) [26, 19], where the EOS is further generalized to

$$p = \gamma\rho - \frac{1}{\rho^{\alpha}}\left[\frac{A}{1+\omega} - 1 + \left(\rho^{1+\alpha} - \frac{A}{1+\omega} + 1\right)^{-\omega}\right]\,. \tag{20}$$

Here A and γ could be both positive or negative constants, and $-l < \omega < 0$ where l is a positive definite constant with values larger than unity. Here also, $0 < \alpha \leq 1$, and the case where $\omega = 0$ gives the equation of state corresponding to the MCG. If we then put $\gamma = 0$, the equation of state corresponding to the GCG is recovered. We can also reach back to the simplest case where $\alpha = 1$, the Chaplygin gas's original equation of state.

Chaplygin gas models have been studied in flat Friedmann models, in terms of the recently proposed "statefinder" parameters [27], dimensionless parameters that allow us to characterise the properties of dark energy in a model-independent manner. It has also been shown that the simple flat Friedmann model with Chaplygin gas can equivalently be described in terms of a homogeneous minimally coupled scalar field ϕ, which has been used in a variety of inflationary models in describing the transition from the quasi-exponential expansion of the early universe to a power law expansion in order to understand the present acceleration of the Universe [2, 28].

The model can be re-expressed as flat Friedman universes containing a scalar field with particular self-interaction potentials [29, 30]; in other words, a very light scalar field ϕ whose effective potential $V(\phi)$ leads to an accelerated phase at the late stages of the Universe [31] by constructing models where the matter responsible for such behaviour is also represented by a scalar field [32, 33].

In [2, 34] a homogeneous scalar field $\phi(t)$ and a potential $V(\phi)$ have been shown to describe Chaplygin cosmology. An extended work is done by [35] with a modified CG. Moreover, the Chaplygin gas is the only gas known to admit a supersymmetric generalisation [36].

2. Chaplygin gas as $f(R)$ gravity?

The main objective of this work is to study models of $f(R)$ gravity which, when we impose the Chaplygin gas equations of state (EoS) to their effective pressure and energy density, produce viable exact solutions that reduce to the ΛCDM scenario in the approximate cosmological limits [37], thus giving a geometric interpretation of Chaplygin gas cosmological models. But first let us briefly discuss $f(R)$ models of gravity. These are models of modified gravity introduced to address the problems and shortcomings of GR, especially in light of the current accelerated epoch of cosmic expansion. They involve modified Einstein-Hilbert action of the form

$$\mathcal{A} = \frac{1}{2}\int d^4x\sqrt{-g}\left[f(R) + 2\mathcal{L}_m\right] . \tag{21}$$

The generalised field equations, using the standard variational principle w.r.t the metric, are given by

$$f'G_{ab} = T^m_{ab} + \frac{1}{2}(f - Rf')g_{ab} + \nabla_b\nabla_a f' - g_{ab}\nabla_c\nabla^c f' , \tag{22}$$

where f', etc. are shorthands derivatives of $f = f(R)$ w.r.t the Ricci scalar R, and T^m_{ab} is the energy-monetum tensor for matter.

These models are among the most widely studied modified theories of gravity [38, 39, 40, 41]. The extra degree of freedom in these models gives us the freedom to explain accelerated cosmic expansion and structure formation of the Universe without adding any extra form of exotic matter.

Assuming the Universe is filled with standard matter and curvature sources, its total energy density, isotropic pressure, anisotropic pressure and heat flux terms are given, respectively, by

$$\rho \equiv \frac{\rho_m}{f'} + \rho_R , \quad p \equiv \frac{p_m}{f'} + p_R , \quad \pi_{ab} \equiv \frac{\pi^m_{ab}}{f'} + \pi^R_{ab} , \quad q_a \equiv \frac{q^m_a}{f'} + q^R_a . \tag{23}$$

Up to first-order perturbations, the curvature fluid components are defined as:

$$\rho_R = \frac{1}{f'}\left[\tfrac{1}{2}(Rf' - f) - \Theta f''\dot{R} + f''\tilde{\nabla}^2 R\right] , \tag{24}$$

$$p_R = \frac{1}{f'}\left[\tfrac{1}{2}(f - Rf') + f''\ddot{R} + f'''\dot{R}^2 + \tfrac{2}{3}\left(\Theta f''\dot{R} - f''\tilde{\nabla}^2 R\right)\right] , \tag{25}$$

$$q^R_a = -\frac{1}{f'}\left[f'''\dot{R}\tilde{\nabla}_a R + f''\tilde{\nabla}_a\dot{R} - \tfrac{1}{3}f''\Theta\tilde{\nabla}_a R\right] , \tag{26}$$

$$\pi^R_{ab} = \frac{f''}{f'}\left[\tilde{\nabla}_{\langle a}\tilde{\nabla}_{b\rangle}R - \sigma_{ab}\dot{R}\right] . \tag{27}$$

For FLRW spacetimes, the Ricci scalar R is given by

$$R = 2\dot{\Theta} + \frac{4}{3}\Theta^2 , \tag{28}$$

where Θ is the cosmic expansion parameter related to the cosmological scale factor $a(t)$ and the Hubble parameter $H(t)$ via the equations

$$\Theta \equiv 3\frac{\dot{a}(t)}{a(t)} = 3H(t) . \tag{29}$$

If the Ricci scalar varies slowly, i.e., if R is almost constant in time, during steady-state exponential expansion in a de Sitter spacetime (such as during inflation or late-time evolution), the approximation $\dot{\Theta} \to 0$ results in

$$R = \frac{4}{3}\Theta^2 = \text{const} . \tag{30}$$

Using this approximation, if we consider the background curvature energy density and isotropic pressure terms defined in Eqs. (24) and (25) above, we obtain the simple relation

$$\rho_R = \frac{1}{4}\left[R(f'+1) - 2f\right] = -p_R\,. \tag{31}$$

This equation of state, with an effective EoS parameter $w_R = -1$, provides the condition for an exponential (accelerated) expansion with a constant Hubble parameter. The energy density ρ_R (with its negative pressure p_R) remains constant and can be interpreted as playing the role of the cosmological constant Λ.

The plan here is, given the different equations of state for the Chaplygin gas models, to come up with models of $f(R)$ gravity that can mimic those models. In other words, we want to do away with these exotic matter forms (Chaplygin gas models) in favor of a modification in the gravitational Lagrangian, i.e., geometry. To solve for the appropriate forms of $f(R)$ corresponding to different Chaplygin gas equations of state, we solve the resulting differential equations using the symbolic computational packages of *Maple*. This is possible because of the nature of the approximations made, namely, the slowly-changing Ricci scalar assumption. Such (analytic) solutions are intractable if this assumption is relaxed, and a future work will involve a fully numerical computation of such solutions, once the appropriate initial conditions are fully understood.

Comparing the original and generalized Chaplygin gas EoS (1) with (31), we obtain

$$p_R = -\rho_R = -\frac{A}{\rho_R^{\alpha}}\,, \tag{32}$$

which, upon using Eq. (24), leads to the o.d.e

$$R\frac{f(R)}{dR} - 2f(R) + R = 4A^{\frac{1}{\alpha+1}}\,. \tag{33}$$

Solving this o.d.e yields

$$f(R) = R + C_1 R^2 - 2A^{\frac{1}{\alpha+1}} \tag{34}$$

for an arbitrary (integration) constant C_1. We note that the ΛCDM solution $f(R) = R - 2\Lambda$ is already a particular solution with $C_1 = 0$ and $A = \Lambda^{\alpha+1}$. In particular, if $\alpha = 0$, then $A = \Lambda$, from which, going back to Eq. (32), one concludes $\rho_R = \Lambda$.

If we include the linearized Laplacian term in Eqs. (24) and (25) and use the eigenvalue $-\frac{k^2}{a^2}$ of the covariantly defined Laplace-Beltrami operator $\tilde{\nabla}^2$ on (almost) FLRW spacetimes

$$\tilde{\nabla}^2 R = -\frac{k^2}{a^2}R \tag{35}$$

for a comoving wavenumber k, we obtain the second-order o.d.e

$$D^2 R f''(R) - Rf'(R) + 2f(R) - R + 4A^{\frac{1}{\alpha+1}} = 0\,, \tag{36}$$

where here we have defined

$$D^2 \equiv \frac{4(2+3\alpha)}{3(1+\alpha)}\frac{k^2}{a^2}\,. \tag{37}$$

The solution of Eq. (36) is given, for arbitrary constants C_2, C_3, by

$$f(R) = R + C_2\left[R^2 - 2RD^2\right] + C_3\left[(R^2 - 2RD^2)Ei\left(1, -\frac{R}{D^2}\right) + (R - D^2)D^2 e^{\frac{R}{D^2}}\right] - 2A^{\frac{1}{\alpha+1}}\,, \tag{38}$$

which should reduce to the quadratic solution (34) for negligible values of D^2, i.e., for small first-order contributions to the energy density and pressure terms.

For the MCG EoS of Eq. (11) , we can write

$$p_R = \gamma \rho_R - \frac{A}{\rho_R^\alpha} \,, \tag{39}$$

and the resulting $f(R)$ model generalizes to

$$f(R) = R + C_4 R^2 - 2\left(\frac{A}{\gamma+1}\right)^{\frac{1}{\alpha+1}} \,, \tag{40}$$

where C_4 is an arbitrary integration constant.

The ΛCDM solution is a limiting case of this generalized model when $C_4 = 0$ and $A = (\gamma+1)\Lambda^{\alpha+1}$. In particular, if $\alpha = 0 = \gamma$, then $A = \Lambda$.

Following similar arguments as in the preceding subsection, if we include the linearized Laplacian contributions to the energy density and pressure, we get Eq. (36) generalized to

$$D^2 R f''(R) - R f'(R) + 2f(R) - R + 4\left(\frac{A}{\gamma+1}\right)^{\frac{1}{\alpha+1}} = 0 \,, \tag{41}$$

the solution of which can be given by

$$f(R) = R + C_5\left[R^2 - 2RD^2\right] + C_6\left[(R^2 - 2RD^2)Ei\left(1, -\frac{R}{D^2}\right) + (R - D^2)D^2 e^{\frac{R}{D^2}}\right] - 2\left(\frac{A}{\gamma+1}\right)^{\frac{1}{\alpha+1}} \,, \tag{42}$$

for an arbitrary integration constants C_5 and C_6. This solution obviously generalizes Solutions (34),(38) and (40) and should reduce to the quadratic solution (34) for vanishingly small B^2 values. In [42], it has been shown that any quadratic Lagrangian leading to an isotropic, homogeneous cosmological model takes the form

$$f(R) = R - 2\Lambda - \frac{1}{6}\beta R^2 \,, \tag{43}$$

where β is an arbitrary, real constant. If we keep only the quadratic solution in (42), i.e., if we set $C_6 = 0$, the Lagrangian (43) corresponds to the choice

$$C_5 = -\frac{1}{6}\beta \,, D = 0 \,, A = (\gamma+1)\Lambda^{\alpha+1} \,. \tag{44}$$

Another interesting fact worth pointing out here is that the condition for the existence of a maximally symmetric vacuum solution in $f(R)$ gravity [42]

$$R_0 f'(R_0) = 2f(R_0) \tag{45}$$

leads to the quadratic solution resulting in the constraint

$$R_0\left(1 - 2C_5 D^2\right) - 4\left(\frac{A}{\gamma+1}\right)^{\frac{1}{\alpha+1}} = 0 \,. \tag{46}$$

The corresponding GR de Sitter, anti-de Sitter and Minkowski solutions $R_0 = 4\Lambda$ (respectively for $\Lambda > 0$, $\Lambda < 0$ and $\Lambda = 0$) are obtained when $C_5 D^2 = 0$ and $A = (\gamma + 1)\Lambda^{\alpha+1}$.

The so-called *modified generalized Chaplygin gas* (mGCG) model is described by a barotropic equation of state of the form [43, 44]

$$p = \beta\rho - (1+\beta)\frac{A}{\rho^\alpha} \,. \tag{47}$$

Models of $f(R)$ gravity that satisfy the condition (31), at the sam time mimicking the mGCG, can be shown to be governed by the same equation as (33) and admit the same solutions (34), provided $\beta \neq -1$. On the other hand, if linearized Laplacian terms are included, then the corresponding differential equation in $f(R)$ generalizes to

$$E^2 R f''(R) - R f'(R) + 2f(R) - R + 4A^{\frac{1}{\alpha+1}} = 0 \,, \tag{48}$$

where we have defined

$$E^2 \equiv \frac{4\left[2 + 3\alpha + 3\beta(1+\alpha)\right]}{3(1+\alpha)(1+\beta)}\frac{k^2}{a^2} \,. \tag{49}$$

Worthy of note is that this equation and its solution

$$f(R) = R + C_7\left[R^2 - 2RE^2\right] + C_8\left[(R^2 - 2RE^2)Ei\left(1, -\frac{R}{E^2}\right) + (R - E^2)E^2 e^{\frac{R}{E^2}}\right] - 2\left(\frac{A}{\gamma+1}\right)^{\frac{1}{\alpha+1}} \,, \tag{50}$$

reduce to their *generalized* counterparts of Eqs. (36) and (38) when $\beta = 0$.

3. Conclusion

To summarize, we have explored exact $f(R)$ gravity solutions that mimic Chaplygin-gas inspired ΛCDM cosmology for the so-called original, generalized, modified and generalized modified Chaplygin gas equations of state. The resulting solutions are generally quadratic in the Ricci scalar, but have appropriate ΛCDM solutions as their limiting cases. These solutions, given appropriate initial conditions, can be potential candidates for scalar field-driven early universe expansion (inflation) and dark energy-driven late-time cosmic acceleration.

The solutions discussed here are based on a slowly-changing Ricci curvature assumption and can generally be obtained using symbolic computing environments such as *Maple*. However, more realistic solutions should relax this assumption, and consider higher-order corrections as well. Such models are generally complex to solve, and advanced numerical computations require more physically motivated initial conditions, currently not fully understood, but which the authors would like to explore further in a future work.

Acknowledgments

ME acknowledges the hospitality of the Department of Physics of North-West University (Mafikeng) where most of this work was conducted. AA acknowledges the Faculty Research Committee of the Faculty of Agriculture, Science and Technology of North-West University for financial support to attend the *28th IUPAP Conference on Computational Physics.*

IOP Conf. Series: Journal of Physics: Conf. Series **905** (2017) 012015 doi:10.1088/1742-6596/905/1/012015

References

[1] Bilić N, Tupper G B and Viollier R D 2002 *Physics Letters B* **535** 17–21
[2] Kamenshchik A, Moschella U and Pasquier V *Physics Letters B* **511**
[3] Gorini V, Kamenshchik A and Moschella U 2003 *Physical Review D* **67** 063509
[4] Gorini V, Kamenshchik A, Moschella U and Pasquier V 2003 **20** 26
[5] Chattopadhyay S and Debnath U 2010 *Astrophysics and Space Science* **326** 155–158
[6] Bento M, Bertolami O and Sen A 2002 *Physical Review D* **66** 043507
[7] Chaplygin S 1944 *Trans. by M. Slud, Brown University*
[8] Xu L, Lu J and Wang Y 2012 *The European Physical Journal C* **72** 1–6
[9] Lu J, Xu L, Wu Y and Liu M 2011 *General Relativity and Gravitation* **43** 819–832
[10] Meng X H, Hu M G and Ren J 2005 *arXiv preprint astro-ph/0510357*
[11] Parker L and Raval A 1999 *Physical Review D* **60** 063512
[12] Fabris J C, Goncurlcalves S V and de Souza P E 2002 *General Relativity and Gravitation* **34** 53–63
[13] Bean R and Dore O 2003 *Physical Review D* **68** 023515
[14] Davidson A, Karasik D and Lederer Y 2001 *arXiv preprint gr-qc/0111107*
[15] Kasuya S 2001 *Physics Letters B* **515** 121–124
[16] Wetterich C 2002 *Physical Review D* **65** 123512
[17] Matos T 2000 *Phys. Rev. D* **63** 63506
[18] Dao-Jun L and Xin-Zhou L 2005 *Chinese Physics Letters* **22** 1600
[19] Pourhassan B 2013 *International Journal of Modern Physics D* **22** 1350061
[20] Bertolami O and Paramos J 2005 *Physical Review D* **72** 123512
[21] Pourhassan B and Kahya E 2014 *Results in Physics* **4** 101–102
[22] Kahya E and Pourhassan B 2015 *Modern Physics Letters A* **30** 1550070
[23] Pourhassan B and Kahya E 2014 *Advances in High Energy Physics* **2014**
[24] Pourhassan B 2015 *arXiv preprint arXiv:1504.04173*
[25] González-Díaz P F 2003 *Physical Review D* **68** 021303
[26] Saadat H and Pourhassan B 2013 *Astrophysics and Space Science* **344** 237–241
[27] Sahni V, Saini T D, Starobinsky A A and Alam U 2003 *Journal of Experimental and Theoretical Physics Letters* **77** 201–206
[28] Kahya E, Pourhassan B and Uraz S 2015 *Physical Review D* **92** 103511
[29] Barrow J D 1988 *Nuclear Physics B* **310** 743–763
[30] Barrow J D 1990 *Physics Letters B* **235** 40–43
[31] Dev A, Alcaniz J and Jain D 2003 *Physical Review D* **67** 023515
[32] Starobinsky A 2000 *Phys. Rev. Lett* **85** 1162
[33] Barrow J D 1987 *Physics Letters B* **183** 285–288
[34] Gorini V, Kamenshchik A, Moschella U and Pasquier V 2004 *Physical Review D* **69** 123512
[35] Debnath U, Banerjee A and Chakraborty S 2004 *Classical and Quantum Gravity* **21** 5609
[36] Jackiw R 2000 *arXiv preprint physics/0010042*
[37] Elmardi M, Abebe A and Tekola A 2016 *International Journal of Geometric Methods in Modern Physics* **13** 1650120
[38] De Felice A and Tsujikawa S 2010 *Living Rev. Rel* **13** 1002–4928
[39] Clifton T, Ferreira P G, Padilla A and Skordis C 2012 *Physics Reports* **513** 1–189
[40] Faraoni V 2008 *arXiv preprint arXiv:0810.2602*
[41] Carloni S, Dunsby P and Troisi A 2008 *Physical Review D* **77** 024024
[42] Barrow J D and Ottewill A C 1983 *Journal of Physics A: Mathematical and General* **16** 2757
[43] Bouhmadi-López M, Brilenkov M, Brilenkov R, Morais J and Zhuk A 2015 *Journal of Cosmology and Astroparticle Physics* **2015** 037
[44] Morais J, Bouhmadi-López M and Capozziello S 2015 *Journal of Cosmology and Astroparticle Physics* **2015** 041

IOP Conf. Series: Journal of Physics: Conf. Series **905** (2017) 012016 doi:10.1088/1742-6596/905/1/012016

Replica Exchange Wang–Landau Simulation of Lattice Protein Folding Funnels

Guangjie Shi[1], **Thomas Wüst**[2] **and David P. Landau**[1]

[1]Center for Simulational Physics, The University of Georgia, Athens, GA 30602, USA
[2]Scientific IT Services, ETH Zürich, 8092 Zürich, Switzerland

E-mail: sgjerry@physast.uga.edu

Abstract. The resolution of Levinthal's paradox concerning the ability of proteins to fold rapidly postulates the existence of a rough "folding funnel" in free energy space that guides the protein to its lowest free energy, native state. To study the folding of the protein ribonuclease A we have mapped it onto a 124 monomer, coarse-grained HP lattice model and onto an H0P model that also includes "neutral" 0-mers in addition to the hydrophobic H-mers and polar P-mers. Using Replica Exchange Wang-Landau sampling, we determined the density of states, g(E), which we then used to calculate the free energy of the protein vs end-to-end distance as a function of temperature. At low temperature the HP model shows a rather shallow and flat free energy minimum, while the H0P model maintains a rough free energy funnel. Unlike the common, schematic figures, we find an asymmetric folding funnel that also changes shape substantially as the temperature decreases. Even the location of the free energy minimum shifts as the temperature decreases.

1. Introduction

Understanding protein folding remains a Grand Challenge problem of modern science. The protein folding funnel concept [1, 2], which is believed to be the explanation of the protein folding problem, has always been portrayed schematically as a function of some unknown reaction coordinate (as seen in Fig. 1) and has never actually been observed. However, due to the complexity of the problem, investigation is only possible through the combinations of coarse-grained models and advanced computational methods. The hydrophobic-polar (HP) lattice protein model [3, 4] is arguably the simplest protein model, yet it has contributed to the understanding of many problems of biological interest [5]-[6] An extension of the HP protein model, the semi-flexible H0P lattice protein model [7], introduced some simple modifications which render the model more realistic with significantly reduced ground-state degeneracy [8].

Even though the HP model is highly simplified, the problem of finding the ground-state structure of a given HP sequence is extremely difficult [9] and ground states are often highly degenerate. Consequently, the HP model represents a significant challenge at the interface between statistical physics and biophysics. The HP model has thus become a testing ground for many advanced algorithms: sequential importance sampling [10], fragment regrowth Monte Carlo [11], chain-growth methods [12, 13], e.g. the pruned-enriched Rosenbluth method (PERM) and its variants [14, 15, 16, 17, 18], multidomain sampler [19], genetic algorithm [20, 21], evolutionary Monte Carlo [22], ant colony models [23] and Wang–Landau sampling [24, 25]. Here we have applied an efficient, parallel version of the Wang–Landau method, i.e. Replica

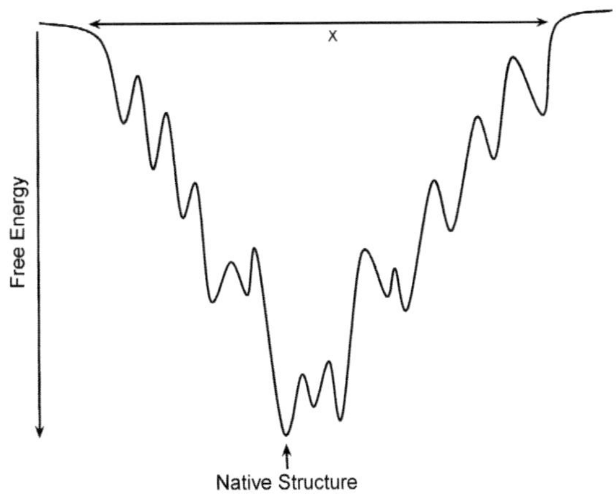

Figure 1. Schematic view of a rough protein folding funnel.

Exchange Wang–Landau (REWL) sampling to both the HP model and the semi-flexible H0P model to uncover protein folding funnels.

2. Model and methods

2.1. HP and semi-flexible H0P lattice protein models

Based on the properties of their side chains, amino acids are classified into hydrophobic (H) and polar (P) in the HP lattice protein model [3, 4]. In this model, a chain of connected monomers represents the amino acid sequence residing on a simple cubic lattice. An effective monomer-monomer coupling $\epsilon_{\rm HH}$ between non-bonded, nearest-neighbor H monomers is used for characterizing the driving force of protein folding: the hydrophobic interaction:

$$\mathcal{H} = -\epsilon_{\rm HH} n_{\rm HH}, \tag{1}$$

where $n_{\rm HH}$ is the number of non-bonded HH contacts.

An improved version of the HP model, i.e. a semi-flexible H0P lattice protein model, introduces a "neutral" type monomer ("0") representing amino acids that are neither hydrophobic nor polar. This model also considers the stiffness of angles formed by bonds connecting monomers, and the general Hamiltonian of the semi-flexible H0P model can be written as

$$\mathcal{H} = -\epsilon_{\rm HH} n_{\rm HH} - \epsilon_{\rm H0} n_{\rm H0} - \epsilon_{\theta} n_{\theta}, \tag{2}$$

where $n_{\rm HH}$ is the number of non-bonded HH contacts ($n_{\rm H0}$ has corresponding meaning), and n_θ represents the number of angles existing in the protein structure (see Fig.2). We used $\epsilon_{\rm HH} = 1$, $\epsilon_{\rm H0} = 0.5$ and $\epsilon_\theta = -0.25$ for this work, but other interaction values or interactions involving "0" monomers (or more distant neighbors) could be easily added. Our intent, however, was to add only as many features as are needed to render the original HP model more realistic while still keeping the model as simple as possible.

2.2. Wang–Landau sampling

Wang–Landau (WL) sampling [26, 27] is an iterative Monte Carlo (MC) simulation method, which performs a random walk over the energy space for estimating the density of states of the system under investigation. This method, using pull moves and bond-rebridging moves to

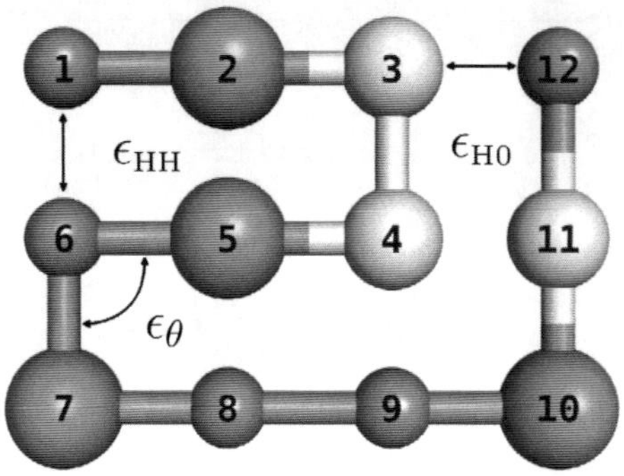

Figure 2. A schematic example of a semi-flexible H0P model. The interaction between monomers 1 and 6 is ϵ_{HH}, and that between monomers 3 and 12 is ϵ_{H0}. The angle constituted by monomers 5, 6 and 7 contributes ϵ_θ energy. In this particular 2-dimensional structure, $n_{\mathrm{HH}} = 1$, $n_{\mathrm{H0}} = 2$ and $n_\theta = 5$. Hydrophobic and "neutral" monomers are colored dark gray and white, respectively, while polar monomers are colored orange. An HP model results if no 0-mers are present and the angle energy is zero. Both models are actually simulated in three dimensions.

generate new, trial states, has been shown to be powerful and efficient in both finding ground state structures and determining density of states ($g(E)$) [24, 25]. During the simulation, a MC trial move, changing the system from energy E_A to energy E_B, will be accepted with probability

$$P(A \to B) = \min\left\{1, \frac{g'(E_A)}{g'(E_B)}\right\}, \tag{3}$$

where $g'(E)$ is an instantaneous estimator for the density of states. This estimator is updated through $g'(E_B) \to f \times g'(E_B)$ at each step, where f is a modification factor. The simulation maintains a histogram of visited energies, which is increased by $H(E_B) \to H(E_B) + 1$. The modification factor will be reduced if the histogram is sufficiently "flat" and at the same time, all the histogram entries are reset to zero. The simulation continues until the modification factor is less than some predefined threshold value f_{final}. In this work, the initial modification factor is set to $f_{init} = e^1$ and decreased via $f \to \sqrt{f}$, until reaching $f_{\mathrm{final}} = 1 \times 10^{-8}$. Initially, $g'(E)$ is set to 1 and we use the "80%" flatness criterion for the histogram, i.e., every entry in the histogram is no less than the 80% of the mean histogram height at each iteration.

2.3. Replica Exchange Wang–Landau sampling

For determining $g(E)$ of systems with the size of interest in this work, it was necessary to employ Replica Exchange Wang–Landau sampling (REWL) [28, 29], a generic, parallel Wang–Landau sampling scheme. As shown in Fig. 3, the energy range of the system is split into multiple, overlapping windows, each of which is simulated by independent processes (random walkers, running serial WL sampling). The replica exchanges between overlapping windows are attempted during the simulation at a fixed time interval, such that each replica can travel through the entire energy space of interest. The simulation terminates only when all random walkers have satisfied the termination criteria independently.

3. Results and discussion

REWL sampling was performed for both the HP sequence (denoted as HP124) [30] and the H0P sequence (denoted as H0P124) based on the mapping of ribonuclease A. (See ref. [7] for more

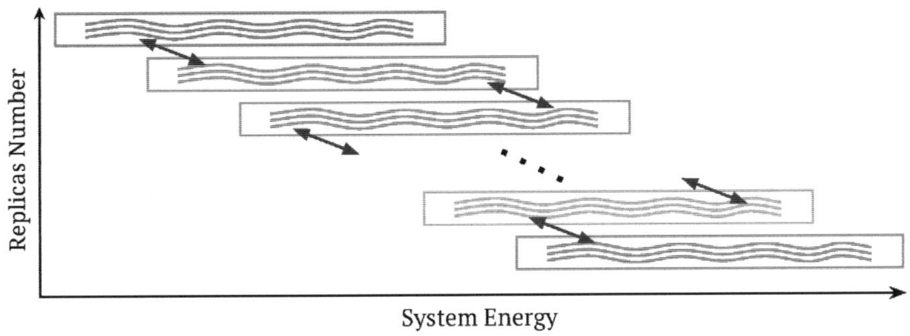

Figure 3. Illustration of the general framework of Replica Exchange Wang–Landau sampling. The system energy range is split into multiple, overlapping energy windows. Each window employs multiple independent processes (or walkers) running serial Wang–Landau sampling. At fixed time intervals, the replica exchange procedure is performed between two neighboring energy windows.

about the mapping.) The density of states was determined to high precision for both models and used to estimate the partition function and both thermodynamic and structural properties. If Q represents a structural quantity which can be used as a reaction coordinate, we can calculate the free energy with respect to temperature and Q based on the joint density of states $g(E, Q)$ as:

$$F\,(T, Q) = -k_B T \, \ln Z(T, Q), \tag{4}$$

where $Z(T, Q)$ is the partition function based on temperature and Q:

$$Z\,(T, Q) = \sum_E g(E, Q)\, e^{-E/k_B T}. \tag{5}$$

After initial exploration, we found that the end-to-end distance:

$$r_{ee} = |\vec{r}_N - \vec{r}_1|$$

was apparently a good candidate for representing the folding funnel, where N is the number of monomers in the chain; $\vec{r}_i$ represents the position of the ith monomer.

In Fig. 4 we present the normalized free energy vs end-to-end distance at two different temperatures for the HP124 lattice protein. The quality of the results was quite high and the statistical errors in the figure are smaller than the size of the symbols. The state, and associated end-to-end distance, with lowest free energy is indicated by a black arrow, while the mean r_{ee} is marked by an orange arrow. At high temperature (as seen in Fig. 4(a)), the free energy shows a shallow, "symmetric" landscape which contains many local maxima and minima. The state with lowest free energy has a different end-to-end distance than the mean end-to-end distance, and there are many minima and maxima between the two states. Upon lowering the temperature we found the free energy curve changes shape and becomes skewed toward the region with low r_{ee} values (as seen in Fig. 4(b)). Moreover, unlike the static schematic portrayals of the protein folding funnel which always has a fixed minimum, the lowest free energy position shifts with the change in temperature, showing the dynamic nature of the folding funnel. Finally, at quite low temperatures as seen in Fig. 4(b), the free energy landscape of HP124 becomes relatively flat near the minimum, and the system can easily move between states with different end-to-end

IOP Conf. Series: Journal of Physics: Conf. Series **905** (2017) 012016 doi:10.1088/1742-6596/905/1/012016

distances. At low temperature, the state with the lowest free energy is then also the state with the mean end-to-end distance. For greater end-to-end distances, the free energy curve assumes a rather odd shape with a flat plateau region.

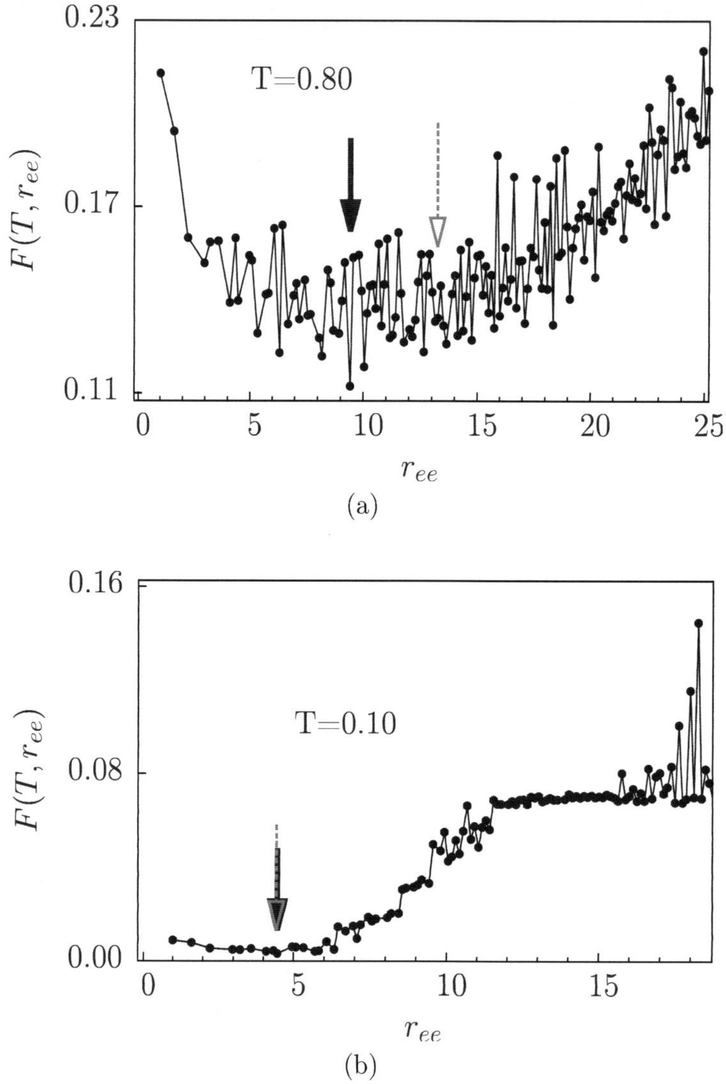

Figure 4. Normalized free energy vs end-to-end distance at (a) $T = 0.8$ and (b) $T = 0.1$ for the HP124 lattice protein. Black arrows indicate the lowest free energy at each temperature (T), orange arrows point to the mean end-to-end distance at that temperature. Error bars are smaller than the data points.

The normalized free energy vs end-to-end distance at two different temperatures for the H0P124 lattice protein shows significant differences from its HP counterpart. Results are shown in Fig. 5. The state with lowest free energy is indicated by a black arrow, while the mean r_{ee} is marked by an orange arrow. At high temperatures (as seen in Fig. 5 (a)) we observed a rugged free energy landscape which is similar to that for HP124, but as the temperature is lowered, substantial differences become apparent. The free energy landscape remains rough but skews towards the region with low r_{ee} values, and the position of the lowest free energy also shifts

IOP Conf. Series: Journal of Physics: Conf. Series **905** (2017) 012016 doi:10.1088/1742-6596/905/1/012016

with temperature. However, at low temperatures, unlike the free energy landscape of HP124, that of H0P124 remains rough even near the bottom (as seen in Fig. 5 (b)) and does not show the curious, plateau region. The inset shows clearly that significant free energy barriers must be overcome to move between free energy minima and the protein can be easily trapped in a metastable state if traditional simulation methods are used.

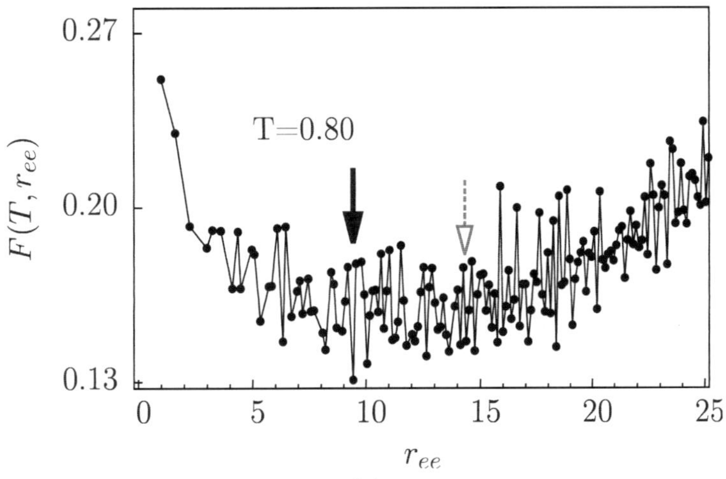

(a)

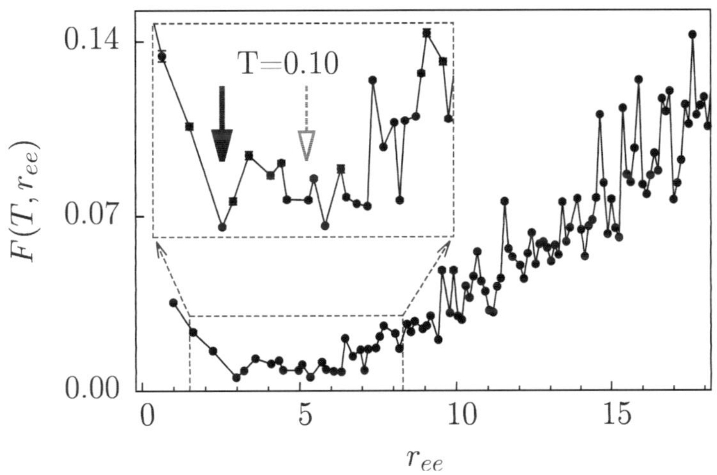

(b)

Figure 5. Normalized free energy vs end-to-end distance at (a) $T = 0.8$ and (b) $T = 0.1$ for the H0P124 lattice protein. Black arrows indicate the lowest free energy at each temperature (T), orange arrows point to the mean end-to-end distance at that temperature. Error bars are smaller than the data points.

4. Conclusion

In summary, we investigated the free energy landscapes for two lattice protein models that are mapped from the protein ribonuclease A. Using the end-to-end distance as the reaction coordinate, we mapped out the folding funnels for these two model proteins. The HP model has a relatively shallow free energy minimum at low temperature, while the H0P model develops a clear, rough free energy funnel, even near the minimum, at all temperatures. In both cases, we find an asymmetric folding funnel that changes shape substantially as the temperature changes. Unlike the schematic figures in the literature, we even observe that the location of the free energy minimum shifts with the change of temperature. Even though both model proteins are highly simplified descriptions of a real one, neither the mapping nor the interactions were tuned to produce a special free energy structure. Moreover, with only a few features added to the HP model, the H0P model already captures some of the essential thermodynamic and folding behavior of real proteins (e.g. rough free energy funnel). Therefore, we believe that the general characteristics of the folding funnels discovered in this work might well persist in a more realistic description of protein folding.

Acknowledgments
This work was supported in part by NSF under Grant No. OCI-0904685. Computing resources were partially provided by the Georgia Advanced Computing Resource Center and the Texas Advanced Computing Center under XSEDE Grant No. PHY130014.

References
[1] Dill K A and Chan H S 1997 *Nat Struct Mol Biol* **4** 10–19
[2] Onuchic J N and Wolynes P G 2004 *Curr. Opin. Struct. Biol.* **14** 70 – 75 ISSN 0959-440X
[3] Dill K A 1985 *Biochemistry* **24** 1501–9
[4] Lau K F and Dill K A 1989 *Macromolecules* **22** 3986–3997
[5] Castells V, Yang S and Van Tassel P 2002 *Phys. Rev. E* **65** 031912
[6] Pattanasiri B, Li Y W, Landau D P, Wüst T and Triampo W 2012 *J. Phys.: Conf. Ser.* **402** 012048
[7] Shi G, Wüst T, Li Y and Landau D 2015 *J. Phys.: Conf. Ser.* **640** 012017
[8] Shi G, Wüst T, Li Y W and Landau D P 2016 *J. Phys.: Conf. Ser.* **686** 012001
[9] Irbäck A and Troein C 2002 *J. Biol. Phys.* **28** 1–15
[10] Zhang J L and Liu J S 2002 *J. Chem. Phys.* **117** 3492
[11] Zhang J, Kou S C and Liu J S 2007 *J. Chem. Phys.* **126** 225101
[12] O'Toole E M and Panagiotopoulos A Z 1992 *J. Chem. Phys.* **97** 8644
[13] Beutler T C and Dill K A 1996 *Protein Sci.* **5** 2037–2043
[14] Grassberger P 1997 *Phys. Rev. E* **56** 3682–3693
[15] Frauenkron H, Bastolla U, Gerstner E, Grassberger P and Nadler W 1998 *Phys. Rev. Lett.* **80** 3149–3152
[16] Bachmann M and Janke W 2003 *Phys. Rev. Lett.* **91** 208105
[17] Hsu H P, Mehra V, Nadler W and Grassberger P 2003 *J. Chem. Phys.* **118** 444
[18] Prellberg T and Krawczyk J 2004 *Phys. Rev. Lett.* **92** 120602
[19] Tang W and Zhou Q 2012 *Phys. Rev. E* **86** 031909
[20] Unger R and Moult J 1993 *J. Mol. Biol.* **231** 75
[21] König R and Dandekar T 1999 *BioSystems* **50** 17–25
[22] Liang F and Wong W H 2001 *J. Chem. Phys.* **115** 3374
[23] Shmygelska A and Hoos H H 2005 *BMC Bioinf.* **6** 30
[24] Wüst T and Landau D P 2009 *Phys. Rev. Lett.* **102**(17) 178101
[25] Wüst T and Landau D P 2012 *J. Chem. Phys.* **137** 064903
[26] Wang F and Landau D P 2001 *Phys. Rev. Lett.* **86** 2050–2053
[27] Wang F and Landau D P 2001 *Phys. Rev. E* **64** 1–16
[28] Vogel T, Li Y W, Wüst T and Landau D P 2013 *Phys. Rev. Lett.* **110** 210603
[29] Vogel T, Li Y W, Wüst T and Landau D P 2014 *Phys. Rev. E* **90** 023302
[30] Lattman E E, Fiebig K M and Dill K A 1994 *Biochemistry* **33** 6158–6166

IOP Conf. Series: Journal of Physics: Conf. Series **905** (2017) 012017 doi:10.1088/1742-6596/905/1/012017

GPUs in a computational physics course

Joan Adler, Gal Nissim and Ahmad Kiswani

Physics Department, Technion -IIT, Haifa, Israel, 32000

E-mail: phr76ja@tx.technion.ac.il

Abstract. In an introductory computational physics class of the type that many of us give, time constraints lead to hard choices on topics. Everyone likes to include their own research in such a class but an overview of many areas is paramount. Parallel programming algorithms using MPI is one important topic. Both the principle and the need to break the "fear barrier" of using a large machine with a queuing system via ssh must be sucessfully passed on. Due to the plateau in chip development and to power considerations future HPC hardware choices will include heavy use of GPUs. Thus the need to introduce these at the level of an introductory course has arisen. Just as for parallel coding, explanation of the benefits and simple examples to guide the hesitant first time user should be selected. Several student projects using GPUs that include how-to pages were proposed at the Technion. Two of the more successful ones were lattice Boltzmann and a finite element code, and we present these in detail.

1. Introduction

Projects are an important part of any Physics education but are especially important for Computational Physics education, [1, 2]. They help students approach realistic problems, learn to present and annotate code and provide material for use in general courses as well as the computational physics one. The course given at the Technion is based on several textbooks [3, 4] as well as own code and is described in [5]. There is a heavy emphasis on visualization, as well as gaining experience in a LINUX and parallel computer environment. Although there have been many changes in computer hardware over time, two of the largest have been the rise and (temporary?) demise of vector programming and the rise in parallel distributed memory architectures concurrent with the use of commodity hardware.

As the computational infrastructure continues to change with time, the most recent innovation in the supercomputer world is the increased adoption of GPUs (Graphical Processing Units) for computation. Thus these algorithms need a place even in an introductory course. Ideally, a senior undergraduate/early graduate course should provide a basis for many years as the student graduates and moves forward. This is especially true in Israel, where students may have a several year break between first and higher degrees due to national service. It is also important for students who take child rearing career breaks.

2. Computational physics course design

Our course has a few weeks of basic algorithms for numerical analysis, with examples taken from undergraduate physics, such as chaos and Schroedinger equation solution, followed by algorithms such as Metropolis simulation and multidimensional integration and Hoshen Kopelman for percolation. Molecular dynamics/simulated annealing approches are discussed. Each week has

IOP Conf. Series: Journal of Physics: Conf. Series **905** (2017) 012017 doi:10.1088/1742-6596/905/1/012017

sample codes with visualization, either in PGPLOT [6] (2D) or with the Technion AViz code [7] for 3D atomistic cases. It takes place in a purpose built classroom with LINUX boxes as well as communication infrastructure for laptops. Many of the FORTRAN routines are adapted from either [3] or [4] with added PGPLOT graphics prepared as students projects from the 1990s.

Once some basic techniques are learnt and familiarity with a LINUX environment and ssh and the compile, link, run paradigm of remote use MPI and queue based submission on our local parallel computer are introduced. Each of the above mentioned examples is reinforced with compulsory individual homework, which the students are encouraged to submit on sourcecode html sites. There is sprited discussion on which language, are MATHEMATICA and MATLAB allowed, but compulsory exercises in C, FORTRAN, and MATHEMATICA, as well as homework with free choice of code ensure the students are exposed to many possibilities. Except for the free choice questions, working code requiring adjustments is supplied to minimise startup pain.

2.1. Choices and tradeoffs

At some time, depending on the class composition and interests, the students vote on the final topics, that range from "big codes" such as QUANTUM ESPRESSO and LAMMPS to computational optics and lattice Boltzmann/finite element methods for fluid dynamics. These topics are presented with the aid of student projects. Unfortunately many of the older ones use X-graphics. This was the best option 20 years ago, and while still good to teach algorithms/physical phenomena are not so suitable for emulation in more modern systems where different options such as QT are more common. This vote is needed because since we only have 13 week semesters, emphasis on any specfic topic has to be at the expense of others. The final weeks need to give the most to the most students in each specific semester.

In order to update, combined with the hardware shift to GPUS, and the interest in python, some fresh codes were needed. Since GPUs are so suited to finite elements, and the current finite element/fluid had the most compilation issues the time had come to prepare new codes, and we describe these below.

During choice-of-code discussions, Technion students always bring up MATLAB, which is much loved locally, and has a site license. The class policy is that it is allowed for projects of students doing experimental research. (Likewise, science education students may use windows software.) Most of the rest of the course is based on public domain or academic-licensed code, to enable portability for the student's future, but for special cases minimal MATLAB/MATHEMATICA use occurs. In the 2016 reality, the changes to FORTRAN compilers (specifically the discontinuation of some FORTRAN 77 commands in GFORTRAN), has meant that this is no longer completely reliable in the very long term. We recompiled our own f77 codes and and adapted our PGPLOT implementations, and an addedum to [5] is in preparation to cover this.

2.2. Fear barriers

Today, even some computer-aware students find it hard to move out of the comfort zone of personal windows, mac or ubuntu desktops and laptops. Since serious computational physics requires access and use of additional servers the first obstacle to overcome is that of compile/link/run to replace some local interface and to ssh to a remote machine. The next fear barrier of remote queue submission follows relatively easily, but introduction to MPI (or OpenMP) is more significant, but harder. Neither of these require a new language, but they do need some introducion to scripting commands which are fairly common to all langauges and for codes like MATHEMATICA/MATLAB. Parallel coding can be done easily for either many canned codes or for the case of trivial parallelization in a suite of simulations where each node receives its own dataset/parameters and communication is minimal. Such situations are useful to build confidence.

Introduction of GPU codes is a new level of change of concept. Unless the GPU use is embedded in the code such as in the case of MATLAB, a GPU suited language has to be selected and rather more understanding of the algorithms required. Hence it is crucial to present examples that show how significant speed up can be.

3. Codes suited to GPUs

The GPU unit has much in common, from an algorithmic viewpoint with vector processors of yore. It can do simple operations very quickly. Thus one needs to choose problems where this is a large component of the total calculation load. However even for these, some effort is needed. It is not that different to teaching parallel coding in that simple well documeted examples give confidence and frameworks. However, there is not the easy way out of trivial parallelization to start students on the path to parallel coding that can be used to begin utilization of the GPUs.

Many excellent GPU codes exist for statistical physics models like the Ising model or molecular dynamics, but many require some sophistication to begin implementing. Within our course framework, these topics were already covered earlier with straightforward codes. Since the finite element codes were based on X-graphics, and this topic was covered at the end of the course only if voted for, the following examples were prepared for this purpose.

3.1. Finite element - water waves

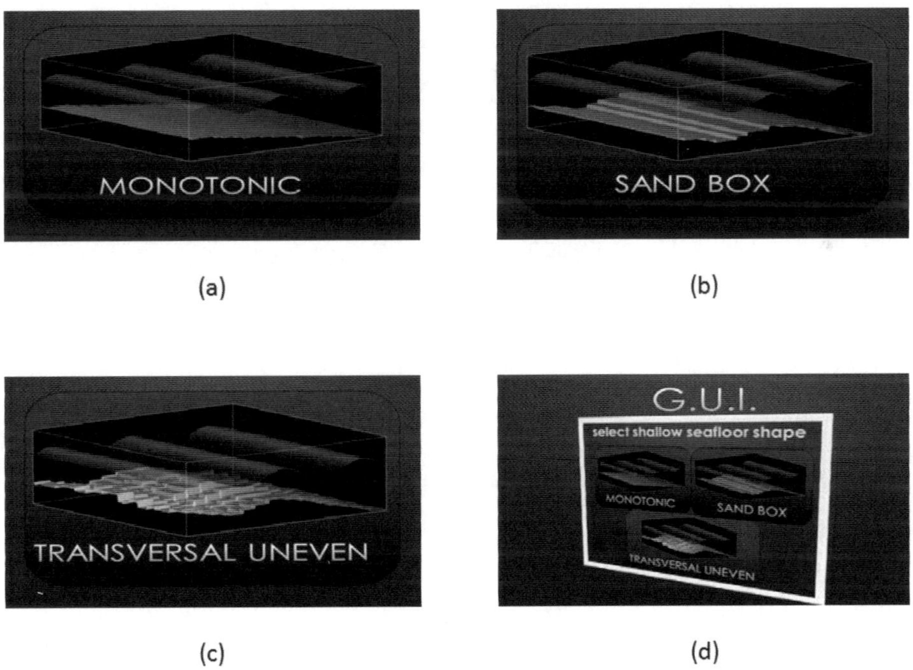

Figure 1. Substrate options (a) monotonic sloped bottom, (b) stepped (sandbox) option, and (c) uneven option. The GUI to select the substrate is shown in (d).

The finite element implementation here takes the easy route of using MATLAB; however it is most probable that some package would be needed for a finite element method implementation so this is a good choice. Full code details and explanations are provided on the project website [8]. At the Technion, there is a site license for MATLAB use. The finite element model simulates

IOP Conf. Series: Journal of Physics: Conf. Series **905** (2017) 012017 doi:10.1088/1742-6596/905/1/012017

water waves with a choice of sand substrates. The sand substrate determines the temporal evolution of the surface waves, which are primarily excited by wind. There are possibilties to move between CPU only and CPU/GPU options. The substrate options are sloped (Figure 1(a)), stepped (Figure 1(b)) or uneven (Figure 1(c)). Theory, results, and downloadable files from the website run as downloaded with relatively recent versions of MATLAB for LINUX or windows, although the GPU options require 2015b and later. The instructions are very clear, with functions and GUIs for each function clearly described, including an explanation of how the GPU parameters can override the setparameter function. An image of the GUI to choose the bottom is given in Figure 1(d).

3.2. Lattice Boltzmann - fluid flow

Another very popular algorithm suited to GPUs is lattice Boltzman. It was previously taught with an earlier lattice Boltzmann FORTRAN90 code and a MATLAB graphics implementation [9] which provides an excellent introduction to the algorithm. An image from this code is shown in Figure 2 for fluid around a spherical object.

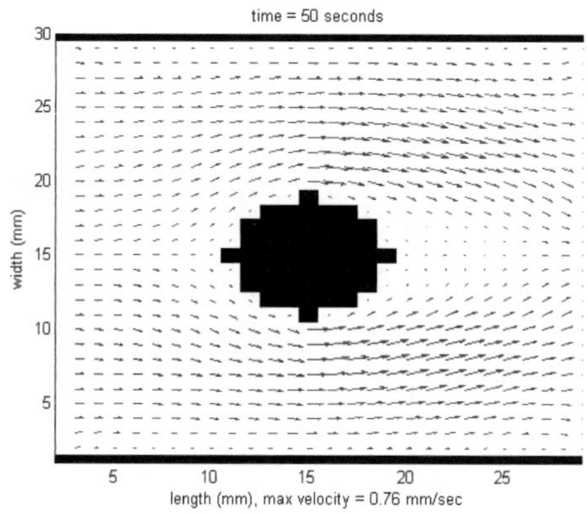

Figure 2. Fluid around a circular object.

However, in order to create a second GPU example a completely new implementation was prepared [10]. The website was prepared as a (downloadable) ipython notebook, with an explanation of the algorithm, source code, compilation instructions and instructions for compiler installation on a unix system. It is a beautiful and quite well documented project, entirely public domain, and we have run it under both CENTOS and UBUNTU, The speedup with the GPU is amazing. The code requires an Nvidia GPU with compute capability of at least 3.0; a version of CMake of at least 2.4 and OpenGL, GLUT and GLEW. The program uses OpenGL to draw a color map of the fluid velocity in the medium (Figures 3 and 4). Although all these work on our classroom server, this code does not have the wider system choice that the one described in the previous section does. This is more than balanced by the excellent speedup characteristics as presented in Table 1 for a Intel i73930k CPU and an Nvidia Geforce GTX 670 GPU. The reader should be warned that root privileges may be needed to install the OpenGL libraries.

4. Conclusions

Two websites that provide downloadable examples of GPU codes suited for an introductory computational physics class have been described, Both include a physical background, discussion of algorithm, downloadable codes with step-by-step instructions and sample results. The level is

Table 1. Speedup for the lattice Boltzman code.

Size	CPU[fps]	GPU[fps]	Speedup
300x300	125	1050	8.4
500x500	32	440	13.7
1000x1000	8	122	15.25
2000x2000	2	33	16.5

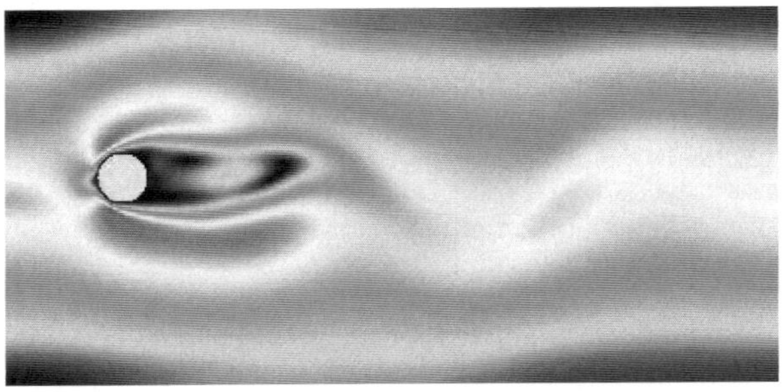

Figure 3. Flow around a single object.

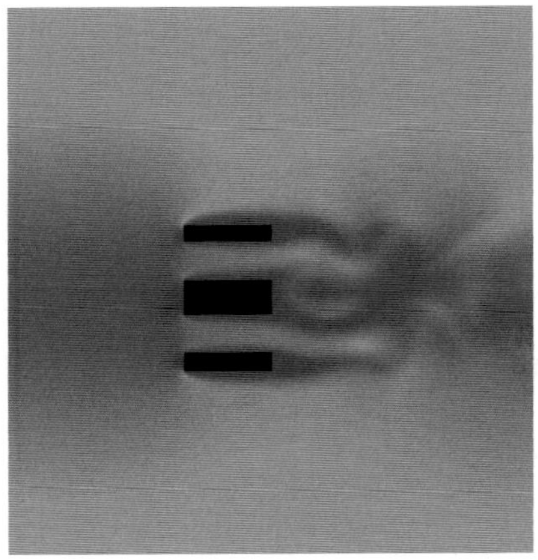

Figure 4. Flow around 3 objects.

suited to senior undergraduate or beginning graduate students wihout prior experience in GPU coding.

Acknowledgments
A very big thank you to the Technion Physics Department staff, especially Israel ben Efraim, Avi Refaeli, Riva Kreizelman and Yan Katz for the renovations and installation of new equipment for the Computer classroom. Thanks to Adam Levi and Igal Rasin for help on selecting GPU hardware, and to research friends who lent me their GPU codes. The educational websites are part of the SimPhoNy project which is funded by FP7 under NMP-2013-1.4-1 call with grant agreement no: 604005.

References

[1] Mazvovsky D, Halioua G and Adler J 2012 *Physics Procedia* vol 34 p.1-5.
[2] Adler J, Artzi Y, ben Bashat L, Izraeli T Y, Kreif M, Lavi I, Leibenzon A, Levi A, Schlesinger I, Toledano E, Peretz U, Weisler Y and Alon Yagil 2014 *Journal of Physics: Conference Series* vol 510, p.012003.
[3] Koonin S C and Meredith D E 1998 "Computational Physics: Fortran version", 2nd edition, Westview Press.
[4] Gould H and Tobochnik J 1998 "An Introduction to Computer Simulation Methods: Application to Physical Systems, Parts I and II" Addison-Wesley.
[5] Adler J 2014 *Physics Procedia* vol 53, p. 2-6.
[6] http://www.astro.caltech.edu/~tjp/pgplot
[7] http://phony1.technion.ac.il/~aviz
[8] http://phelafel.technion.ac.il/~gn/project
[9] http://phelafel.technion.ac.il/~drorden/project
[10] http://phony1.technion.ac.il/~ahmad

IOP Conf. Series: Journal of Physics: Conf. Series **905** (2017) 012018 doi:10.1088/1742-6596/905/1/012018

Coding considerations for standalone molecular dynamics simulations of atomistic structures

R O Ocaya and J J Terblans

Department of Physics, University of the Free State, South Africa.

E-mail: ocayaro@ufs.ac.za

Abstract. The laws of Newtonian mechanics allow ab-initio molecular dynamics to model and simulate particle trajectories in material science by defining a differentiable potential function. This paper discusses some considerations for the coding of ab-initio programs for simulation on a standalone computer and illustrates the approach by C language codes in the context of embedded metallic atoms in the face-centred cubic structure. The algorithms use velocity-time integration to determine particle parameter evolution for up to several thousands of particles in a thermodynamical ensemble. Such functions are reusable and can be placed in a redistributable header library file. While there are both commercial and free packages available, their heuristic nature prevents dissection. In addition, developing own codes has the obvious advantage of teaching techniques applicable to new problems.

1. Introduction

The last decades have seen rapid growth of atomistic/molecular simulation (AMS) as a powerful tool in material science. The method can postulate material properties that are difficult or costly to study experimentally [1, 2, 3, 4, 5]. In AMS, macroscopic material properties are aggregations of atomic interactions and dynamics. A wide span of properties, such as structural changes due to temperature and applied forces, are better understood through parametric variations effected in the method. Examples include energetics of cluster formation [6, 7, 8], diffusion [9, 10], crack formation and propagation [11], the impact of vacancy or adatom defects, the load-bearing capacity, and so on.

There are two approaches to MD [12, 13, 14]: classical MD and ab-initio MD [15]. In ab-initio MD, the starting point is an incremental and iterative numerical basis arising from a concisely bounded mathematical model i.e. the forces can be deduced by differentiation of a potential energy function. The temporal and spatial evolution of each particle of the ensemble then permits the determination of the trajectories. While in principle simple, the analytical complexity increases with the number of system particles, the efficacy of the mathematical model and the required level of computational accuracy. Complexity bears directly on overall computational time and storage. Much effort is being expended to discover better potential energy functions and more efficient computational algorithms for them. Ongoing developments in internet and computer technology make it ever easier to simulate these systems. The current trend is to discover new efficient, high-speed interconnects of several multiple-core central or graphical processing units into super-computing clusters that collaborate to solve a complex problem [5]. This approach defines a clear futuristic research niche that is relevant to all areas of research. The single, standalone machine, however, is still an essential part in most institutional supercomputer set ups. Its participation is therefore vital in the computational scheme since, as the starting point, algorithms are presumed to run on each node in some form or other. More complex systems are then expected to integrate several such nodes through generalization and scaling.

Modern computer architectures are by their nature scalable i.e. programs can readily be extended to bigger systems with minimal modification. Thus, larger particle ensembles can exploit distributed processing to realize computational time and storage cost savings. There is also the ease of porting codes to run on more heterogeneous nodes that suit specific users. Typical ensembles are defined in terms of the total number of individual particles in the system in addition to macroscopic material properties such as pressure, volume and temperature. Perturbations then ultimately give an idea of the system dynamics i.e. the trajectories of individual particles, their interactions through collisions, repulsions, attractions, adhesion, steady state and equilibrium distances, angles and similar geometric quantities can be computed and over time. Deductions of connected macroscopic parameters such as kinetic and potential energy, diffusion and other transport coefficients, structure factors, spectral density functions, distribution functions, then follow. Numerical integration algorithms rely on the computation of the system differential equations at well-defined, linearly incremented instants of time t. There are a variety of integration schemes, each depending on the specific variables of interest and what is known or needs to be known at each time step δt.

This paper implements all integrations using the Störmer-Verlet (SV) technique for pairwise particle interactions. At each iteration, the computational end point per particle is said to be determined when the force F_i acting on the i-th particle is calculated. It is understood that all pre-models are based on approximations. Therefore, the outputs are accepted to be accurate to a specified degree for a set number of iterations. These trade-offs are usually made to reduce execution speed. Although the potential of modeling and simulation for research is clear, it remains largely in the domain of organizations that have large computational facilities with direct vested interests in materials modeling. This does not imply that a standalone computer typically found today is not up to many of the tasks. The algorithms can be ported to run on distributed clusters by implementing freely available message passing interfaces (MPI).

2. Motivation and significance

While computational materials science has grown substantially, the bulk of research output still remains largely experimental, inevitably leaving the avenue of computation largely unexplored. There are a number of likely reasons for this. One could be a misconception of computation as solely the domain of computer science. Another could be that the real, general shortage of programming skills required to code the underlying functions places a high starting inertia on most researchers. Interestingly, most established undergraduate physical and applied science curricula have some aspects of computer programming. It is likely that many researchers today have either some programming exposure or access to a computer science department, thus permitting collaboration. The aim of this article is therefore twofold. Firstly, it presents a toolkit of functions and examples that were developed and tested over a period of several months to solve AMS equations of face-centred cubic (fcc) crystalline structures using embedded atom methods (EAM) on a standalone computer. The algorithms are coded using Microsoft Visual C/C++ language and are easily ported to GNU compilers such as `gcc`. The software has been used by the authors to calculate for instance fcc metal bulk and surface energetics. The potential to use the codes for scientific visualization and discovery therefore exists. Secondly, it provides repeatable information to peers who have some basic knowledge of C-programming and an interest in materials modeling and simulation. The aim is to facilitate their progression through first simulations without the detractions of deep coding. The nature of materials science today is such that the work will be ongoing as newer functions are implemented and the package updated. The article also outlines the mathematical basis of the algorithms for parallel reading. Finally, an example of the usage of the package is given.

3. Software description

The illustration package implements a three-dimensional (3D) Sutton-Chen (SC) form of the Finnis-Sinclair (FS) EAM potential [16, 17, 18, 19] in the canonical ensemble. EAM methods are less accurate than first-order principles that consider all particle interactions individually but are computationally cheaper and can handle larger time scales and more particles [20]. For a monoatomic N-particle ensemble, the energy EAM Hamiltonian is expressed as [18, 20, 21]

$$E_{tot} = \frac{1}{2}\sum_{ij}^{N} V(r_{ij}) + \sum_{i}^{N} F(\bar{\rho}_i), \tag{1}$$

where $V{=}V(r_{ij})$ is the pairwise interaction potential energy between particles i and j separated as distance r_{ij}, and $F{=}F(\bar{\rho}_i)$ is the embedding energy of atom i as a function of the host electron density.

According to SC, the FS total potential energy for fcc metals is written in the translation and rotation invariant form [11, 16, 23, 24]

$$V = \varepsilon \sum_{i=1}^{N}\left(\sum_{j=i+1}^{N} \left(\frac{\sigma}{r_{ij}}\right)^{n} - c\sqrt{S_i}\right), \tag{2}$$

where

$$S_i = \sum_{j=1, j\neq i}^{N} \left(\frac{\sigma}{r_{ij}}\right)^{m}, \tag{3}$$

and σ is a length parameter e.g. lattice constant of the fcc unit cell, ε is an energy scaling factor for the system. The constants c, m and n, where $m < n$, are fitting parameters. Force is gradient of potential, i.e.

$$\overline{F} = -\overline{\nabla} V(r). \tag{4}$$

Hence the force on the i-th particle can be expressed as

$$\overline{F}_i = \varepsilon \sum_{j=1,\ j\neq i}^{N} \left[n\left(\frac{\sigma}{r_{ij}}\right)^{n} - \frac{cm}{2}\left(\frac{1}{\sqrt{S_i}} + \frac{1}{\sqrt{S_j}}\right)\left(\frac{\sigma}{r_{ij}}\right)^{m}\right] \frac{\bar{r}_{ij}}{r_{ij}^{2}}, \tag{5}$$

where $\bar{r}_{ij}$ is the vector between particles i and j and is given in terms of the positions of each particle i.e.

$$\bar{r}_{ij} = \bar{x}_j - \bar{x}_i. \tag{6}$$

Discretization of the continuous domain into a finite set of points achieves numerical approximations that are useful to solve a variety of difficult differential equations. In MD such equations are derived from the Schrödinger equation under specific boundary conditions [12, 13, 14]. The continuous closed-form solutions may be extremely difficult to solve analytically so that numerical approximations may be the only recourse. An necessary requirement in MD discretization is energy conservation i.e. the Hamiltonian (H) must not change over discrete steps [25]. However, discretization leads to unavoidable approximation errors. Effort must then be spent to keep the variations in H within acceptable bounds. A commonly used approach to discretize MD systems is the so-called velocity Störmer-Verlet method. It is a discretization of

IOP Conf. Series: Journal of Physics: Conf. Series **905** (2017) 012018 doi:10.1088/1742-6596/905/1/012018

Newtonian equations of motion [11, 26, 27]. In the variant of the method that is used in this software, using the discretized notation $\bar{p}_i^n := \bar{p}_i(t_n)$ which describes the vector $\bar{p}$ sampled at the discrete n-th time interval, where $\bar{p}$={$\bar{x}$, $\bar{v}$, $\bar{F}$}, it follows that the approximation of $\bar{p}_i^{n+1}$, i.e. on the next or $(n+1)$-st time interval or $(t+\delta t)$ gives the new positions

$$\bar{x}_i^{n+1} = \bar{x}^n + \delta t \bar{v}_i^{n+1} + \frac{\bar{F}_i^n}{2m_i}\delta t^2. \tag{7}$$

Similarly, the next velocities are

$$\bar{v}_i^{n+1} = \bar{v}_i^n + \left(\frac{\bar{F}_i^n + \bar{F}_i^{n+1}}{2m_i}\right)\delta t. \tag{8}$$

The discretization error in the method is $O(\delta t^2)$. The method requires starting values of positions and velocities for each particle. Therefore in the software the user-created ASCII text file e.g. `data.txt` contains initial particle data in the 3D format:

```
mass x_pos y_pos z_pos v_x v_y v_z
```

It is necessary to scale all variables to reference values and dimensionless equations. The user then interprets the output in that context [11].

The canonical ensemble is a statistical arrangements of particles and their states in a mechanical system that is in thermal equilibrium with a heat bath at fixed temperature [20]. For a system of N particles, volume V and temperature T, the system is said to be a NVT canonical ensemble. Specifying T facilitates the estimation of the system energies using the energy equipartition theorem (EPT) of thermodynamics. Conversely, a knowledge of the energies allows an estimate of the system temperature [11, 28]. Thus, the evolution of the macroscopic variables T, V and P that portray system phases may be followed by considering particle interactions. According to EPT, the kinetic energy E_k of a particle with three degrees of freedom is

$$E_k = \frac{3}{2}Nk_BT, \tag{9}$$

where k_B is Boltzmann constant.

3.1. Software Architecture

The package combines a compiled executable and C-style library file. The aforementioned MD functions are in the `VSV_02.h` header whose functions are called from the main program `main.c`. The particle structure is defined, initial particle boundary conditions are loaded from a user-edited ASCII disk data file containing unambiguous masses, 3D coordinates and initial velocity components of each particle. The 3D-space particle structure is

```
typedef struct{
double m;
double x[DIM];
double v[DIM];
double F[DIM];
double F_old[DIM];
} Particle;  // 72 bytes

// ParticleList structure
typedef struct ParticleList{
Particle p;
struct ParticleList *next;
} ParticleList;
```

The central coding theme employed is function reusability and a by-reference passable RAM-based particle array for speedier computations. Reusable codes are readily adaptable to new MD problems.

3.2. Software Functionalities
Fig. 1 shows the main functionalities of the software in pictorial form. Further information about the most current build of the program code and user manuals is available for download under the GNU General Public License v3.0 [22]. The code has evolved to include the calculation of ergodics such as vacancy energy, cohesion energy, deterministic thermostats and Monte-Carlo thermostats.

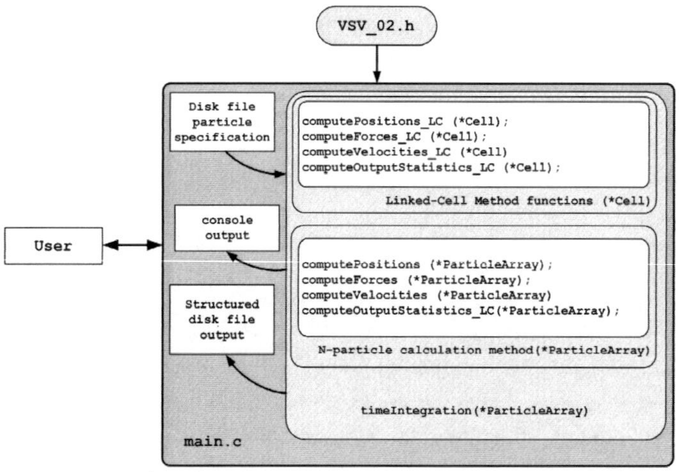

Figure 1. Pictorial depiction of the functionalities of the program.

3.3. Main code snippets
All included functions (`func`) are declared in the form

```
return_type func(arrangement *particle, types other_variables)
```

where `arrangement` can be either particle, particlelist or cell array, the latter two being dedicated to the LCM method. The `other_variables` can be summation indices, etc. The 3D inter-particle distance r_{ij} in Eq.(6) for all MD processes is calculated using

```
Particle *p;
double r_ij;
for (d=0; d<3; d++)
  r_ij += sqr(p[j].x[d]-p[i].x[d]);
r_ij = pow(r,0.5);     // 3D distance
```

Stating the condition: $r_{ij}>r_{cut}$, where r_{cut} is a cut-off radius within this snippet will improve calculation speed by ignoring interactions beyond r_{cut}.

4. Illustrative Example

Doye & Wales [23] have investigated global minima in transition metal SC clusters described by various (m-n) pairs i.e. (12-6), (9-6) and (10-8) for clusters up to $N = 80$ atoms. Their published work provide sufficient data to illustrate the calculation of energy minima using the present software. For instance, specifying the data points in Table 1 for $N = 3$ for copper with normalized energy and lattice parameters $\varepsilon = 1$, $\sigma = 1$, $m = 6$, $n = 9$, $m_o = 63$ and $c = 39.432$

Table 1. Sample data for fcc copper obtained from [23].

m	x	y	z	v_x	v_y	v_z
63	-0.5264206369	0.1019469049	0.0252070114	0	0	0
63	-0.8377547075	-0.1828574677	-0.4214018482	0	0	0
63	-0.2345929501	-0.0832085752	-0.4827850722	0	0	0

Doye and Wales [23] developed a method to improve structural identifications of metallic atomic clusters based on experimental data. Their Monte Carlo based method deforms an arbitrarily constructed hypersurface of the atomic clusters for up to 80 atoms. By searching for a global potential energy minimum they are able to suggest the most stable structure of the collection of atoms. A difficulty of their method, hence the need for the globalized search, is that the global minimum of adjacent surfaces are not necessarily related and therefore a newer local surface is needed for the atoms on the two surfaces. Therefore, the difficulty with the method is a complexity of $\approx\mathcal{O}(N^2)$. In contrast, the method of the present paper is deterministic, has a complexity of $\approx\mathcal{O}(N \log N)$ and relies on the complete solution of the potential energy of the atomic cluster as is. Thus, we make no assumptions about the stability or structure of the cluster, and consequently we can apply it to a much larger number of atoms. Table 2 shows the output of global minima calculation for a stable copper atom cluster. The output energy is scaled by ε. While Table 2 only compares 3-atom clusters with those in Doye and Wales [23], we verify their outputs for all their 80 atoms with no noticeable increase in calculation time.

Table 2. Program output for Table 1 input data.

N	12-6	9-6	10-8
3	-1704.6905	-480.8560	-633.7771

Our present interest is to show how initial experiments in MD simulations for fcc metals, for instance, may be facilitated in a standalone setting. It has been written from personal experiences that confirm how first computer-based experiments can be difficult to initiate. By coding a dissectable software and presenting alongside it the mathematical basis for the algorithms the progression from simple to more advanced MD systems is easier. FCC metal properties were correctly calculated such as the bulk and surface formation energetics down to single atoms, forces on individual particles, diffusion coefficients, cluster formation mechanisms, that successfully reproduced published data in similar simulation programs.

5. Conclusions

This article emphasizes that the software program tool, like any tool, has to be coded and then applied correctly for any measure of effectiveness. The ease of interpreting the outputs of the program can lead to postulations and predictions about new configurations of the material. With increasing interest in the use of software to investigate materials, more researchers are increasingly having to code material behavior. In this work, the authors make no assumptions about the C-compilers, nor presume to recommend any particular programming environment as being more suitable for a given purpose. Rather, the emphasize is on underlying methods and pathways to successful coding. Having a software that can be updated easily, ported over many platforms and scalable are advantageous for wider applicability. The programs codes presented here have been tested successfully on Microsoft C/C++ compiler (Windows 7) and gcc in Ubuntu Linux with little modification for various fcc metal parameters. The outputs compare favorably with published literature sources and confirm the applicability of the software.

Acknowledgments

The authors hereby acknowledge the National Research Foundation (NRF) for funding support through KIC 2016, Grant No. 104078.

References

[1] Mendelev M I, Han S, Srolovitz D J, Ackland G J, Sun D Y and Asta M 2003 *Philos. Mag.* **83** 3977–3994 doi: 10.1080/14786430310001613264.
[2] Sutton A P and Chen J 1990 *Philos. Mag. Lett.* **61** 139–156
[3] Das A and Ghosh M M 2015 *Comput. Mater. Sci.* **101** 88–95 doi: 10.1016/j.commatsci.2015.01.008
[4] van der Walt C, Terblans J J and Swart H C 2014 *Comput. Mater. Sci.* **83** 70–77 doi: 10.1016/j.commatsci.2013.10.039
[5] Abraham M J, Murtolad T, Schulz R, Pálla S, Smith J C, Hessa B and Lindahl E 2015 *SoftwareX* doi: 10.1016/j.softx.2015.06.001
[6] Smirnov B M 1995 *Zh. Eksp. Teor. Fiz.* **107** 2080–2091
[7] Terblans J J 2002 *Surf. Interface Anal.* **33** 767–770 doi: 10.1002/sia.1451
[8] Mattsson T R and Mattsson A E 2002 *Phys. Rev. B* **66** 214110
[9] Sebastian I S, Aldazabal J, Capdevila C and Garcia-Mateo C 2008 *Phys. Status Solidi A* **205** 6 1337–1342 doi: 10.1002/pssa.200778124
[10] Jian-Min Z, Fei M and Ke-Wei X 2004 *Appl. Surf. Sci.* **13** 7 1009-1963/2004/13(07)/1082-09
[11] Griebel M, Knapek S and Zumbusch G 2007 *in Barth TJ et al. (Eds.), Numerical Simulation in Molecular Dynamics, in: Texts in Computational Science and Engineering 5* (Berlin: Berlin)
[12] Car R and Parrinello M 1985 *Phys. Rev. Lett* **55** 22 2471–2474
[13] Car R and Parrinello M 1989 *The Unified Approach for Molecular Dynamics and Density Functional Theory, in Simple Molecular Systems at Very High Density* **186** *of NATO ASI Series B, Loubeyre PP and Boccara N (eds.)* (New York: Plenum Press) 455–476
[14] Remler D K and Madden P A 1990 *Mol. Phys.* **70** 6 921–966
[15] Tuckerman M E 2002 *J. Phys.: Condens. Matter* **14** R1297–R1355 Online at stacks.iop.org/JPhysCM/14/R1297
[16] Finnis M W and Sinclair J E 1984 *Philos. Mag. A* **50** 45
[17] Daw M S, Foiles S M and Baskes M I 1993 *Mater. Sci. Rep.* **9** 251
[18] Daw M S and Baskes M I *Phys. Rev. B* **29** 6443
[19] Todd B D and Lynden-Bell R M 1993 *Surface Science* **281** 191–206
[20] Chamati H, Papanicolaou N I, Mishin Y and Papaconstantopoulos D A 2006 *Surface Science* 1–11 doi: 10.1016/j.susc.2006.02.10
[21] Mishin Y 2005 *in: Yip S (Ed.), Handbook of Materials Modeling* (Springer: The Netherlands)
[22] Ocaya R O and Terblans J J 2016 *SoftwareX* doi: 10.1016/j.softx.2016.05.005
[23] Doye J P K and Wales D J 1998 *New J. Chem.* 733–744
[24] Swope W, Andersen H, Berens P and Wilson K 1982 *J. Chem. Phys.* **76** 637–649
[25] Landau L and Lifschitz E 1976 *Mechanics, Course of Theoretical Physics, Vol. 1* (Pergamon Press: Oxford)
[26] Hockney R 1970 *Methods Comp. Phys.* **9** 136–211
[27] Verlet L 1967 *Phys. Rev.* **159** 98–103
[28] Chandler D 1987 *Introduction to modern statistical mechanics* (Oxford University Press: New York)

IOP Conf. Series: Journal of Physics: Conf. Series **905** (2017) 012019 doi:10.1088/1742-6596/905/1/012019

Learning problem-solving skills in a distance education physics course

G J Rampho[1] and M Z Ramorola[2]

[1] Department of Physics, University of South Africa, P.O. Box 392, Pretoria 0003, South Africa

[2] Department of Science and Technology Education, University of South Africa, P.O. Box 392, Pretoria 0003, South Africa

E-mail: ramphjg@unisa.ac.za

Abstract. In this paper we present the results of a study on the effectiveness of combinations of delivery modes of distance education in learning problem-solving skills in a distance education introductory physics course. A problem-solving instruction with the explicit teaching of a problem-solving strategy and worked-out examples were implemented in the course. The study used the *ex post facto* research design with stratified sampling to investigate the effect of the learning of a problem-solving strategy on the problem-solving performance. The number of problems attempted and the mean frequency of using a strategy in solving problems in the three course presentation modes were compared. The finding of the study indicated that combining the different course presentation modes had no statistically significant effect in the learning of problem-solving skills in the distance education course.

1. Introduction

The role of computational physics in the quest to understand and explain the physical world has risen to equal that of physics theory and experiment [1]. This explains the increasing need to introduce computational physics skills to physics learners at undergraduate level [2], if not as early as experimentation is. Almost all the studies on the teaching and learning of computational physics and problem-solving skills were conducted and focused in the traditional face-to-face education [3, 4]. In this work the worked-out example instructional approach was implemented to teach problem-solving skills in a distance education introductory physics course. The aim of the present research was to assess the effectiveness of the worked-out example instructional approach in a distance education environment. The results of this study can assist in the development of the worked-out example instructional approach for effective use in distance education settings.

One of the aims of physics courses is to teach problem-solving skills [5]. This aim is reflected in the learning activities and assessment instruments in physics courses being dominated by problem solving tasks. Research [3] has shown that the explicit teaching of a problem-solving strategy in science courses can also increase success rates in the courses. The present study implemented the explicit teaching of a problem-solving strategy with worked-out examples to teach problem-solving skills in a distance education physics course. The implemented instructional approach had not been tested for effectiveness, however, the components (problem-solving strategy and worked-out examples) were shown to be effective in this regard. The study was conducted on an actual distance education course delivery. There are practically no authentic studies reported on the learning of problem-solving skills in distance education. Moreover, most of the studies reported, including those focusing on the traditional face-to-face education, are conducted under ideal experimental settings [6, 7].

In distance education the normal teaching and the required learning take place in different locations, and the teaching is facilitated through technology [8]. Distance education delivery modes include (i) correspondence where print as well as information and communication technologies are used, (ii) online where the Internet and computer technologies are used and (iii) face-to-face classes where traditional classroom instructional technologies are used [9]. The effective use of these technologies in the delivery of distance education relies on the sound knowledge of the technology affordances and the appropriate pedagogy for the content delivered [10]. This study investigated the effect of using combinations of presentation modes of a distance education physics course in learning a problem-solving strategy. The research questions addressed in this study are:

(i) Is there any statistically significant difference in the problem-solving performance between students who use different combinations of the three presentation modes of a distance education course?
(ii) Is there any statistically significant difference in the frequency of strategy use between students who use different combinations of the three presentation modes of a distance education course?

These questions were also investigated in the study by Selçuk *et al.* [7] in a traditional face-to-face environment using an experimental research design. In the present study the questions are investigated in a distance education environment. The same course content was treated in the three course presentation modes. The research method of the study is presented in the next section.

2. Research method

The study was conducted at the University of South Africa (UNISA) which is a single-mode distance education institution situated in South Africa, Africa. The physics course considered for the study was presented through correspondence, online, and face-to-face modes.

(i) Correspondence: This was the traditional tuition mode at UNISA. In this learning mode tuition took place through sending (receiving) the course material (study guide and tutorial letters) to (from) the university using postal services, e-mail and telephone.
(ii) Online: The *myUnisa* learning management system [11] was used for the online presentation of the course material (study guide and tutorial letters). Weekly online course discussions were scheduled on the discussion forum. Participation in the online interactions required access to computer and internet facilities, which could be accessed at the university regional centers.
(iii) Face-to-face: Two face-to-face discussion classes in the semester were scheduled for the course and the students were informed of the date, time and venue at least one month before the class. The classes were held at different regional centers in the Gauteng Province. The classes focused on problem-solving and students were working individually during the problem-solving activity.

The same course materials were treated in all the three course-presentation modes [12]. The research design, sampling and data collection are discussed next.

2.1. Research Design

The present study investigated the effect of using combinations of correspondence, face-to-face and online presentation modes of a distance education physics course in learning a problem-solving strategy. The study used a quantitative research method to formulate an answer to the research questions. The *ex post facto* research design was used in the study because it was not possible to either control for extraneous variables or randomly assign subjects to groups [13]. The following three subsections summarize the instructional approach, population and sampling, and data collection, respectively.

2.2. Instructional Approach

The course considered for this study was a distance education semester course. The course was the calculus-based first-level physics mechanics course. The course was presented through correspondence as well as online using a learning management system. Two face-to-face classes were organized as a learner-support measure. The course was divided into four learning units with each unit scheduled to be treated over three weeks. A pacer was set for the course completion. The instructional approach for the course involved the (i) direct explanation of the prescribed problem-solving strategy, (ii) modeling of the problem-solving strategy through worked-out examples, (iii) independent practice of the strategy through formative exercise problems and (iv) provision of immediate feedback for attempts to formative exercise problems [7]. For each course unit students were required to first study the summary of the mechanics concepts, laws and principles covered. Three worked-out examples and three formative exercise problems were assigned for each course unit. The format of the solutions of the worked-out examples was consistent with the prescribed problem-solving strategy and the solution rubric. The example and exercise problems were taken from standard mechanics textbooks [14, 15]. The summaries of the mechanics concepts, laws and principles were discussed with the instructor, or tutor. However, the solutions to the exercise problems were constructed by the students without assistance.

Heller *et al.* [3] have designed a problem-solution rubric based on a sound problem-solving strategy. This solution rubric has been well research and is adopted and adapted for use in this study. These authors also designed a tool for determining the difficulty level of a physics problem based on the familiarity of problem context, cues and information given, explicitness of the question, number of possible approaches as well as number of equations to solve. According to this tool the most difficult physics problem has a difficulty rating of six (6) while the least difficult has a difficulty rating of one (1). The example and exercise problems were chosen to be of difficulty rating 3 or less.

2.3. Population and Sampling

The target population was all the students registered for the first-level mechanics physics course. Research [16] has shown that demographic and situational factors can affect learning, including in distance education. In this study the target population was treated as a homogeneous group. However, it was noted that the target student population consisted of males and females with ages between 18 years and 50 years, the majority of whom resided in South Africa. Some of the students were employed in various economic sectors while others were not employed. All the students were familiar with information and communication technologies (computer, cellphone) and had used instant-messaging and e-mail services. Also, based on the admission requirements, the students were familiar with distance education technologies and were proficient in the language used in the course presentation. However, the students were used to the traditional face-to-face teaching, which differed from the distance teaching and learning in this course.

All the students registered for the course were invited to participate in the course by taking part in the online course activities as well as attend the face-to-face class of the course, in addition to the traditional correspondence tuition. Three main subgroups of students in the course were identified based on whether the students participated in the online part and/or face-to-face class of the course. As a result, a stratified random sampling technique [17] was used to identify groups of students depending on the additional participation mode in the learning activities of the course. Few students participated in the face-to-face classes as well as the online components of the course [18] which resulted in all the participants in these learning modes being selected for the study. However, a simple random sampling technique [17] was used to select a proportional number of students using the correspondence learning mode only.

2.4. Data Collection

The research data were obtained using researcher-designed criterion-referenced tests with ten (10) closed problems as test items. The problems were testing the knowledge and understanding of physics concepts at various levels of Blooms taxonomy of learning objectives [19]. Primary and secondary data were collected to answer the research questions. Primary data consisted of (i) solutions to the formative practice problems, (ii) the number of problems solved, and (iii) the frequency of strategy use in solving problems in the summative examination. Secondary data consisted of (i) participation in the correspondence, online and face-to-face learning activities. All necessary permissions where obtained before collecting the data. Several measures were taken to increase the validity and reliability of the data collection instruments as well as the results.

The problems to be solved were selected from a standard internationally published physics textbook [15] used in previous official formative and summative assessment tests in the course. In addition, a test consisting of problems with deep similarity characteristics to the problems in the assessment tests was given to students registered for the course in the previous semester. This helped improve the clarity and comprehensibility of the problems in the tests. The same rubric used to assess solutions to the formative practice problems was used to assess solutions to problems in the formative and summative assessment tests. The tests were moderated by an internal physics expert for face validity and content validity. Also, the tests were marked by an external marker and the marked tests were moderated by two internal physics expert. The results are presented and discussed in the next section.

3. Results and discussion

The target population consisted of 179 active students in the course. Only 19 students legitimately participated in the online and 19 students attended the face-to-face class. After trimming the face-to-face class group was reduced to 15 students. Therefore, the sample size was restricted to 45 students, 15 in each group. No matching in the sample was done because this would have further reduced the already small sample size. The summative assessment test consisted of ten problems and was to be completed in two hours. The solution rubric of Heller *et al.* [3] was adopted in this study to assess solutions to problems from the formative and summative assessment tests. The rubric has been widely researched and modified over the years [20]. This rubric divides the problem solving process into five categories and provides simple grading mechanism for each category. The rubric is in line with the assessment policy of UNISA and is consistent with best assessment practices. Only problems of difficulty level between 2 and 4 [3] were chosen for the domain-referenced researcher-designed tests. To simplify the analysis of the results the number of problems with complete solutions (out of ten) and the frequency of strategy use where categorized as shown in Table 1.

The summative test was evaluated by two physics experts and the evaluated test was further assessed by the researcher. The inter-rater ratio [13]

$$R = \frac{\text{Number of actual agreements}}{\text{Number of possible agreements}} \tag{1}$$

can be used as a measure of consistency in the evaluation of the test. A value of $R = 0.65$ was obtained for the number of problems with complete solutions in the summative assessment

IOP Conf. Series: Journal of Physics: Conf. Series **905** (2017) 012019 doi:10.1088/1742-6596/905/1/012019

test. The results for the problem-solving performance are given and discussed in the next two subsections.

Table 1. Categories of problem-solving performance measures considered.

Category	Number of Problems	Strategy use
A	8 or more	6 or more
B	5, 6 or 7	3, 4 or 5
C	4 or less	2 or less

3.1. Number of problems solved

The results for the number of problems attempted in the summative examination are presented in Table 2 for the three groups. The number of problems with complete solutions N_p as well as the mean score M_p for each category and each group is given in the table. It can be seen from this table that students in the correspondence group generated more Category A solutions (53% with a mean of 52) than students in the face-to-face and online groups. On the other hand, students in the face-to-face and online groups generated more Category B solutions, 73% and 60%, respectively, with the corresponding mean of about 40. In general the performance of the face-to-face and the online group are similar but lower than that of the correspondence group. The results for the correspondence group follow the same trend as in the formative assignments (not shown) while the face-to-face and online results follow a different trend. This indicates that combining correspondence with face-to-face or online can increase the learning of problem-solving skills in a distance education course.

Table 2. The number of problems solved (N_p) and the mean score (M_p) for the course-presentation modes.

Category	Correspondence		Face-to-face		Online	
	N_p	M_p	N_p	M_p	N_p	M_p
A	8	52.4	3	39.7	4	59.5
B	3	33.0	11	39.1	9	42.7
C	4	34.2	1	21.0	2	20.0

One-way analysis of variance was used to further investigate the performance of the students presented in Table 2 for the number of problems solved. The F statistic and p-value for the three groups were calculated with a 95% confidence level ($\alpha = 0.05$) assuming equal variance in the groups. The results are given in Table 3. The calculated p-value is 0.583, and the F statistic value is 0.547 with a critical value $F_c = 3.2199$. Therefore, the analysis of the descriptive data indicate that combining the learning modes had no statistically significant effect on the number of problems solved ($F < F_c$, $p > 0.05$). However, this finding should be treated with caution since there was no control for extraneous variables.

Table 3. Inferential statistics for the number of problems solved.

Source	df	SS	MS	F	p	F_c
Between groups	2	4.044	2.022	0.547	0.583	3.2199
Within groups	42	155.2	3.695			
Total	44	159.2				

3.2. Frequency of strategy use

The frequency of strategy use was determined from the number of problems with complete solutions. The results obtained from the summative assessment test for the three categories of the three groups show that the face-to-face group had a higher frequency of strategy use, Category B solutions (87%), than the correspondence and the online groups, which had more Category C solutions (80% and 53%, respectively). However, unlike the correspondence and the online groups, the face-to-face group had no Category A solutions. The results for the correspondence and online groups appear very similar. For further analysis of these results, descriptive statistics for the frequency of strategy use were calculated and the results are shown in Table 4. In this table the sample size (N_s), the maximum value (Max), the range (*Range*), the mean (Mf), standard deviation (SD) and standard error (SE) are shown. It can be seen from the table that of the three groups, the face-to-face group had the lowest *Range* (3), the lowest SD (0.9) and the lowest SE (0.2). The results for the correspondence and online groups also appear similar.

Table 4. Descriptive statistics for the frequency of strategy use.

Group	N_s	Max	$Range$	Mf	SD	SE
Correspondence	15	9	9	1.9	2.2	0.6
Face-to-face	15	5	3	3.3	0.9	0.2
Online	15	10	10	3.2	2.8	0.7

Table 5. Inferential statistics for the frequency of strategy use.

Source	df	SS	MS	F	p	F_c
Between groups	2	19.733	9.867	2.187	0.125	3.2199
Within groups	42	189.467	4.511			
Total	44	209.2				

One-way analysis of variance was used to further investigate the descriptive statistics of the frequency of strategy use presented in Table 4. The F statistic and p-value for the three groups were calculated with a 95% confidence level ($\alpha = 0.05$) assuming equal variance in the groups. The results are given in Table 5. The calculated p-value is 0.125, and the F statistic value is 2.187

with a critical value $F_c = 3.2199$. Therefore, the analysis of the descriptive data indicate that the use of the blended-learning approach had no statistically significant effect on the frequency of strategy use ($F < F_c$, $p > 0.05$). However, this finding should be accepted with caution since there was no control for extraneous variables. Conclusions are presented in the next section.

4. Conclusion

The objective of this research was to investigate the effectiveness of the three delivery modes of distance education in learning problem-solving skills in a distance education physics course. A problem-solving instructional approach that involved the explicit teaching of a problem-solving strategy and worked-out examples was implemented in the course. The *ex post facto* research design with stratified sampling was used for the study. A sample of 45 students with 15 in each stratum was considered. The number of problems solved and the frequency of strategy use in solving problems were assessed for each of the course presentation modes. These problem-solving performance measures were compared between the three modes. Small differences in the means were observed in the problem-solving performance of the three groups of students. However, no significant difference was found in the effectiveness of the three presentation modes. It should be noted that extraneous variables were not controlled for in the study. Therefore, the results must be treated with caution. However, the finding of the study is consistent with the findings of [21]. Nonetheless, further investigations are required to verify the results. The finding of this study suggests that learning may be affected less by the type of technology used in teaching and learning in distance education.

Acknowledgments

This work is part of a dissertation submitted to the University of South Africa for the degree of Master of Education.

References

[1] Landau D P and Binder K 2009 *A Guide to Monte-Carlo Simulations in Statistical Physics* 3rd ed (New York: Cambridge University Press)
[2] Chetty N 2011 *Comp. Phys. Comm.* **182** 2065–70
[3] Heller P, Keith R and Anderson S 1992 *Am. J. Phys.* **60** 627–36
[4] Salagaram T and Chetty N 2013 *J. Phys. Conf. Series* **454** 012075
[5] Maloney D P 2011 *Rev. Phys. Educ. Res.* **2** 1–33
[6] Gök T and Silay İ 2010 *Lat. Am. J. Phys. Educ.* **4** 7–21
[7] Selçuk G S, Çalişkan S and Erol M 2008 *Lat. Am. J. Phys. Educ.* **2** 151–66
[8] Moore M G and Kearsley G 2005 *Distance education: A systems view* 2nd ed (Belmont: Wadsworth)
[9] Peters O 1998 *Learning and Teaching in Distance Education: Pedagogical analyses and interpretations in an international perspective* (New York: Routledge)
[10] Mishra P and Koehler M J 2006 *Teachers College Record* **108** 1017–54
[11] myUnisa 2014 URL `http://www.unisa.ac.za/sites/myunisa/default/`
[12] Rampho G J 2015 *Teaching problem-solving skills in a distance education programme using a blended-learning approach* MEd dissertation (University of South Africa: Unpublished)
[13] Cohen L, Manion L and Morrison K 2007 *Research methods in education* 6th ed (New York: Routledge)
[14] Halliday D, Resnick R and Walker J 2011 *Principles of Physics* 9th ed (New York: Wiley)
[15] Knight R D 2013 *Physics for scientists and engineers: A strategic approach* 3rd ed (San Francisco: Pearson)
[16] Vlachopoulos D and Tsokkas G 2005 *Int. J. Soc. Behav. Educ. Econ. Manag. Eng.* **9** 264–67
[17] Creswell J W 2012 *Educational Research: Planning, Conducting, and Evaluating Quantitative and Qualitative research* 4th ed (Boston: Pearson)
[18] Gulati S 2008 *Int. Rev. Res. ODL* **9** 1–16
[19] Krathwohl D R 2002 *Theor. Pract.* **41** 212–8
[20] Hull M M, Kuo E, Gupta A and Elby A 2013 *Phys. Rev. ST Phys. Educ. Res.* **9** 010105
[21] Russell T L 2001 *The No Significant Difference Phenomenon: as reported in 355 research reports, summaries and papers* (Raleigh: North Carolina State University)

IOP Conf. Series: Journal of Physics: Conf. Series **905** (2017) 012020 doi:10.1088/1742-6596/905/1/012020

Asymmetric Shock Wave Generation in a Microwave Rocket Using a Magnetic Field

Masayuki Takahashi

Department of Aeronautics and Astronautics, The University of Tokyo, Bunkyo-ku 113-8656, Japan

E-mail: m.takahashi@al.t.u-tokyo.ac.jp

Abstract. A plasma pattern is reproduced by coupling simulations between a particle-in-cell with Monte Carlo collisions model and a finite-difference time-domain simulation for an electromagnetic wave propagation when an external magnetic field is applied to the breakdown volume inside a microwave-rocket nozzle. The propagation speed and energy-absorption rate of the plasma are estimated based on the breakdown simulation, and these are utilized to reproduce shock wave propagation, which provides impulsive thrust for the microwave rocket. The shock wave propagation is numerically reproduced by solving the compressible Euler equation with an energy source of the microwave heating. The shock wave is asymmetrically generated inside the nozzle when the electron cyclotron resonance region has a lateral offset, which generates lateral and angular impulses for postural control of the vehicle. It is possible to develop an integrated device to maintain beaming flight of the microwave rocket, achieving both axial thrust improvement and postural control, by controlling the spatial distribution of the external magnetic field.

1. Introduction

It is necessary to develop innovative launch schemes to promote utilization of small satellites because the launch timings and orbits of small satellites are restricted in conventional launch schemes when sharing a chemical rocket with a main satellite. A microwave propulsion system has been proposed to realize frequent and flexible launches of small satellites while reducing the launch cost through repetitive launches of inexpensive vehicles [1–5]. Intense microwaves are radiated from a ground-based oscillator to a thruster equipped with a parabolic mirror, and the incident beam is focused by the mirror to induce air breakdown by the strong fields in the focal region. A dense plasma is generated by the gas breakdown and propagates toward the nozzle exit while forming strong shock waves because the energy absorbed by the plasma is transferred to neutral particles through the collision process. The vehicle obtains an impulsive thrust from interactions between the thruster wall and the shock waves. Chemical fuel aboard the vehicle can be removed or reduced because ambient air acts as the fuel, which decreases the launch cost.

Pressure dependencies of the breakdown structure and the thrust performance of a microwave rocket were assessed previously by placing a parabolic thruster in a vacuum chamber [2, 3, 6]. A discrete plasma was formed at atmospheric pressure because a standing wave was induced in front of the plasma due to waves reflected by the overcritical plasma. On the other hand, the breakdown pattern became diffusive at lower pressures because rapid electron diffusion smeared out the plasma pattern. The propagation speed of the ionization front was found to increase with

IOP Conf. Series: Journal of Physics: Conf. Series **905** (2017) 012020 doi:10.1088/1742-6596/905/1/012020

decreasing ambient pressure due to the rapid diffusion of electrons. The thrust performance of the microwave rocket degraded at lower pressures because the propagation speed of the ionization front became faster and the energy-absorption rate of the plasma decreased. It is necessary to suppress the propagation speed of the ionization front and improve the energy-absorption rate in order to enhance the generated shock wave inside the rocket nozzle at lower pressures. The breakdown pattern of the microwave plasma was numerically reproduced by coupling fluid or particle models of plasma transport with electromagnetic wave propagation [7–9]. Based on a fully kinetic simulation, it was proposed to apply an external magnetic field to the breakdown volume in order to confine the plasma inside the rocket nozzle [9]. Axial thrust performance was improved at lower pressures because electron cyclotron resonance (ECR) heating was combined with plasma confinement [10, 11].

In addition to improving axial thrust, it is necessary to obtain postural control impulses in order to maintain beam-riding flight [12–14]. The generated thrust is drastically degraded if the incident beam has lateral and angular misalignments with respect to the vehicle axis. The vehicle finally deviates from the incident beam axis if restoring forces for diminishing incidental misalignments are not attained during flight. The capturing ability of the incident beam, which is called beam-riding performance, is crucial to maintaining higher thrust and stable flight without deviation from the beam line. The beam-riding performance is evaluated entirely by the vehicle's feedback abilities against lateral and angular offsets from the incident beam line. The "lightcraft" vehicle, which has an annular shroud in order to generate the centering-feedback force, was proposed by Myrabo [12], and flight demonstrations of the lightcraft were conducted using high-averaged-power repetitive-pulses. The lightcraft was passively pushed back toward the center of the incident beam because the blast wave interacted with the ring-shaped shroud when the incident beam encountered lateral misalignment. However, the lightcraft, weighting 50 g, deviated from the incident beam line because of insufficient angular feedback performance, and its flight altitude was only 71-m altitude in 2001. It is necessary to propose an active control scheme for beam riding because passive control such as that of the lightcraft cannot maintain stable flight in a repetitive-pulse launch.

The objective of this study is propose an active control technique for maintaining stable flight of a beamed-energy propulsion vehicle. To actively control the lateral and angular impulses acting on the vehicle, it is necessary to deform the generated shock wave by controlling the spatial distribution of energy absorption by the plasma. The asymmetric energy-absorption region can be obtained when the magnetic field is asymmetrically applied to the breakdown volume inside the rocket nozzle. The lateral and angular impulses are evaluated using CFD code when the asymmetric shock wave interacts with the thruster.

2. Numerical Methods

A one-dimensional (1D) particle-in-cell with Monte Carlo collisions (PIC-MCC) model is coupled with a finite-difference time-domain (FDTD) simulation to reproduce the breakdown structure induced by microwave irradiation when the external magnetic field is applied to the discharge volume. The propagation speed and energy-absorption rate of the plasma are evaluated in the discharge simulation. The axial, lateral, and angular impulses are assessed using computational fluid dynamics (CFD) code by reproducing shock wave generation and reflection by the thruster wall.

2.1. Electromagnetic wave propagation

Electromagnetic wave propagation and interaction with the plasma are reproduced using the 1D FDTD method to examine the breakdown pattern, propagation speed of the ionization front, and energy-absorption rate of the plasma. Maxwell's equations are numerically solved to simulate electromagnetic wave propagation. For the boundary conditions of the FDTD simulation, a Mur absorption boundary [15] is utilized to model the open boundary by damping the electromagnetic waves. A total field-scattered field implementation is employed to inject microwaves into the computational domain. The current density **J** in the Ampere-Maxwell law is evaluated in the 1D PIC-MCC simulation to introduce interactions between the motion of the charged particles and the microwave oscillations.

2.2. Fully kinetic simulation with collision reactions

The external magnetic field is applied to the breakdown volume to enhance electron heating by satisfying the ECR condition. The PIC method is utilized to reproduce interactions between the cyclotron motions of the charged particles and microwave oscillations during microwave breakdown. The 1D simulation is performed with three velocity (3V) components to reduce the computational load for simulating the collisional plasma. In the 1D3V PIC module, the equations of motion for the electron and ion particles are temporally integrated using the Buneman-Boris method [16] to predict the charged-particle motions. The current density $\mathbf{J} = qn\mathbf{v}$ is assessed on computational grids using linear weighting of super-particles in the PIC module and is fed back into Maxwell's equations in the FDTD simulation. The microwave reflection and damping caused by the plasma are reproduced by the feedback of the current density to Maxwell's equations. Collision reactions between the electron and neutral particles are introduced by combining the MCC model with the PIC method. The Nanbu method [17] is used to evaluate the existence or non-existence and the reaction types of the collisions. Eighteen reactions (e.g., elastic collision, vibrational excitation, rotational excitation, electron-impact excitation, and ionization) are introduced in the MCC model based on Phelps's cross section for nitrogen gas [18]. The MCC model is coupled with the PIC model using the Birdsall method [19].

2.3. Shock wave propagation induced by gas heating

A shock wave is induced inside a microwave-rocket nozzle when the microwave energy absorbed by the plasma is transferred to the energy of the neutral particles. The energy absorbed by the plasma is evaluated by the PIC-MCC and FDTD coupling simulation and deposited into a source term in the fluid simulation to reproduce the propagation of shock waves. The two-dimensional (2D) Euler equations are numerically integrated to reproduce the shock wave propagation around the microwave rocket as follows:

$$\frac{\partial \mathbf{Q}}{\partial t} + \frac{\partial \mathbf{E}}{\partial x} + \frac{\partial \mathbf{F}}{\partial y} = \mathbf{S}, \tag{1}$$

where **Q** is a conservative vector, **E** and **F** are inviscid flux vectors in the x- and y-directions, and **S** is the heating term estimated by the PIC-MCC calculation. Discretization is performed in a cell-centered finite-volume manner [20]. The AUSM-DV method [21] is employed for numerical fluxes with the second-order MUSCL method [22]. In the MUSCL method, characteristic variables are selected as interpolation quantities and a minmod limiter is used to fulfill the TVD condition. Time integration for unsteady shock wave propagation is performed using the first-order explicit Euler method.

3. Simulation Conditions

3.1. FDTD and PIC-MCC coupling simulation

The maximum intensity of the incident field E_y is fixed at 5 MV/m, and the microwave frequency is 110 GHz following Hidaka's experiments [6]. The 1D simulation domain has 2.25λ (the microwave wavelength is $\lambda = 0.27$ cm), and the number of grid points used is 40,000 in order to resolve the Debye length of the electrons. The initial plasma spot, which has a Gaussian distribution, is set on the computational domain to simplify the first gas breakdown under microwave focusing. The microwaves are injected from the left boundary ($x = 0$ cm) to reproduce microwave-plasma interaction. An external magnetic field B_z is applied to the computational domain to confine electron transport while the incident electric field E_y oscillates in the y-direction. ECR heating occurs when the external magnetic field is chosen as 3.93 T because oscillation of the electric field E_y is coupled with the electron cyclotron motion twinning around the magnetic field B_z. The ambient pressure is set as 0.003 atm to obtain a large Hall parameter.

3.2. CFD simulation to reproduce shock wave propagation

A cylinder-type microwave rocket is numerically modeled to estimate the lateral and angular impulses obtained by microwave irradiation using CFD code. The nozzle radius is 3 cm and its length L is 18 cm following Oda's experiments [2,3]. The distance between the center of gravity and the nozzle head is 7.2 cm. The absorption energy and propagation speed of the ionization front are calculated by the PIC-MCC and FDTD coupling simulation in order to introduce the energy source into the CFD simulation. An external magnetic field has a lateral offset (Fig. 1); however, the incident beam has no offset in this simulation. The magnetic field satisfying the ECR condition is applied into only 253 cm $\leq y \leq$ 256 cm (the nozzle axis exists at $y = 253$ cm). However, there is no magnetic field at 250 cm $\leq y <$ 253 cm to generate the asymmetric shock wave. The heating region moves from the nozzle head toward the nozzle exit ($x = 193$ cm) at the propagation speed of the ionization front. Microwave irradiation is terminated when the heating region at the ECR condition reaches the nozzle exit. The slip boundary condition is used on the thruster wall, and the other outer boundaries are treated as zeroth-order extrapolations. The axial, lateral, and angular impulses are calculated by integrating the surface pressures on the thruster walls. The lateral force is defined in the positive y-direction, and the clockwise direction is taken to be positive for the angular impulse.

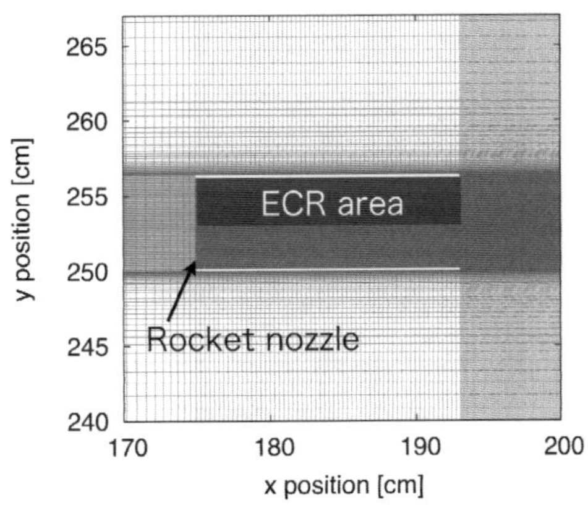

Figure 1. Computational grid of rocket nozzle. ECR magnetic field is applied with a lateral offset against the nozzle axis.

4. Results

4.1. Propagation speed and energy-absorption rate of plasma

5-MV/m, 110-GHz microwaves are irradiated onto an initial plasma spot at an ambient pressure of 0.003 atm. An external magnetic field of 3.93 T is applied to the discharge volume to induce ECR heating by satisfying $\omega = \omega_c$, where ω is the incident microwave frequency and ω_c is the electron cyclotron frequency. The breakdown structure, propagation speed of the ionization front, and energy-absorption rate of the plasma are assessed using the PIC-MCC and FDTD coupling simulation to model an energy source in the CFD simulation.

At 0 T, the electron density is small at an ambient pressure of 0.003 atm because of infrequent ionization collisions. The incident microwaves penetrate into the plasma spot because the electron density does not exceed the cut-off density for microwave reflection (Fig. 2(a)). However, the incident microwaves are reflected by overcritical plasma when a magnetic field of 3.93 T is applied to the discharge volume because electron-impact ionization rapidly occurs due to high-energy electrons generated by ECR heating (Fig. 2(b)). The ionization front propagates toward the microwave source by repeated microwave reflection, electron diffusion, and ionization at the plasma front. By tracing the position of the ionization front, the propagation speeds are evaluated as 1,430 km/s and 1,250 km/s at magnetic fields of 0 T and 3.93 T, respectively, at an ambient pressure of 0.003 atm. The energy-absorption rate of the plasma is estimated by tracing the electron energy at the plasma front in order to model the source term in the shock wave simulation. At 0 T and 3.93 T, the energy-absorption rates by the plasma are assessed as 3.3×10^{11} W/m^3 and 5.0×10^{13} W/m^3, respectively. The shock wave induced by the microwave irradiation is reproduced based on the propagation speed and the energy-absorption rate evaluated by the PIC-MCC and FDTD coupling simulation.

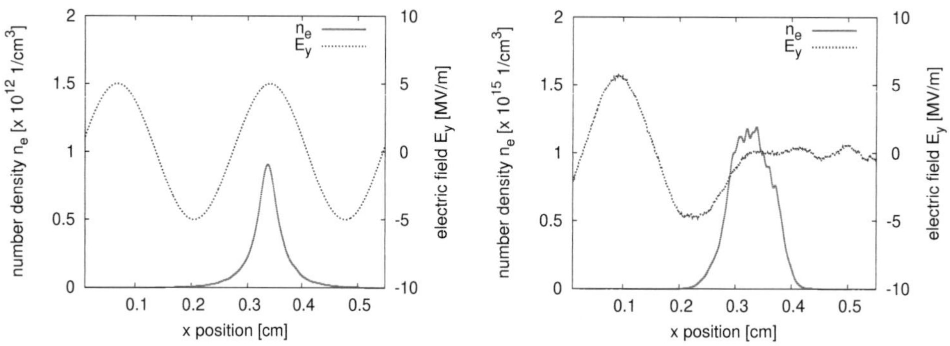

(a) 0 T, 1.6 ns (b) 3.93 T, 1.6 ns

Figure 2. Electron number density and electric field at 0.003 atm. 5-MV/m, 110-GHz microwaves are irradiated onto the initial plasma spot, and microwave plasma propagates toward the microwave source (left boundary). Electrons and ions diffuse in the x-direction, and the incident microwave field oscillates in the y-direction. External magnetic fields B_z of 0 T and 3.93 T are applied in the z-direction.

4.2. Lateral and angular impulses driven by asymmetric shock waves

A shock wave is induced inside the microwave-rocket nozzle when microwaves are irradiated onto the vehicle because the microwave energy absorbed by the plasma is transferred to the energy of neutral particles through collisional reactions. The shock wave is enhanced when a magnetic field satisfying the ECR condition is applied to the breakdown volume because the energy-absorption rate is improved by resonant heating. An asymmetric shock wave is induced when the applied magnetic field has a lateral offset relative to the vehicle axis (the ECR condition is satisfied at 253 cm $\leq y \leq$ 256 cm, while the vehicle axis exits at y = 253 cm), which provides lateral and angular impulses to the thruster for achieving postural control.

The shock wave is asymmetrically generated inside the rocket nozzle because the energy-absorption region caused by the ECR heating is asymmetric due to the lateral offset of the external magnetic field (Fig. 3(a)). Lateral and angular impulses are not obtained when the lateral offset of the magnetic field does not exist; however, the thruster obtains lateral and angular impulses with a lateral offset because the asymmetric shock wave interacts with the rocket nozzle (Fig. 3(b)). Negative lift is generated because the asymmetric shock wave strongly interacts with the side wall at $y = 250$ cm. An anti-clockwise moment is obtained because the high pressure of the shock wave interacts with the nozzle corner at $(x,\ y) = (175\ \text{cm},\ 250\ \text{cm})$. The amplitude and direction of the lateral and angular impulses can be controlled if the offset of the magnetic field is changed by controlling the current density of an external coil. The axial thrust is also improved because the shock wave enhanced by ECR heating interacts with the nozzle head. It is possible to develop an integrated device that can achieve axial thrust improvement and postural control by applying an external magnetic field onto a microwave-rocket nozzle.

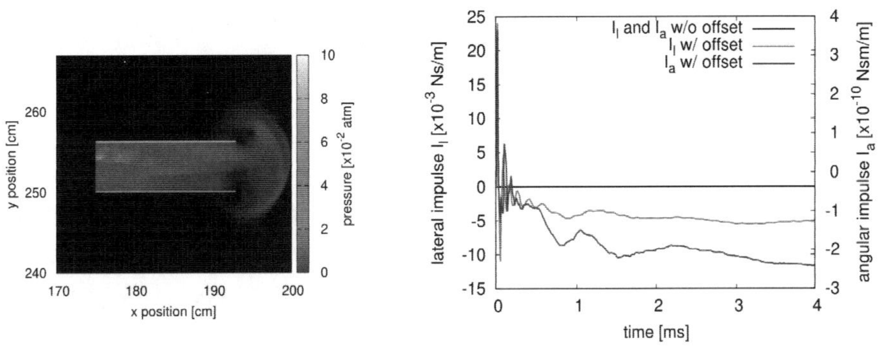

(a) pressure distribution (b) lateral and angular impulses

Figure 3. Pressure distribution around the microwave rocket and lateral and angular impulses. The axis of microwave irradiation is aligned with the vehicle axis. An external magnetic field is applied to the breakdown volume with and without a lateral offset.

5. Conclusion

By coupling the PIC-MCC model with FDTD simulation, the propagation speed and the energy-absorption rate of plasma were estimated for the case where an external magnetic field is applied into the discharge volume induced by microwave irradiation. The breakdown pattern was reproduced and shock wave propagation was reproduced using CFD code based on the propagation speed and the energy-absorption rate of the plasma. An asymmetric shock wave was induced inside the microwave-rocket nozzle when the ECR region had a lateral offset, providing lateral and angular impulses for postural control of the vehicle. In addition to axial thrust enhancement, postural control can be conducted by utilizing an external magnetic field satisfying the ECR condition.

Acknowledgments

The fully kinetic simulations in this work were performed on a Silicon Graphics International (SGI) Altix UV1000 at the Advanced Fluid Information Research Center, Institute of Fluid Science, Tohoku University. The fluid simulations in this work were performed on a FUJITSU PRIMEHPC FX100 at the Japan Aerospace Exploration Agency. This work was supported by JSPS KAKENHI Grant Number JP16J09910.

References

[1] Nakagawa T, Mihara Y and Komurasaki K 2004 *J. Spacecraft and Rockets* **41** (1) 151–153
[2] Oda Y and Komurasaki K 2006 *J. Appl. Phys.* **100** 113307
[3] Oda Y 2008 *Doctor Thesis* (Tokyo: The University of Tokyo)

IOP Conf. Series: Journal of Physics: Conf. Series **905** (2017) 012020 doi:10.1088/1742-6596/905/1/012020

[4] Fukunari M, Komatsu R, Arnault A, Yamaguchi T, Komurasaki K and Arakawa Y 2013 *Vacuum* **88** 155–159
[5] Fukunari M, Arnault A, Yamaguchi T and Komurasaki K 2014 *Appl. Optics* **53** (31) 16–22
[6] Hidaka Y, Choi E M, Mastovsky I, Shapiro M A, Sirigiri J R and Temkin R J 2009 *Phys. Plasmas* **16** 055702
[7] Boeuf J P, Chaudhury B and Zhu G Q 2010 *Phys. Rev. Lett.* **104** 015002
[8] Kourtzanidis K, Boeuf J P and Rogier F 2014 *Phys. Plasmas* **21** 123513
[9] Takahashi M and Ohnishi N 2014 *Appl. Phys. Lett.* **105** 223504
[10] Takahashi M and Ohnishi N *Trans. Japan Soc. Aero. Space Sci., Aero. Tech. Japan* (accepted)
[11] Takahashi M and Ohnishi N 2016 *J. Appl. Phys* **120** 063303
[12] Myrabo L N, Messitt D G and Mead F B Jr 1998 *AIAA Paper* 98-1001
[13] Kenoyer D A, Anderson K S and Myrabo L N 2008 *Beamed Energy Propulsion: Fifth International Symposium on Beamed Energy Propulsion, AIP Conference Proceedings* vol 997 (New York: American Inst. of Physics) pp 325–337
[14] Scharring S, Eckel H A and Röser H P 2011 *Int. J. Aero. Innovations* **3** (1) 15–31
[15] Mur G 1981 *IEEE Trans. EMC* **EMC-23** (4) 377–382
[16] Buneman O 1967 *J. Comput. Phys.* **1** 517–535
[17] Nanbu K 1994 *Jap. J. Appl. Phys.* **33** 4752–4753
[18] Phelps A V 2008 *Compilation of Electron Cross Sections* (University of Colorado) [http://jila.colorado.edu/~avp/collision_data/electronneutral/ELECTRON.TXT]
[19] Birdsall C K 1991 *IEEE Trans. on Plasma Sci.* **27** 65–85
[20] Hirsch C 1984 *Numerical Computation of Internal and External Flows* vol 1 (New York: John Wiley and Sons) pp 237–266
[21] Wada Y and Liou M S 2006 *AIAA Paper* 2006-1358
[22] van Leer B 1979 *J. Comput. Phys.* **32** (1) 101–136

IOP Conf. Series: Journal of Physics: Conf. Series **905** (2017) 012021 doi:10.1088/1742-6596/905/1/012021

Formalism of compound particles for simulation of the heavy ions in a stationary nonequilibrium warm dense matter

Zh A Moldabekov[1], T S Ramazanov[1], M T Gabdullin[1], A Tikhonov[2], K Baigarin[2] and M Kaikanov[3]

[1]Institute of Experimental and Theoretical Physics, al-Farabi Kazakh National University, al-Farabi Avenue 71, Almaty 050040, Kazakhstan

[2]Nazarbayev University, Kabanbay batyr ave. 53, Astana 010000, Kazakhstan

[3]National Laboratory Astana, Kabanbay batyr ave. 53, Astana 010000, Kazakhstan

E-mail: `zhandos@physics.kz`

Abstract. The screened interaction potential between two compound particles, compound particle and charged particle (ion or electron) in multipole approximation for simulation of nonequilibrium warm dense matter are discussed. The density and temperature range has been considered at which the formalism of compound particles can be used. It is proposed that the presented screened potential can be useful for the simulation of the heavy ions in the presence of streaming electrons. Discussions about the implication of a compound particle picture for consideration of the dynamics of the beam of the charged particles in plasmas are given. The proposed model of interaction between heavy ions consists of dipole terms and the short range repulsion due to the Pauli exclusion principle.

1. Introduction

Plasma consists of different particles types both charged and neutral. Usually, different species of particles have a large difference in physical properties such as mass and charge. The difference in the mass between particles of different species, e.g between electron and ion in usual two component plasma, leads to the different characteristic time scales of plasma components. In this regard, the subsystem of particles with similar values of the mass and charge can be considered within so called one component plasma model with neutralizing background. In the case of the weak electron-ion coupling, the electrons can be treated within linear response theory and the ions by molecular dynamics or Monte-Carlo simulation as has been proposed by Ludwig et al. [1]. This approach is also useful for the simulation of so called complex or dusty plasma (see, for example, [5] and references therein) where the subsystem of strongly coupled nano- or micro-sized dust particles with ideal electrons and ions as a background can be investigated in the framework of the one component plasma model. The idea of Ludwig et al. [1] has been used by Graziani et al. to decouple the electron kinetic equation using the Singwi-Tosi-Land-Sjölander scheme [2] or a recently derived extension [3, 4].

For simulation of the heavy component of the plasma, which can be nonideal, the collective screening effect due to mobile component (electrons) of plasma has to be taken into account properly in the effective screened interaction potential [6, 7]. The effective screened interaction potentials can be obtained by methods based on the kinetic equations and the formalism of the dielectric response function. However, these methods are not useful for derivation of the screened potential if considered particles have a dipole moment, as a dipole field does not have spherical symmetry. In the recent work [8] it has been shown that the multipole expansion method allows to find screened interaction potentials between compound particles with dipole moments, if one knows the screened potential of single charged particle at a given plasma parameters. In this case the problem of finding the effective interaction potential is reduced to the problem of finding derivatives.

In this paper some problems related to the calculation of the screened interaction potential between particles with non zero dipole moment by standard way are discussed and the formalism of the compound particles for the simulation of the heavy component of the plasma in presence of the streaming of particles of mobile component relatively to the inert ones are proposed. We consider interaction model of heavy ions in the presence of streaming electrons in warm dense matter and in the case when ion beam penetrates into the plasma. The temperature and density range is discussed at which the formalism of compound particles is applicable.

2. Interaction potential between compound particles in nonequilibrium plasmas

The key of the multiscale approach is to absorb the fast particle kinetics into the effective screened interaction potential Φ of the heavy particles with charge Q where the screening is provided by the mobile particles (electrons in two component plasma) via a dielectric function ϵ, e.g. [1]

$$\Phi(\mathbf{r}) = \int \frac{\mathrm{d}^3k}{2\pi^2} \frac{Q^2}{k^2\epsilon(\mathbf{k}, \omega = 0)} e^{i\mathbf{k}\cdot\mathbf{r}} \quad , \tag{1}$$

taken in the static limit. The effect of streaming electrons (ions) is important in warm dense matter. Then one have dynamically screened potential and the frequency argument becomes $\omega = \mathbf{ku}$ which leads to wake effects, that are well investigated theoretically and experimentally for dusty plasmas, e.g. [9–11] and quantum plasmas, e.g. [12] and references therein.

In the case of quantum plasmas, an accurate calculation of dynamically screened potential of an ion takes from few minutes for zero temperature limit up to several hours for finite temperature case. Therefore it is important to have simple analytical model which takes into account wake effect for simulation of the inert particles in nonequilibrium plasma. Such interaction model can be proposed knowing properties of the dynamically screened potential.

2.1. Dynamically screened ion potential in two component quantum plasmas

Recently, we have done detailed analysis of the dynamically screened ion potential in warm dense matter and dense plasmas [12–14] as a function of the electrons degeneracy parameter and so called streaming parameter (Mach number) which is in fact dimensionless velocity of streaming electrons or of the ion penetrating into plasma. In particular, the transition from the high temperature regime (classical) to the low temperature regime (quantum) was investigated. Strong deviations from the static Yukawa type potential [6] as well as a possibility of an attraction between like-charged ions were shown. The conditions have been found at which streaming of the plasma background is giving rise to a dynamically screened charge field with a negative trailing potential minimum behind considering ion, which may lead to the attraction between like-charged ions. The physical mechanism is that the streaming fast particles gives rise to a deflection of the latter and, eventually, to an excess density behind the test particle. This interpretation was directly confirmed by computing the electron density perturbation which reveals a clear enhancement behind the ion [12]. In figure 1 the summary of the analysis of dynamically screened ion potential is presented at $(u/u_F; \theta)$ plane in the form of the diagram, where u_F is the Fermi velocity, u electrons streaming velocity (or projectile velocity), and

$\theta = k_BT/E_F$ is the degeneracy parameter of electrons. In figure 1 the characteristic quantum velocity reads:

$$v_Q^2 = \frac{2\langle K \rangle_Q}{m} = \frac{3}{2}v_F^2\theta^{5/2}\int_0^\infty dy \frac{y^{3/2}}{\exp(y-\beta\mu)+1}. \tag{2}$$

This diagram was recently used with a special focus on anomalous wake effects [13]. Here we use this diagram to present at which parameters the formalism of compound particles can be used. In the regions II and III, the dynamically screened potential has oscillations with more than one negative minima. In these regions the dynamically screened potential probably can not be presented by a relatively simple analytical formula. In the region IV, the dynamically screened potential can have negative minimum, but the distance between the ion and this negative minimum is more large than the mean interparticle distance (if neglect Friedel oscillations [12]). In this region, the screening length behind and ahead of the ion different from each other and rescaling of the screening length can be used [15–18]. In the region I the ion wake field possesses only one main potential minimum. *At these plasma parameters we have a positively charged point like ion and a negatively charged cloud of the focused electrons. One can consider this as one compound particle with non zero total charge and dipole moment.* Therefore, in the region I, the dynamically screened ion potential can be replaced by the statically screened potential of a compound particle.

In figure 2 the distance between an ion and a negative trailing potential minimum behind considering ion for different values of the mean interparticles distance of electrons $a = (4\pi n_e/3)^{-1/3}$ as a function of the streaming velocity (or projectile velocity) at $\theta = 0.01$ is shown. With increase of the degeneracy parameter up to $\theta \simeq 0.1$ this distance remains nearly constant [12]. Further increase in the degeneracy parameter up to ~ 0.5 can increase the distance from an ion to the negative minimum of the dynamically screened ion potential no more than 1.5 times [12, 13].

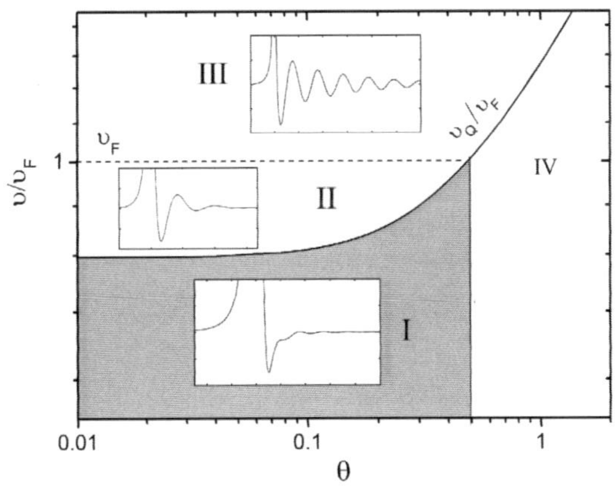

Figure 1. In the region I the ion wake field possesses only one main potential minimum and the dynamically screened ion potential can be replaced by statically screened potential of the compound particle.

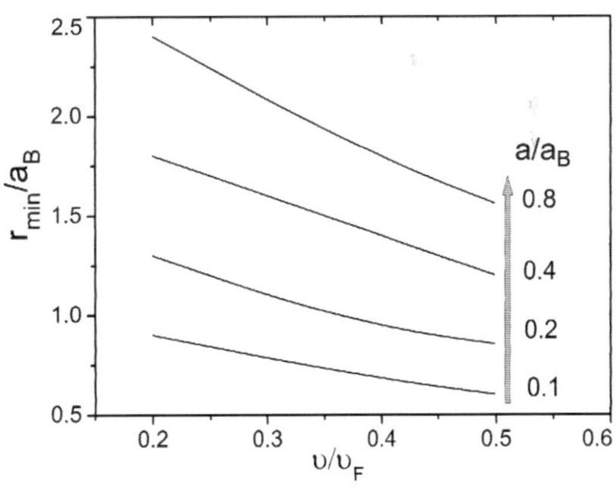

Figure 2. Distance from ion to the negative minimum of the dynamically screened ion potential for different values of the ratio of the mean interparticle distance of the electrons to the first Bohr radius at $\theta = 0.01$.

From quasi-neutrality condition, for density of ions we have $n_i = n_e/Z_i$, where Z_i is the ion charge in units of the electron charge (ion charge number). For applicability of the compound particle formalism within multipole approximation the mean distance between ions has to be larger than the linear size of the compound particle. Therefore, we have condition:

$$Z_i \gg \left(r_{\min}[v,\theta]/a\right)^3, \tag{3}$$

here it is emphasized that r_{min} depends on both streaming velocity v and the degeneracy parameter of the electrons θ.

In figure 3, minimal values of the heavy ion charge $Z_i^{\min}$ obtained from equation (3) are presented. As it was mentioned, the formalism of compound particles can be used when $Z \gg Z_i^{\min}$. In experiments on warm dense matter, the ion charge can be of order ten. Therefore, as it is seen from figure 3, the formalism of compound particles is justified if $a/a_B \gtrsim 0.8$ (here a is the electrons mean interparticle distance). This corresponds to the number density of electrons $n_e \lesssim 3 \times 10^{23} cm^{-3}$. The temperature of the electrons related to the degeneracy parameter and mean interparticle distance as $T \simeq \frac{\theta}{(a/a_B)^2}\, 0.58 \times 10^6 (K)$. Using this relation one can obtain the maximum value of the temperature $T_{\max} \sim 4.5 \times 10^5 K$ up to which the formalism of compound particles can be used.

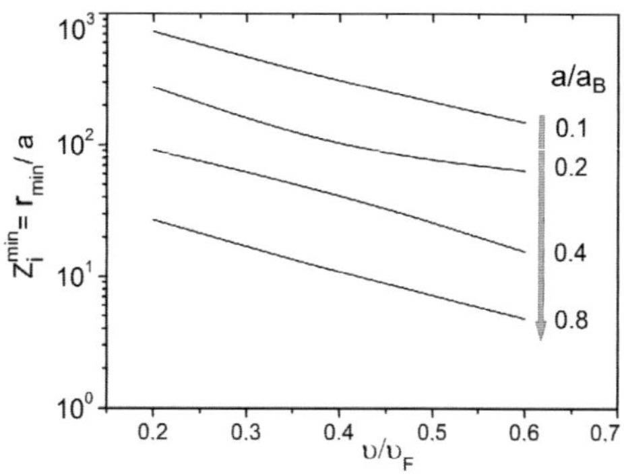

Figure 3. The minimal value of the heavy ion charge $Z_i^{\min}$ obtained from equation (3) at different values of a/a_B.

Up to now, the interaction between ions in the presence of streaming electrons has been considered. In the case of the ion beam – plasma interaction, the density of the beam ions $n_i^{\rm beam}$ much less than that of plasma n_e (electrons). As the result, the formalism of compound particles can be used at higher densities of the plasma for simulation of the interaction of beam ions with each other and with the plasma ions. In this case the following condition must be satisfied :

$$\frac{n_i^{\rm beam}}{n_e} \ll \left(r_{\min}[v,\theta]/a\right)^3. \tag{4}$$

In the high energy ion beam experiments, the ion beam density can be up to $n_i^{\rm beam} \sim 10^{14} cm^{-3}$ [20]. From data presented in figure 3 it is clearly seen that in the case of the ion beam – plasma interaction the formalism of compound particles can be used almost at any plasma density if $\theta = k_B T/E_F \leq 0.5$.

2.2. Screened interaction potential between compound particles with the strong repulsion at short distance

After consideration plasma parameters at which the formalism of compound particles can be used, we turn to the model of interaction between compound particles with nonzero total charge and dipole moment. As it has been discussed, the formalism of compound particles can be used for the simulation of the heavy ion component with $Z_i > 1$. In this case a strong additional repulsion between ions due to bound electrons must be taken into account [19,21]. This repulsion between ions appears at short distance as a result of the Pauli exclusion principle when shells of ions start to overlap. In [21] the following screened ion-ion interaction potential with a strong repulsion at short distance for the simulation of the ion component of an equilibrium warm dense matter was proposed:

$$\Phi_Y(r;n,T) = \frac{Z_i e}{r} e^{-k_Y r} + \frac{a}{r^6}, \tag{5}$$

here a defines the strength of short range repulsion and can be understood as a radial extension of the ion, $k_Y^2 = k_{TF}^2 \theta^{1/2} I_{-1/2}(\eta)/2$ is the screening length which interpolates between Debye and Thomas-Fermi expansions, I_ν is the Fermi integral of order ν, $\eta = \mu/k_B T$ is the chemical potential of electrons, and $\theta = k_B T/E_F$ is the degeneracy parameter which defines whether plasma degenerate or classical.

In the case of nonequilibrium warm dense matter, the interaction potential between compound particles has additional terms which responsible for charge-dipole and dipole-dipole interactions. Screened interaction potential between compound particles with additional dipole terms was derived in recent work [8]. Using this potential and adding the short range repulsion we have:

$$\Phi(\mathbf{R}) = \frac{Q_1 Q_2}{R}\exp(-Rk_Y) + \frac{(Q_2\mathbf{d}_1 - Q_1\mathbf{d}_2)\cdot\mathbf{R}}{R^3}(1+Rk_Y)\exp(-Rk_Y) + \\ + \frac{\left[(\mathbf{d}_1\mathbf{d}_2)R^2 - (\mathbf{d}_1\mathbf{R})(\mathbf{d}_2\mathbf{R})\left(3+Rk_Y\right)\right](1+Rk_Y) + (\mathbf{d}_1\mathbf{R})(\mathbf{d}_2\mathbf{R})Rk_Y}{R^5}\exp(-Rk_Y) + \frac{a}{r^6}. \tag{6}$$

here Q_1 and Q_2 are charges of ions under consideration.

In the case of parallel dipole moments of compound particles (expected in the case of stationary nonequilibrium state), additional dipole-dipole interaction leads the stronger repulsion in perpendicular to dipole moment direction in comparison with the Yukawa potential without dipole terms [22]. In parallel to dipole moment direction, the screened interaction potential has region with attraction between compound particles [8]. In this direction, the short range repulsion $\sim 1/r^6$ prevents collapse due to dipole-dipole attraction.

The repulsion constant a was determined fitting the structure factor obtained from density functional theory by hypernetted chain calculation or molecular dynamics simulation on the basis of potential (5). The value of a obtained for equilibrium case can be used for nonequilibrium warm dense matter as a repulsion due to quantum symmetry effect does not affected by dynamic screening. The value of the dipole moment depends on the electron stream velocity or projectile velocity and can be obtained by fitting the depth and position of the negative minimum of the dynamically screened ion potential. The maximum absolute value of the depth of the negative minimum for considering parameters (region I in figure 1) is of order $\sim Z_i^2 \times 0.1\,[Ha]$ in atomic units (see figure 5 (a) in [12]).

3. Conclusion

We have proposed the interaction model between heavy ions within formalism of compound particles for simulation of stationary nonequilibrium warm dense matter. This model can be used, for example, in the simulation of the ions using molecular dynamics. The proposed model has two parameters the dipole moment d of the compound particle and the radial extension a of the heavy ion. The radial extension for a silicon ion was considered in [19, 21]. The dipole moment can be varied taking into account the maximal absolute value of the negative minimum obtained from calculation of the dynamically screened ion potential. The density and temperature range at which the formalism of compound particles can be used has been discussed. It has been shown that the density and temperature have to be $n_e \precsim 3\times 10^{23} cm^{-3}$ and $T \precsim 4.5\times 10^5 K$, respectively. Additionally, the degeneracy parameter has to be $\theta \leq 0.5$. These parameters corresponds to the warm dense matter regime.

The formalism of compound particles was used successfully for understanding properties of the charged dust particles in complex plasmas [23–29]. In the case of the complex plasma, the negatively charged dust particle injected into the plasma with streaming ions creates a focused ion cloud [24, 25] and the dust particle together with the ion cloud can be considered as one compound particle [26]. It was shown that additional charge-dipole and dipole-dipole interaction has strong impact on both structural and dynamical properties of the dusty plasma [22, 27–29]. Therefore, it is expected that the formalism of compound particles can be useful for the investigation of the warm dense matter.

Acknowledgments
This research was funded under the program number 011503029 "NU-Berkeley strategic initiative in warm-dense matter, advanced materials and energy sources for 2014-2018" from the Ministry of Education and Science of the Republic of Kazakhstan

References

[1] Ludwig P, Bonitz M and Kählert H 2010 *J. Phys.: Conf. Series* **220** 012003
[2] Michta D, Surh D and Graziani F 2013 *High Energy Density Physics* **9** 696
[3] Kählert H, Kalman G and Bonitz M 2014 *Phys. Rev. E* **90** 011101
[4] Kählert H, Kalman G and Bonitz M 2015 *Contrib. Plasma Phys.* **55** 352
[5] Bonitz M, Henning C and Block D 2010 *Rep. Prog. Phys.* **73(6)** 066501
[6] Moldabekov Zh, Schoof T, Ludwig P, Bonitz M and Ramazanov T 2015 *Phys. Plasmas* **22** 102104
[7] Ramazanov T S, Moldabekov Zh A and Gabdullin M T 2015 *Phys. Rev. E* **92** 023104
[8] Ramazanov T, Moldabekov Zh and Gabdullin M 2016 *Phys. Rev. E* **93** 053204
[9] Tsytovich V N, Morfill G E, Vladimirov S V and Thomas H M 2008 *Lecture Notes in Physics* 731 (Berlin Springer: Heidelberg)
[10] Block D, Carstensen J, Ludwig P, Miloch W J, Greiner F, Piel A, Bonitz M and Melzer A 2012 *Contrib. Plasma Phys.* **52** 804
[11] Ludwig P, Kählert H and Bonitz M 2012 *Plasma Phys. Control. Fusion* **54** 045011
[12] Moldabekov Zh A, Ludwig P, Bonitz M and Ramazanov T S 2015 *Phys. Rev. E* **91** 023102
[13] Moldabekov Zh A , Ludwig P, Bonitz M and Ramazanov T S *Contrib. Plasma Phys.* **56** 442
[14] Moldabekov Zh A, Ludwig P, Joost J P, Bonitz M and Ramazanov T S 2015 *Contrib. Plasma Phys.* **55** 186
[15] Grabowski P E, Surh M P, Richards D F, Graziani F R and Murillo M S 2013 *Phys. Rev. Lett.* **111** 215002
[16] Issanova M K, Kodanova S K, Ramazanov T S, Bastykova N Kh, Moldabekov Zh A and Meister C V 2016 *Laser and Particle Beams* **34** 457
[17] Ramazanov T S, Kodanova S K, Moldabekov Zh A and Issanova M K 2013 *Phys. Plasmas* **20**, 112702
[18] Song M Y and Jung Y D 2006 *Phys. Letters A* **360** 149
[19] Vorberger J, Donko Z, Tkachenko I M and Gericke D O 2012 *Phys. Rev. Letters* **109** 225001
[20] Kawata S, Karino T and Ogoyski A I 2016 *Matter and Radiation at Extremes* **1** 89
[21] Wünsch K, Vorberger J, and Gericke D O 2009 *Phys. Rev. E* **79** 010201
[22] Ramazanov T S, Gabdulin A Zh and Moldabekov Zh A 2015 *IEEE Transactions on Plasma Science* **43** 4187
[23] Lee MJ and Jung YD 2016 *Phys. Plasmas* **23** 033702
[24] Sukhinin G I and Fedoseev A V 2010 *IEEE Transactions on Plasma Science* **38** 2345
[25] Sukhinin G I, Fedoseev A V and Salnikov M V 2016 *Contrib. Plasma Phys.* **56** 397
[26] Lapenta G 1995 *Physical Review Letters* **75** 4409
[27] Kodanova S K, Ramazanov T S, Bastykova N Kh and Moldabekov Zh A 2015 *Phys. Plasmas* **22** 063703
[28] Ramazanov T S, Gabdulin A Zh and Moldabekov Zh A 2016 *Contrib.Plasma.Phys.* **56** 391
[29] Kodanova S K, Ramazanov T S, Issanova M K, Nigmetova G N and Moldabekov Zh A 2015 *Contrib. Plasma Phys.* **55** 271

IOP Conf. Series: Journal of Physics: Conf. Series **905** (2017) 012022 doi:10.1088/1742-6596/905/1/012022

Sustainable numerical scheme for molecular dynamics simulation of the dusty plasmas in an external magnetic field

K N Dzhumagulova and T S Ramazanov

IETP, al Farabi Kazakh National University, 71, al Farabi av., Almaty, 050040, Kazakhstan

dzhumagulova.karlygash@gmail.com

Abstract. The method, which allows one to carry out computer simulation of a system of charged particles in a strong external homogeneous magnetic field with a time step that is independent on the Larmor oscillation time, was generalized for the case of the presence of the surrounding background for the moving particles. Correctly taking into account a strong magnetic field and friction force, which both depend on the particles velocities, we obtained solution resistant to a change in the time step within the second-order Velocity Verlet propagation scheme.

1. Introduction

Classical molecular dynamics is a problem of numerical solving of the Newton equations of motion for the many particle systems. The particles interact with each other and often turn up under an influence of the external fields. Additionally, the particles can be immersed in any external stationary environment, for example, one may consider the suspended Brownian particles in a liquid or the charged particles in a neutral gas (weakly ionized plasma), and finally, the charged particles of the condensed material in the background plasma (dusty plasma) [1,2]. If the concentration of the particles of the background medium itself is much higher than the concentration of the immersed particles, the environment can be considered as a continuum which makes some averaged effect on the immersed particles. This effect on the particle dynamics can be described by the friction force:

$$F_{fr}(t) = -\nu\vartheta(t) \ , \qquad (1)$$

where ν is the friction coefficient, and $\vartheta(t)$ is the particle velocity, which is relative to the stationary background.

In this work, we considered a system of the charged particles exposed in a static uniform external magnetic field. Also the effect of the friction force was taken into account. Implementation of the external magnetic field and the friction force in the MD simulation is not a very difficult task. In Refs.[3-6] the different properties of a system of charged particles in a magnetic field were investigated on the basis of the molecular dynamics. In Refs.[7-10] the simulations of the dusty plasmas with taking into account of the background plasma and random force were performed on the basis of the Langevin dynamics.

In Ref.[11] authors described the way for obtaining of the sustainable numerical scheme for simulation of the particle system in the presence of some forces, depending on the particle velocity. There also second-order Velocity Verlet scheme, taking into account the external uniform magnetic field, was presented. Since MD investigates the trajectory of each particle, time step Δt is chosen small enough to have the necessary number of steps in the period of the Larmor oscillation in order to properly monitor the movement of particles in a spiral. Under the influence of the magnetic field B particle with specific charge q/m performs the rotation with the Larmor frequency $\Omega = qB/m$. One of the conditions for selection of Δt is related to the magnetic field strength:

$$\Omega \Delta t < 2\pi \tag{2}$$

When the magnetic field is weak, the condition (2) is performed quite easily. In a strong magnetic field for the condition (2) one should select very small time steps, which leads to a sharp increase in the amount of calculations, sometimes critical. In work [11] the stable numerical scheme was obtained on the basis of the expansion of $\vec{r}(t+\Delta t)$ and $\vec{\vartheta}(t+\Delta t)$ in the Taylor series. This scheme as was shown is resistant to a change in the time step at large external magnetic fields. Time step in this scheme is independent of the Larmor period of oscillation. In this paper, we introduce the frictional force in the Velocity Verlet scheme, performing all steps described in [11] for obtaining of the stable scheme for the case with background.

2. The Velocity Verlet scheme for simulation of the charged particles in an external magnetic field and background medium

2.1 Explicit and implicit Velocity Verlet schemes

Lets consider the system of N charged particles. The position, velocity, and acceleration of the i - th particle at time t are given by the three-dimensional vectors $\vec{r}_i(t), \vec{\vartheta}_i(t), \vec{a}_i(t)$. The components of the vectors are *x, y,* and *z,* respectively. We can write Newton's equations of the all N particles as a system of 6N first-order differential equations:

$$\frac{d\vec{r}_i}{dt} = \vec{\vartheta}_i(t) \quad , \tag{3}$$

$$\frac{d\vec{\vartheta}_i}{dt} = \vec{a}_i(t) = \frac{\vec{F}_i}{m_i} \quad , \tag{4}$$

here m_i is the mass of the particle *i,* $\vec{F}_i$ is the force acting on the particle i and $i=1,2..N$. In general case, $\vec{a}_i$ can be a function of the coordinates and velocities of all particles.
To solve first- order differential equations (3)-(4), there are many numerical algorithms, including the Velocity Verlet scheme (VV), which is one of the most popular for MD simulations.

$$\vec{r}(t+\Delta t) = \vec{r}(t) + \Delta t \vec{\vartheta}(t) + \frac{1}{2}(\Delta t)^2 \vec{a}(t) + O\left((\Delta t)^3\right), \tag{5}$$

$$\vec{a}(t+\Delta t) = \vec{a}(\vec{r}_1(t+\Delta t),\ldots,\vec{r}_N(t+\Delta t),\vec{\vartheta}_1(t+\Delta t),\ldots,\vec{\vartheta}_N(t+\Delta t),t+\Delta t) \ , \tag{6}$$

$$\vec{\vartheta}(t+\Delta t) = \vec{\vartheta}(t) + \frac{1}{2}\Delta t(\vec{a}(t) + \vec{a}(t+\Delta t)) + O\left((\Delta t)^3\right) . \tag{7}$$

In scheme (5) - (7) one can see $t+\Delta t$ on both sides of Eqs. (6) and (7), so it appears as implicit scheme. In many simple tasks when a magnetic field and friction are absent particle acceleration does not depend on the particle velocity and the equation (6) can be written as follows:

$$\vec{a}(t+\Delta t) = \vec{a}(\vec{r}_1(t+\Delta t),\ldots,\vec{r}_N(t+\Delta t),t+\Delta t) \tag{8}$$

From equation (5) one can calculate $\vec{r}(t+\Delta t)$ at time t. Using (8) $\vec{a}(t+\Delta t)$ can be found, and finally $\vec{\vartheta}(t+\Delta t)$ via (7). Thus, implicit form of equations (5) - (8) disappears. In this case, VV scheme is a second- order integration scheme, i.e., the error term is $O((\Delta t)^2)$.

Obviously, when the magnetic field affects on the charged particles, and (or) the particles undergo the friction force in the background medium, the acceleration depends on the velocity, and (8) is not correct, we have an implicit scheme (5) - (7). As was shown in Ref.[11], in the case of a static homogeneous external magnetic field acting on the system, VV can be modified in order to repair an explicit form. We performed this taking into account the friction force. In a homogeneous magnetic field $\vec{B}(0,0,B)$ directed along the axis z, the acceleration of each particle, also experiencing the friction in the background gas, can be written as follows:

$$\vec{a}(t) = \vec{a}^C(t) - \Omega\vec{e}_z \times \vec{\vartheta}(t) - \nu\vec{\vartheta}(t), \tag{9}$$

here $\vec{a}^C(t)$ is part of acceleration which does not depend on the velocity:

$$\vec{a}^C(t) = \vec{a}^C(\vec{r}_1(t),...,\vec{r}_N(t);t), \tag{10}$$

$\Omega = qB/m$ is the Larmor frequency, $\vec{e}_z = (0,0,1)$ is the unit vector directed along the z axis. To simplify the notation, we considered a system where all the particles have the same charge- to-mass ratio. In subsections 2.2 and 2.3, we described two different approaches, taking into account the magnetic field and the friction force in the VV scheme, which we in agreement with Ref.[11] called as "inversion $\vec{e}_z \times \vec{\vartheta}$" and "Taylor expansion".

2.2 *Inversion* $\vec{e}_z \times \vec{\vartheta}$

In expression (9), we have the cross product $\vec{e}_z \times \vec{\vartheta}$, which after substitution in Eqs. (5) - (7) mixes the ϑ_x and ϑ_y components of the each individual particle. Rewriting Eq.(7) in an explicit form for $\vec{\vartheta}(t+\Delta t)$, we have the following equations of the explicit VV scheme with the acceleration given by (9):

$$r_x(t+\Delta t) = r_x(t) + \Delta t\vartheta_x(t) + \frac{1}{2}(\Delta t)^2\left[a_x^C(t) - \nu\vartheta_x(t) + \Omega\vartheta_y(t)\right] + O((\Delta t)^3), \tag{11}$$

$r_y(t+\Delta t)$ can be obtained from (11) by replacing $x \rightarrow y$ and $\Omega \rightarrow -\Omega$

$$r_z(t+\Delta t) = r_z(t) + \Delta t\vartheta_z(t) + \frac{1}{2}(\Delta t)^2\left[a_z^C(t) - \nu\vartheta_z(t)\right] + O((\Delta t)^3), \tag{12}$$

$$\vartheta_x(t+\Delta t) = \frac{1}{K}\left[\vartheta_x(t) + \frac{1}{2}\Delta t\left(\begin{array}{c} a_x^C(t) + a_x^C(t+\Delta t) + \Omega\vartheta_y(t) - \nu\vartheta_x(t) + \\ \dfrac{\Omega}{1+\nu\Delta t/2}\left(\begin{array}{c}\vartheta_y(t) + \dfrac{\Delta t}{2}a_y(t) + \dfrac{\Delta t}{2}a_y(t+\Delta t) - \\ \dfrac{\Delta t}{2}\Omega\vartheta_x(t) - \dfrac{\Delta t}{2}\nu\vartheta_y(t)\end{array}\right)\end{array}\right)\right] + O((\Delta t)^3), \tag{13}$$

where $K = 1 + \dfrac{\Omega^2(\Delta t)^2}{4+2\nu\Delta t} + \dfrac{\nu\Delta t}{2}$, $\vartheta_y(t+\Delta t)$ can be obtained from (13) by replacing $x \rightarrow y$ and $\Omega \rightarrow -\Omega$

$$\vartheta_z(t+\Delta t) = \frac{1}{1+\nu\Delta t/2}\left[\vartheta_z(t) + \frac{1}{2}\Delta t\left(a_z^C(t) + a_z^C(t+\Delta t)\right)\right] + O((\Delta t)^3) \tag{14}$$

It is shown below the scheme (11) - (14) is stable only in the case of weak magnetic fields. When a strong magnetic field is applied, it becomes unstable with respect to the time step and requires a large amount of calculations with a small time step.

2.3 *The Taylor expansion*

In work [11] for an arbitrary value of magnetic field the authors developed the robust numerical scheme based on the Taylor expansion of the particle acceleration and velocity, followed by the correct choice of all the terms that are not higher than $O((\Delta t)^2)$. This scheme has been successfully used in many studies, for example in Ref.[12]. Applying the same technique for the case when the particles are immersed in a homogeneous stationary environment we obtained the following equations for the positions and velocities of the particles:

$$r_x(t+\Delta t) = r_x(t) - \frac{1}{(\Omega^2+\nu^2)}\begin{bmatrix}\left(\nu\vartheta_x(t)+\Omega\vartheta_y(t)\right)\cdot\left(\exp(-\nu\Delta t)\cos(\Omega\Delta t)-1\right)+ \\ \left(\nu\vartheta_y(t)-\Omega\vartheta_x(t)\right)\exp(-\nu\Delta t)\sin(\Omega\Delta t)\end{bmatrix}$$
$$+\frac{1}{(\Omega^2+\nu^2)^2}\begin{bmatrix}C(\Omega\Delta t)\left((\nu^2-\Omega^2)a_x^C(t)+2\nu\Omega a_y^C(t)\right)+ \\ S(\Omega\Delta t)\left((\nu^2-\Omega^2)a_y^C(t)-2\nu\Omega a_x^C(t)\right)\end{bmatrix}+O((\Delta t)^3) \qquad (15)$$

$r_y(t+\Delta t)$ can be obtained from (15) by replacing $x \to y$ and $\Omega \to -\Omega$)

$$r_z(t+\Delta t) = r_z(t) + \Delta t\vartheta_z(t) + \frac{1}{2}(\Delta t)^2\left[a_z^C(t) - \nu\vartheta_z(t)\right] + O((\Delta t)^3), \qquad (16)$$

where we have defined:

$$S(\Omega\Delta t) \equiv \exp(-\nu\Delta t)\sin(\Omega\Delta t) - \Omega\Delta t \qquad (17)$$
$$C(\Omega\Delta t) \equiv \exp(-\nu\Delta t)\cos(\Omega\Delta t) - 1 + \nu\Delta t \qquad (18)$$

$$\vartheta_x(t+\Delta t) = \exp(-\nu\Delta t)\left(\vartheta_x(t)\cos(\Omega\Delta t)+\vartheta_y(t)\sin(\Omega\Delta t)\right)+$$
$$\frac{1}{\Omega^2+\nu^2}\begin{bmatrix}\exp(-\nu\Delta t)\left(\Omega\sin(\Omega\Delta t)-\nu\cos(\Omega\Delta t)\right)a_x^C(t)+\nu a_x^C(t)- \\ \exp(-\nu\Delta t)\left(\Omega\cos(\Omega\Delta t)+\nu\sin(\Omega\Delta t)\right)a_y^C(t)+\Omega a_y^C(t)\end{bmatrix}+$$
$$\frac{1}{\left(\Omega^2+\nu^2\right)^2}\begin{bmatrix}\left\{\exp(-\nu\Delta t)\left((\nu^2-\Omega^2)\cos(\Omega\Delta t)-2\nu\Omega\sin(\Omega\Delta t)\right)+(\Omega^2-\nu^2)+(\Omega^2+\nu^2)\nu\Delta t\right\}\frac{d}{dt}a_x^C(t)+ \\ \left\{\exp(-\nu\Delta t)\left((\nu^2-\Omega^2)\sin(\Omega\Delta t)+2\nu\Omega\cos(\Omega\Delta t)\right)-2\nu\Omega+(\Omega^2+\nu^2)\nu\Delta t\right\}\frac{d}{dt}a_y^C(t)\end{bmatrix}+ \qquad (19)$$
$$O((\Delta t)^3)$$

$\vartheta_y(t+\Delta t)$ can be obtained from (19) by replacing $x \to y$ и $\Omega \to -\Omega$

$$\vartheta_z(t+\Delta t) = \frac{1}{1+\nu\Delta t/2}\left[\vartheta_z(t)+\frac{1}{2}\Delta t\left(a_z^C(t)+a_z^C(t+\Delta t)\right)\right]+O((\Delta t)^3) \qquad (20)$$

The system of Eqs. (15) - (20) is a sustainable second-order numerical scheme for simulation of the charged particles in an external homogeneous stationary magnetic field and background environment. The choice of the integration step for it is not limited to a condition $\Omega\Delta t << 2\pi$. In these equations, there are not references to the acceleration caused by the Lorentz force and by frictional force, they are fully integrated into these equations. There are references only on $\vec{a}^C$ caused by the interaction of the particles, as well as by the external forces beyond the control of the particles velocities. The choice of the time step is now depends only on the time scale conditioned by $\vec{a}^C$. At $\nu = 0$ Eqs.(15) - (20) transform to the corresponding equations presented in [11] for the case without the background.

3. Numerical example

Lets consider the following example. We calculated the trajectory of the first charged particle (charge and mass are $q_1 = -1, m_1 = 1$), moving in a gaseous environment with a friction coefficient ν in the Coulomb field of the second stationary charged particle ($q_2 = 1$), and in a static homogeneous external magnetic field. At the initial time $t = 0$ position and velocity of the first particle are determined by the vectors $\vec{r}(0) = (-1,0,0)$ and $\vec{\vartheta}(0) = (0,1,0)$, respectively. The second particle is at the origin (0,0,0). Then, the acceleration of the first particle is given by:

$$\vec{a}(t) = -\frac{\vec{r}}{|\vec{r}|^3} - \Omega\vec{e}_z \times \vec{\vartheta}(t) - \nu\vec{\vartheta}(t) \tag{21}$$

We numerically calculated the equations of motion from $t = 0$ to $t = 20$ in a wide range of variation of the parameters $\Omega, \Delta t$ and ν on the basis of both methods, by an algorithm (11) - (14) and the algorithm of the Taylor expansion (15) - (20). Fig. 1 shows the trajectories calculated for different values Ω and ν on the basis of the equations (15) - (20), the trajectories calculated by the scheme of the Taylor series expansion [11], which does not take into account the friction force ($\nu = 0$), are also given. As can be seen from these graphs at $\nu = 0.001$ the trajectories, calculated by (15) - (20) and [11], are almost the same. With an increase in ν noticeable differences occur. It is also shown that with an increase in the magnetic field the Larmor rotations occur in addition to the rotation around of the force center of the electric field.

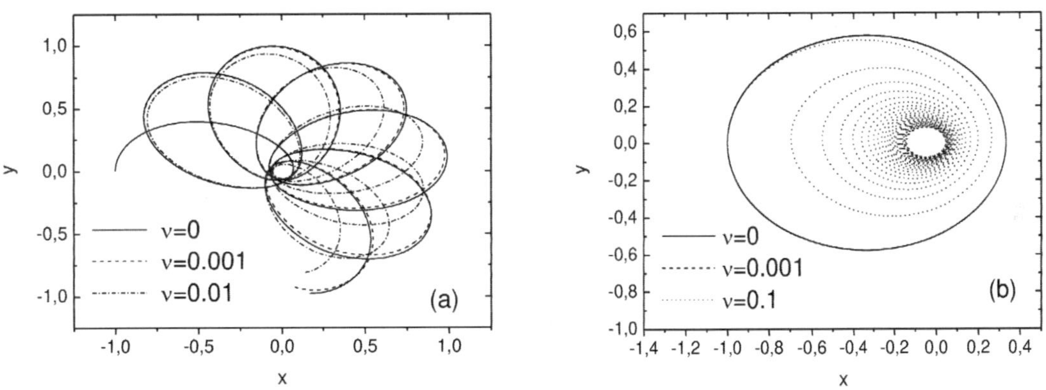

Figure 1. Trajectories of charged particle with acceleration (21). (a) - $\Omega = 1$ (b) - $\Omega = 0.001$

Fig. 2a shows the trajectories calculated for different values of the time step Δt on the basis of Eqs. (11) - (14), as well as Fig. 2b displays these calculated by a scheme of the Taylor expansion (15) - (20). In the first case on Fig. 2a it is visible that an increase in the time step results in the instability of the solution of the particle motion equations. Conversely, Fig. 2b shows that an algorithm of the Taylor series expansion (15) - (20) is resistant to a change in the time step. At $\Delta t = 0.1$ the trajectory becomes more polygonal in connection with a decrease in the number of the calculation points on the Larmor spiral, but the deviation of the solutions from the trajectory calculated at $\Delta t = 0.001$ remains within the margin of error. Thus solutions on the basis of the algorithm (11) - (14) depend strongly on Δt, while an algorithm of the Taylor expansion (15) - (20) is more stable relatively to a change in the time step, including range when $\Delta t > 0.1$. Recommended step is $\Delta t = 0.1$, since at its bigger values points of position still lie on the "exact" trajectory, but their number, attributable to a single coil, decreases.

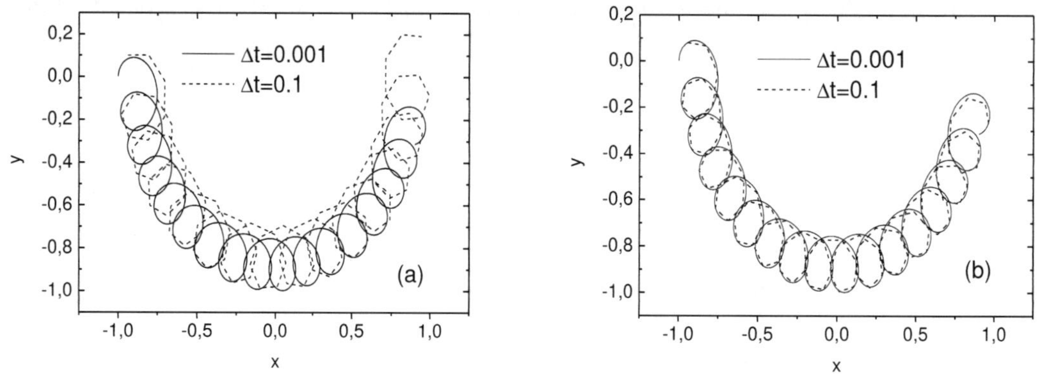

Figure 2. Trajectories of charged particle with acceleration (21). $\Omega = 10, \nu = 0.001$ (a) – algorithm (11)-(14), (b) - algorithm (15)-(20)

4. Conclusion

We presented a stable numerical scheme for solving of the equations of motion of charged particles in a background medium, as well as in a strong external static homogeneous magnetic field. To obtain it the method of the expansion in the Taylor series has been used. The Lorentz force and friction force depending on the particle velocity have been taken into account. In the resulting scheme the choice of the time step does not depend on the magnetic field.

References

[1] Fortov V E, Ivlev A V, Khrapak S A, Khrapak A G and Morfill G E 2005 *Phys. Rep.* **421** 1
[2] Complex and Dusty Plasmas: From Laboratory to Space / Ed. By V.E. Fortov, G.E. Morfill. – CRC Press 2009
[3] Ott T, Bonitz M, Hartmann P and Donkò Z 2011 *Phys. Rev.* E **83** 046403
[4] Ott T, Loewen H, Bonitz M 2013 *Phys. Rev. Lett.*. **111** 065001
[5] Uchida G, Konopka U and Morfill G 2004 *Phys. Rev. Lett.* **93** 155002
[6] Ott T, Loewen H, Bonitz M 2014 *Phys. Rev.* E **89** 013105
[7] Baimbetov F B, Ramazanov T S, Dzhumagulova K N, Kadyrsizov E R, Petrov O F and Gavrikov A V 2006 *J. Phys. Math. Gen.* **39** 4521
[8] Dzhumagulova K N, Ramazanov T S, Ussenov Y A, Dosbolayev M K and Masheyeva R U 2013 *Contrib. Plasma Phys.* **53** 419
[9] Dzhumagulova K N, Ramazanov T S, Ussenov Y A, Dosbolayev M K and Masheyeva R U 2013 Contrib. Plasma Phys. **53** 419
[10] Dzhumagulova K N, Ramazanov T S and Masheyeva R U 2013 *Phys. Plasmas* **20** 113702
[11] Spreiter Q and Walter M 1999 *J. Comput. Phys.* **152** 102
[12] Dzhumagulova K N, Masheyeva R U, Ramazanov T S and Donkò Z 2014 *Phys. Rev.* E **89** 033104

Software development for the calculation of dynamic properties of dense plasmas: Coulomb logarithm, relaxation and transport properties

S K Kodanova, T S Ramazanov, M K Issanova and Zh A Moldabekov

Institute of Experimental and Theoretical Physics, Al-Farabi Kazakh National University, Al-Farabi av. 71, Almaty 050040, Kazakhstan

E-mail: kodanova@physics.kz

Abstract. Progress in the theory of effective interaction potentials allows us to calculate the Coulomb logarithm, relaxation and transport properties of weakly coupled as well as nonideal plasmas taking into account both collective screening and quantum diffraction effects. As a result, the fast calculation of the mentioned properties of the classical and semiclassical plasmas is possible without time-consuming simulation methods such as molecular dynamics simulation. For this purpose, user-friendly software with a wide range of functions for simulation of dynamic properties of dense plasmas using the theory of effective potentials is needed. The range of applicability of the effective interaction potentials theory is discussed and part of the software for the simulation of the Coulomb logarithm, relaxation and transport properties of dense plasmas are presented.

1. Introduction

At the present time, the dense plasma is the subject of active theoretical investigations due to its relevance to the inertial confinement fusion. In particular, these investigations were triggered by the experiments at the National Ignition Facility [1] and magnetized Z-pinch experiments at Sandia [2]. To obtain a thermonuclear reaction in the above-mentioned facilities it is necessary to study transport properties and relaxation times of the temperature of dense plasma. Computer simulation can answer many important questions. It should be noted that it is especially important to study relaxation times of electrons and ions. In particular, during compression of a target by the flow of high-energy electrons the non-isothermal plasma with heated electrons and cold ions is created. Non-isothermal plasma also appears during interaction of heavy ion beams with a target [3,4]. The temperature equalizes much faster within subsystems of electrons and ions than between electrons and ions. This is explained by a large difference between masses of ions and electrons. Therefore it is necessary to develop new technology and make information about the properties of the plasma which are available for users.

Programs for simulation and calculation of inertial confinement fusion targets are very complicated and require a lot of computation time. The complexity of calculations is caused by necessity of taking into account a large number of different physical processes simultaneously: transport of particles; plasma thermodynamics; energy absorption; radiation transport; stopping power etc. Each part of the program related to the description of one or more of these processes is already quite complicated.

Recent progress in the theory of effective interaction potentials [5–8, 10–28] allows us to make fast and accurate calculations Coulomb logarithm, temperature relaxation time and such transport properties as viscosity and diffusion coefficient of strongly coupled dense plasma with parameters relevant to inertial confinement fusion plasma and warm dense matter. In the next section, the current state and the applicability range of the effective interaction potential method are discussed. After that the software package for the calculation of the Coulomb logarithm, temperature relaxation time and transport properties of matter under extreme conditions is presented.

2. Effective interaction potentials

2.1. Dimensionless parameters

The method of effective interaction potentials based on the pair interaction approximation with the collective (many-particle) effects and quantum effects are incorporated into the pair interaction potential. For further consideration of the applicability range of the effective interaction potentials method, we introduce the following dimensionless parameters widely used as characterization the plasma state:

i. the electron degeneracy parameter, $\theta = k_B T_e / E_F$, here E_F is the Fermi energy of electrons,

ii. the density parameter (Brueckner or quantum coupling parameter), $r_s = a/a_B$, where $a = (4/3 \times \pi n_e)^{-1/3}$ is the mean interparticle distance for electrons,

iii. the coupling parameters of the ions $\Gamma_i = Q_i^2/(a_i k_B T_i)$ and electrons $\Gamma_e = e^2/(a_e K)$, where the ion charge is $Q_i = -Ze$, K is the characteristics kinetic energy of electrons which should be taken equal to $k_B T_e$ if $\theta > 0.5$ and E_F if $\theta < 0.5$, and $a_i = (4/3 \times \pi n_i)^{-1/3}$ is the average distance between ions,

iv. the classical non-ideality parameter $\gamma = e^2/(r_D k_B T_e)$, where $r_D = (r_{\mathrm{De}}^2 + r_{\mathrm{Di}}^2)^{1/2}$ is the Debye screening length with $r_{\mathrm{De(i)}} = \left(k_B T_{\mathrm{e(i)}}/4\pi n_{\mathrm{e(i)}} Q_{\mathrm{e(i)}^2}\right)^{1/2}$.

The classical coupling parameter of electrons can be found if the degeneracy and density parameters are known. Thus, the plasma electrons in equilibrium are characterized by the two parameters θ and r_s. The temperature in Kelvin or eV is obtained from $T \simeq \frac{\theta}{r_s^2} \cdot 0.58 \times 10^6\,[K]$ and $k_B T \simeq \frac{\theta}{r_s^2} \cdot 50.12\,[eV]$.

2.2. Plasma with weakly coupled electrons and ions

In the case of classical plasma ($\theta > 1$) with $\gamma << 1$, the screened Debye potential can be used

$$\Phi_{\alpha\beta}(\mathbf{r}) = \frac{Q_\alpha Q_\beta}{r} exp(-r/r_D), \tag{1}$$

where α, β indicate the type of the considered particle (electron or ion).

In the case of classical ($\theta > 1$) non-ideal plasma with $\gamma < 4$, the following screened pair interaction potential can be used [20, 21]:

$$\Phi(r) = \frac{\gamma}{r} exp(-r) \frac{1 + \gamma f(r)/2}{1 + c(\gamma)}, \tag{2}$$

where

$$f(r) = (exp(-\sqrt{\gamma} r) - 1)(1 - exp(-2r))/5, \tag{3}$$

and $r = r/r_D$,

$$c(\gamma) = -0.008617 + 0.455861\gamma - 0.108389\gamma^2 + 0.009377\gamma^3. \tag{4}$$

In the case of partially degenerate plasma $0.5 < \theta \leq 1$ with weakly coupled electrons $\Gamma_e < 1$ and ions $\Gamma_i < 1$, the following effective interaction potential is applicable [5, 6, 12]:

$$\Phi_{\alpha\beta}(\mathbf{r}) = \frac{Q_\alpha Q_\beta}{r\sqrt{1-(2r_D^{-1}/\lambda_{\rm ee}\gamma^2)^2}}\left(\left(\frac{1/\lambda_{\rm ee}^2 - B^2}{1-B^2\lambda_{\alpha\beta}^2}\right)exp(-Br) - \left(\frac{1/\lambda_{\rm ee}^2 - A^2}{1-A^2\lambda_{\alpha\beta}^2}\right)exp(-Ar)\right) - \\ - \frac{Q_\alpha Q_\beta}{r}\frac{(1-\delta_{\alpha\beta})}{1+C}exp(-r/\lambda_{\alpha\beta}), \tag{5}$$

$\lambda_{\alpha\beta} = \hbar/\sqrt{4\pi m_{\alpha\beta}k_B T_{\alpha\beta}}$, $m_{\alpha\beta} = m_\alpha m_\beta/(m_\alpha + m_\beta)$, $T_{\alpha\beta} = \sqrt{T_\alpha T_\beta}$, $\gamma^2 = r_{\rm Di}^{-2} + 1/\lambda_{\rm ee}^2$ and $C = (k_D\lambda_{\rm ei}^2 - k_i\lambda_{\rm ee}^2)\,/\,(\lambda_{\rm ee}^2/\lambda_{\rm ei}^2 - 1)$, $A^2 = \gamma^2/2\left(1+\sqrt{1-(2r_D^{-1}/\lambda_{\rm ee}\gamma^2)^2}\right)$, $B^2 = \gamma^2/2\left(1-\sqrt{1-(2r_D^{-1}/\lambda_{\rm ee}\gamma^2)^2}\right)$.

If both ions and electrons are weakly coupled, the method of effective potentials can be used to consider properties of both ions and electrons.

2.3. Plasma with non-ideal ions and weakly coupled electrons

If ions are strongly coupled, the screening effect due to weakly coupled electrons can be included into the ion-ion pair interaction potential.

In the case of classical plasma ($\theta > 1$) with non-ideal ions $\Gamma_i > 1$ and weakly coupled electrons $\Gamma_e < 1$, the ions can be considered within one component plasma model (OCP) with interparticle screened Coulomb interaction potential [25]:

$$\Phi(\mathbf{r}) = \frac{Q_i^2}{r}exp(-r/r_{Di}), \tag{6}$$

In the case of quantum or partially degenerate plasma ($\theta \leq 1$) with nonideal ions $\Gamma_i > 1$ and weakly coupled electrons $r_S \leq 1$ the following screened potential can be used [22–25]

$$\Phi(r) = \frac{Q_i^2}{2r}\left[(1+b)\,e^{-k_+ r} + (1-b)\,e^{-k_- r}\right], \tag{7}$$

with $\alpha = 3\sqrt{8\beta}\lambda I'_{-1/2}(\eta_0)/\pi$, $\lambda = 1/9$, where $b = 1/\sqrt{1-\alpha}$, $\beta = (k_B T_e)^{-1}$ and $k_\pm = k_{TF}(1\mp\sqrt{1-\alpha})^{1/2}/\sqrt{\alpha/2}$. $I_p(\eta) = \int_0^\infty dx\, x^p/(1+e^{x-\eta})$ denotes the Fermi integral and $I'_p(\eta)$ its derivative with respect to η. η_0 is determined by the normalization, $n_0 = \sqrt{2}I_{1/2}(\eta_0)/\pi^2\beta^{3/2}$ with the inverse temperature β. The inverse Thomas-Fermi screening length for finite temperatures is given as $k_{TF} = (4I_{-1/2}(\eta_0)/\pi\sqrt{2\beta})^{1/2}$.

Once the ion-ion pair interaction potential is chosen depending on plasma parameters, the screened potential (6) or (7) can be used for calculation of the pair distribution function $g(r)$ by HNC. Now, the pair distribution function allows us to obtain numerical values the effective ion-ion potential from the relation $\Phi_{\rm ii} = -k_B T_i ln(g(r))$ [26–28].

To define the conditions which must be satisfied for applicability of the method of effective potentials, we present the diagram in figure 1. It is indicated that if the electrons are strongly coupled then electrons and ions must be considered by molecular dynamics simulation. If, additionally, the electrons are degenerate then quantum molecular dynamics has to be used.

When the effective interaction potentials for the given plasma parameters are determined, the relaxation, transport properties and Coulomb logarithm can be obtained by solving the classical scattering problem on the basis of the effective potential [5, 12, 26, 28].

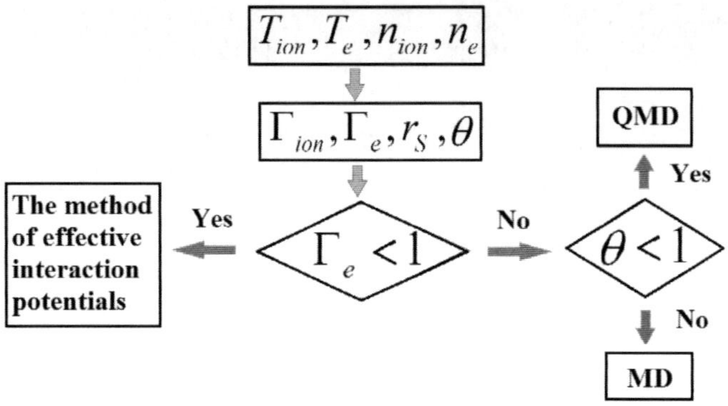

Figure 1. The diagram shows the case when the effective interaction potentials can be used for straightforward calculation of the scattering cross sections and transport properties of the plasma. In the case of plasma with non-ideal electrons molecular dynamics simulation (classical or quantum) of two component plasma should must in order to determine dynamical properties of plasma.

In the case of the plasma with non-ideal electrons, the so-called quantum potential approach [29,30] at $r_S > 3$ makes molecular dynamics simulation significantly faster as the exact quantum consideration of electrons can be avoided. Quantum potentials do not take into account the collective effect, but correctly reproduce quantum effects at short interparticle distances. At higher densities, the orbital free molecular dynamics simulation can be used [31].

2.4. Transport and relaxation properties of dense ICF plasmas

The transport and relaxation properties are obtained on the basis of Coulomb logarithm using the effective potentials for ICF plasma. The Coulomb logarithm based on the effective interaction potential of particles is determined by the scattering angle of pair Coulomb collisions. Introducing the center of mass in the collision process, the Coulomb logarithm is written as [9, 10, 13]:

$$\Lambda_{ei} = \frac{1}{b_\perp^2} \int_0^{b_{\max}} \sin^2\left(\frac{\theta_c}{2}\right) b\, db\,, \tag{8}$$

The center-of-mass scattering angle θ_c can be obtained from the formula [9]:

$$\theta_c = \pi - 2b \int_{r_0}^{\infty} \frac{dr}{r^2}\left(1 - \frac{\Phi_{\alpha\beta}(r)}{E_c} - \frac{b^2}{r^2}\right)^{1/2}, \tag{9}$$

here $E_c = \frac{1}{2} m_{\alpha\beta} v^2$ is the energy of the center of mass, $m_{\alpha\beta} = m_\alpha m_\beta/(m_\alpha + m_\beta)$ is the reduced mass of the particles of kinds α and β (ion or electron); $b_\perp = Z_\alpha Z_\beta e^2/(m_{\alpha\beta} v^2)$, $b_{\min} = \max b_\perp, \lambda_{\alpha\beta}$ describes the minimum impact parameter, where $\lambda_{\alpha\beta}$ is the thermal de Broglie wave length.

The diffusion coefficient, viscosity of plasma are connected with the effective collision frequency by the equations:

$$D = \frac{k_B T}{m_e \nu_{eff}}, \tag{10}$$

$$\eta = \frac{5}{4}\sqrt{\frac{m}{\pi}}\frac{(k_BT)^{5/2}}{e^4\Lambda}, \tag{11}$$

where e is the electron charge, m_e is the mass of the electron, n is the density of plasma particles, and

$$\nu_{eff} = (4/3)\sqrt{2\pi}e^4\Lambda/\sqrt{m_e}(k_BT)^{3/2} \tag{12}$$

is the effective collision frequency directly proportional to the Coulomb logarithm.

The relaxation rate of the electron-ion temperature, i.e., the rate of energy exchange, is determined by the difference of the average energy or temperature:

$$\frac{dT_e}{dt} = \frac{T_i - T_e}{\tau_{ei}}, \frac{dT_i}{dt} = \frac{T_e - T_i}{\tau_{ie}}, \tag{13}$$

$$\tau_{ei} = \frac{3m_em_i}{8\sqrt{2\pi}n_ie^2\Lambda}(\frac{k_BT_e}{m_e} + \frac{k_BT_i}{m_i})^{3/2}. \tag{14}$$

where τ_{ei} is relaxation times of the temperature in the plasma.

3. Software package

In this work we present a software package based on modern information technologies that allows us to get rapid analysis and visualization of properties of dense plasmas. We calculate dynamic properties of plasma by using the Coulomb Logarithm on the basis of effective potentials taking into account quantum diffraction and screening effects at short and large distances. The developed software for calculation and analysis of relaxation and transport properties of dense plasma has been created on the basis of object-oriented language C#. For graphical representation of the results a set of components ZedGraph was used. C# - type-safe object-oriented programming language was designed for the development of a variety of high-power applications realized in the .NET Framevork. Using the C# language, one can create standard Windows applications, the XML-web services, distributed components, the application client-server, database applications, etc. The Visual C# provides a developed code editor, designers with a convenient user interface, built-in debugger, and many other tools that simplify the development of applications based on C# language and the .NET Framework. ZedGraph allows its users to build graphs of high complexity. This software package is designed to calculate dense plasma relaxation and transport properties with a function of visualization and display of the results as graphs in real time.

The software package consists of three namespaces: GEOTOOLS, PHYTOOLS, SOLUTIONS (see figure 2). Each namespace contains realized and abstractly inherited classes kept in a separate folder with identical names. Such a partition ensures the flexibility of the program code and the ability to improve the program code without any dramatic changes in the existing one. Figure 2 shows a scheme of the software package.

The namespace GEOTOOLS (Geometry tools) contains classes designed to control geometrical parameters and tasks of visualization (visualization realized by using OPENGL library). GEOTOOLS consists of the following classes: GCOLOR3f, GPOINT3f, GVECTOR3f, GDISTANCE3f, GBOX3f an abstract class GOBJECT3f and a static class UNIQUENUMBER, which defines a unique number for each object created in the above classes.

The namespace PHYTOOLS (Physics tools) contains classes designed to solve physical problems. This namespace has the enumerator (enum) POTENTIAL containing the names of potentials, and a static class TENSION containing single method GETVECTORVALUE3f which returning the vector quantity field on a single charge at a given point are implemented.

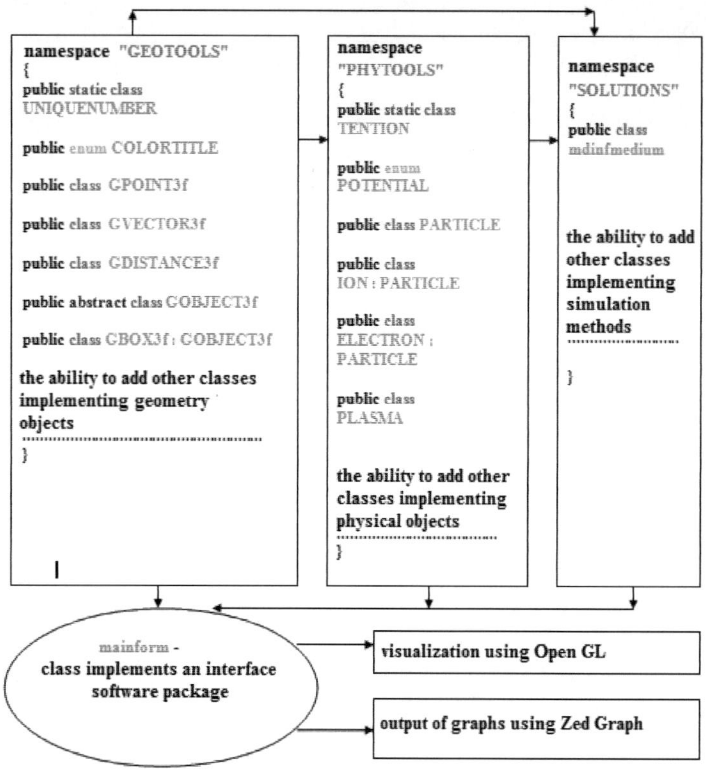

Figure 2. Scheme of the software package.

The main classes of the namespace PHYTOOLS are PARTICLE, its heirs ELECTRON and ION as well as class PLASMA.

The class PARTICLE consists of parameters of particles such as mass, charge, wavelength and magnitude of the velocity vector, the force acting on the particle and its position in space. Classes ELECTRON and ION are heirs of the class PARTICLE differing in mass, charge, and wavelength.

Class PLASMA contains an array of classes ELECTRON and ION, and also physical parameters such as the plasma coupling parameter, concentration, plasma temperature, etc.

Namespace SOLUTIONS is for storage of classes that realize certain modeling (calculation) methods. In this case, the namespace SOLUTIONS keeps the only class mdinfmedium. The object class mdinfmedium has to be started by the user in a separate thread by calling the main method of the object class mdinfmedium.

The software allows us to set the parameters of the task, to control the procedure of the calculations (see figures 3, 4, 5). The results are displayed during calculations in the form of a chart, which enables the user to stop calculations pressing the button when any deviations are found. There is the panel Parameters in the left part of the software interfaces. It displays constant parameters, as well as the data calculated by the system itself. The program provides an opportunity to compare the results of calculations with theoretical results of other authors and allows us to make the calculations in a wide range of dense plasma parameters.

The calculation time of the Coulomb logarithm on the basis of the effective interaction potentials is present in the figure 6. For comparison the calculation time of the relaxation time or of the some transport characteristics using MD simulations needs from few hours up

IOP Conf. Series: Journal of Physics: Conf. Series **905** (2017) 012023 doi:10.1088/1742-6596/905/1/012023

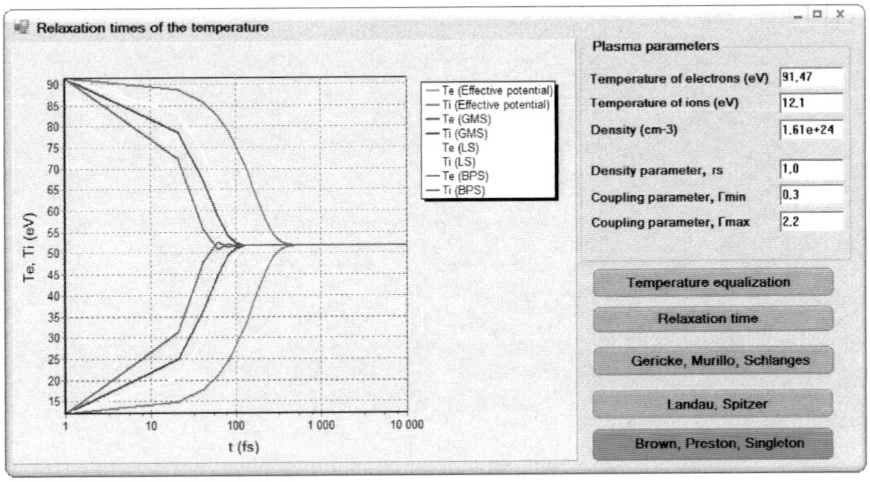

Figure 3. The interface of the software package for simulation of the relaxation processes.

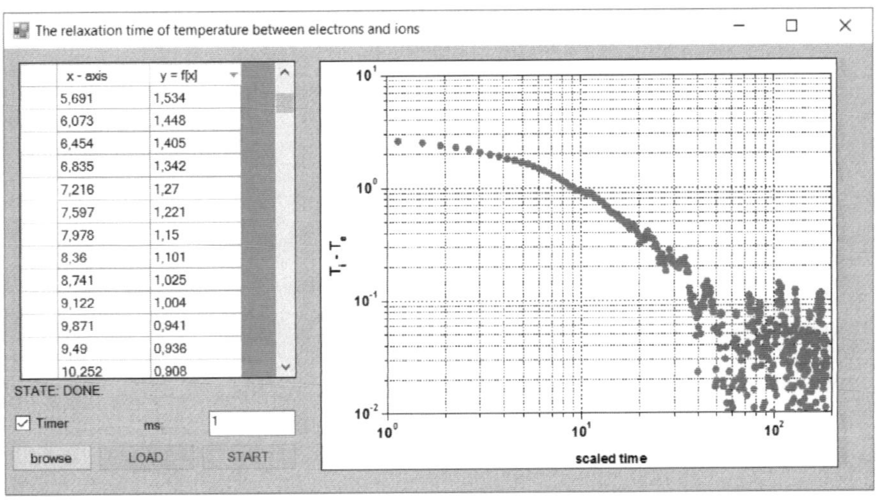

Figure 4. The view of the relaxation time for temperature between electrons and ions.

to few month depending on plasma parameters and the computational power. Therefore, if the considering plasma parameters correspond to the range of applicability of the known and well checked effective interaction potential, fast calculation of the relaxation time and transport properties can be done on the basis of it.

4. Conclusion

Development and application of information systems, providing reliable information on a wide range of properties of the dense plasma are one of the modern forms of scientific research. The program module combines a user interface package with powerful algorithms and numerical methods of high level. This module helps scientists, engineers and experimentalists to study the processes in plasma. The software package includes a set of tools that can be used both for training and as a basis for future independent work.

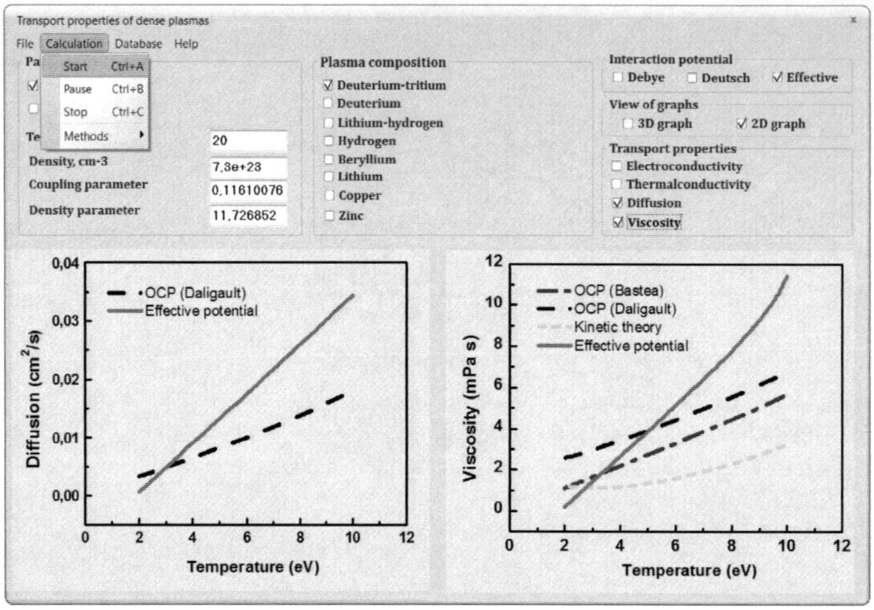

Figure 5. Software package for the simulation of the transport properties.

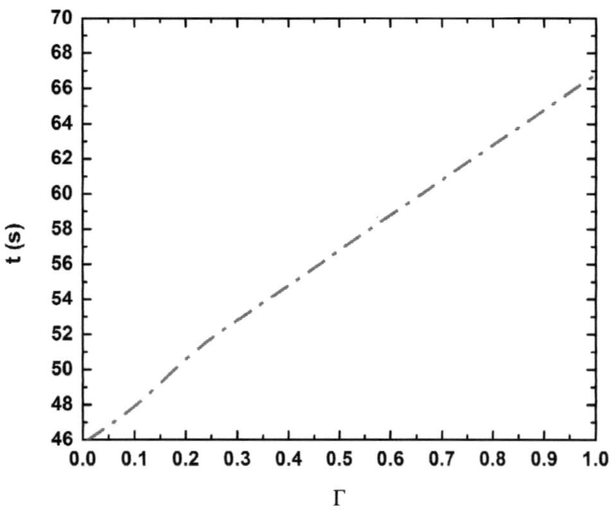

Figure 6. Calculation time of the Coulomb Logarithm on the basis of effective interaction potentials.

The software which realizes fast calculation of transport properties and relaxation time on the basis of Coulomb logarithm using effective potential theory of dense ICF plasma are presented. The choice of programming language was based on such criteria as the possibility of creation of a friendly reliable interface, the possibility of including databases by means of language itself, widened mathematical apparatus, effective compilation, possibility of development of the software system and compilers. The package gives the possibility to set the parameters and to monitor the process of calculations.

Acknowledgments
This work has been supported by the Ministry of Education and Science of the Republic of Kazakhstan under Grant No. 3083/GF4 (2017) aimed at the to develop software package for studying of the transport and dynamic properties of dense ICF plasmas.

References

[1] Hurricane O A, et al 2014 *Nature* **506** 343
[2] Gomez M R, et al 2014 *Phys. Rev. Lett.* **113** 155003; Schmit P F, et al 2014 *Phys. Rev. Lett.* **113** 155004
[3] Tahir N A, et al 2010 *New Journal of Physics* **12** 073022
[4] Minaev S, Alexeev N, Golubev A, Hoffmann D H H, Kulevoy T, Sharkov B, Sitnikov A, Tahir N A and Varentsov D 2010 *Nucl. Inst. Meth.A* **620** 99104
[5] Issanova M K, Kodanova S K, Ramazanov T S, and Hoffmann D H H 2016 *Contrib. Plasma Phys.* **56** (5) 425
[6] Ramazanov T S, Moldabekov Zh A and Gabdullin M T 2015 *Phys.Rev. E* **92** 023104
[7] Ramazanov T, Moldabekov Zh and Gabdullin M 2016 *Phys. Rev. E* textbf93 053204
[8] Ramazanov T S, Moldabekov Zh A, Gabdullin M T and Ismagambetova T N 2014 *Phys. Plasmas* **21** 012706
[9] Ordonez C A, Molina M I 1994 *Phys. Plasmas* **1** 2515
[10] Ramazanov T S and Kodanova S K 2001 *Phys. Plasmas* **8** 5049
[11] Kodanova S K, Ramazanov T S, Issanova M K, Nigmetova G N and Moldabekov Zh A 2015 *Contrib. Plasma Phys.* **55** 271
[12] Issanova M K, Kodanova S K, Ramazanov T S, Bastykova N Kh, Moldabekov Zh A and Meister C-V 2016 *Laser and Particle Beams* **34** 457-466
[13] Ramazanov T S, Kodanova S K, Moldabekov Zh A and Issanova M K 2013 *Phys.Plasmas* **20** 11270
[14] Moldabekov Zh A, Ludwig P, Joost J P, Bonitz M and Ramazanov T S 2015 *Contrib. Plasma Phys.* **55** 186
[15] Moldabekov Zh A, Ludwig P, Bonitz M and Ramazanov T S 2016 *Contrib. Plasma Phys.* **56** 442
[16] Ramazanov T S, Dzhumagulova K N, and Akbarov A Zh 2006 *J. Phys. A: Math. Gen.* **39** 4335
[17] Song M-Y and Jung Y-D 2003 *Phys. Plasmas* **10**, 3051 (2003)
[18] Kim H-M and Jung Y-D 2007 *Phys. Plasmas* **14** 074501
[19] Ki D-H and Jung Y-D 2010 *Phys. Plasmas* **17** 074506
[20] Baimbetov F B, Nurekenov Kh T, and Ramazanov T S 1995 *Phys. Lett. A* **202** 211
[21] Baimbetov F B, Nurekenov Kh T, Ramazanov T S 1996 *Physica A* **226** 181
[22] Stanton L G, and Murillo M S 2015 *Phys. Rev E* **91** 033104; *Pulisher's Note ibid:* **91** 049901
[23] Akbari-Moghanjoughi 2015 *Phys. Plasmas* **22** 022103; *Erratum* 2015 ibid: **22** 039904
[24] Moldabekov Zh, Schoof T, Ludwig P, Bonitz M and Ramazanov T 2015 *Phys. Plasmas* **22** 102104
[25] Ludwig P, Bonitz M and Kählert H 2010 *J. Phys.: Conf. Series* **220** 012003
[26] Baalrud S D and Daligault J 2013 *Phys. Rev. Lett.* textbf110 235001
[27] Daligault J 2012 *Phys. Rev. E.* textbf86 047401
[28] Baalrud S D 2010 *Phys. Plasmas* **19** 030701
[29] Filinov A V, Golubnychiy V O, Bonitz M, Ebeling W, and Dufty J W 2004 *Phys. Rev. E* **70** 046411
[30] Filinov A V, Bonitz M, and Ebeling W 2003 *J. Phys. A: Math. Gen.* **36** 5957
[31] Sjostrom T and Daligault J 2014 *Phys. Rev. Lett.* **113** 155006

Development of Plasma Fluid Model for a Microwave Rocket Supported by a Magnetic Field

Masayuki Takahashi

Department of Aeronautics and Astronautics, The University of Tokyo, Bunkyo-ku 113-8656, Japan

E-mail: m.takahashi@al.t.u-tokyo.ac.jp

Abstract. A fluid model of plasma transport is developed to reproduce a plasma pattern induced by microwave irradiation when an external magnetic field is applied to the breakdown volume. Transport coefficients in the fluid model are evaluated using a fully kinetic simulation under a magnetic field to maintain consistency of electron transport between the particle and fluid models. The electron-density profile and propagation speed of the ionization front obtained by the fluid model agree with those of the particle model. Multidimensional or longer time-scale simulations can be conducted using the fluid model in the case of the application of an external magnetic field, with the simulation reducing computational cost compared to the fully kinetic model.

1. Introduction

Microwave rockets have been proposed to conduct frequent and flexible operations for small satellites to accelerate several areas of space utilization, particularly in the fields of business, scientific research, and national defense. It can achieve this by reducing the launch cost of a payload mass [1–5]. During the launch of a microwave rocket, intense microwaves are transmitted from a ground-based oscillator to the vehicle, which has a parabolic mirror to focus the incident beam. A dense plasma is generated inside the microwave-rocket nozzle as the intense field induced by mirror focusing breaks down the ambient air. A strong shock wave is generated when the plasma absorbs the energy of the incident microwave, providing an impulsive thrust to the vehicle due to interactions between the shock wave and the thruster wall. The launch cost of small satellites is decreased as the fuel needed on the vehicle can be reduced or removed by transmitting energy for propulsion from the ground. There is a large cost required to develop the ground-based microwave oscillator; however, this initial burden will become cost effective over time with repetitive launches of inexpensive rockets [5].

Flight demonstrations of microwave rockets has been conducted by irradiating a microwave beam using a high-power gyrotron [1–5]. A parabolic thruster was placed in a chamber to evaluate the ambient pressure dependence of thrust performance as the microwave rocket travels in rarefied gas conditions at higher altitudes. The higher thrust performance was obtained at atmospheric pressure, while forming a filamentary plasma. The filamentary plasma was caused by a standing wave, which was induced by wave reflection of the overcritical plasma. The thrust performance degraded with a decrease in the ambient pressure, and the plasma pattern transitioned from the discrete plasma to a diffusive breakdown pattern due to infrequent elastic collision. One-dimensional (1D) particle simulations for the charged-particle transport were conducted to identify the mechanism behind the decrease in the thrust performance at lower pressures. An increase in propagation speed of the ionization front and a decrease of the energy-absorption rate by the plasma lowered the thrust performance at rarefied gas conditions [6]. To improve the thrust performance, the propagation speed of the ionization front must be suppressed by applying an external magnetic field to the breakdown region of the microwave rocket. Electron cyclotron resonance (ECR) heating was induced by satisfying the resonant condition to increase the energy-absorption rate of the plasma. This improved thrust performance due to an enhanced shock wave inside the rocket nozzle [7,8]. Previous simulations did not address multidimensionality of the plasma structure when an external magnetic field was applied, as the computational requirements of such a particle simulation were huge. In addition, the largely increased time-scale physics of a microwave plasma cannot be completely captured using the fully kinetic model when a magnetic field is applied. A fluid model of the plasma transports, which has a smaller computational load, was coupled with the electromagnetic wave propagation in past studies [9–14]; however, the breakdown physics under the magnetic field was not examined using the fluid model. It is necessary to develop the fluid model under the magnetic field to examine the multidimensional plasma structure and the breakdown physics with a longer time scale.

The objective of this study is to construct a fluid model, which has simulations that require a smaller computational load than the particle model, so as to reproduce the breakdown physics induced by microwave irradiation when a magnetic field is applied. The transport coefficients under the magnetic field are evaluated using the fully kinetic model, and these coefficients are introduced in the fluid model to maintain consistency between the particle and fluid models. The 1D plasma structure simulated by the fluid model is compared with the 1D breakdown structure obtained by the particle model to validate the developed fluid model.

2. Numerical methods

2.1. Electromagnetic field calculation

An electromagnetic wave simulation is coupled with the fluid model of the plasma transport via the current density to examine the breakdown physics induced by microwave irradiation as interactions between the incident microwave and plasma play an important role for formation of the plasma pattern. Maxwell's equations are numerically integrated using the finite-difference time-domain (FDTD) method with Mur's boundary conditions [15] to reproduce the propagation of the incident microwave;

$$\nabla \times \mathbf{E} = -\frac{\partial \mathbf{B}}{\partial t}, \tag{1}$$

$$\nabla \times \mathbf{H} = \epsilon_0 \frac{\partial \mathbf{E}}{\partial t} + \mathbf{J}, \tag{2}$$

where $\mathbf{E}$ is the electric field, $\mathbf{B}$ is the magnetic flux density, $\mathbf{H}$ is the magnetic field, t is the time, ϵ_0 is the electric permittivity in vacuum, and $\mathbf{J}$ is the current density. Here, $\mathbf{B} = \mu \mathbf{H}$ with the magnetic permeability μ. The current density $\mathbf{J}$ in Eq. (2) is evaluated using the fluid description of plasma transport to model interactions between the dense plasma and propagation of the incident microwave. The time step of the FDTD simulation is determined by the Courant-Friedrichs-Lewy condition for light speed.

2.2. Fluid model of plasma transport

A precise plasma structure is obtained if the particle model for the charged-particle transport is coupled with electromagnetic wave propagation; however, the particle model uses substantial computational resources. In this study, the fluid description of plasma transport is made based

IOP Conf. Series: Journal of Physics: Conf. Series **905** (2017) 012024 doi:10.1088/1742-6596/905/1/012024

on a quasi-neutral assumption to reduce the computational cost. A simple diffusion equation for electrons is numerically integrated based on past studies [11–14];

$$\frac{\partial n_e}{\partial t} - \nabla \cdot (D_{eff} \nabla n_e) = \nu_i n_e, \tag{3}$$

where n_e is the electron density, D_{eff} is the effective diffusion coefficient, and ν_i is the ionization frequency. The effective diffusion coefficient is written as $D_{eff} = (\alpha D_e + D_a)/(\alpha + 1)$ with the ambipolar diffusion coefficient $D_a = (\mu_i D_e + \mu_e D_i)/(\mu_i + \mu_e)$ and $\alpha = \tau_M \nu_i$. Here, $\tau_M = \epsilon_0/[en(\mu_i + \mu_e)]$ is Maxwell's relaxation time, where μ_e is the electron mobility, μ_i is the ion mobility, D_e is the electron diffusion coefficient, and D_i is the ion diffusion coefficient. The diffusion coefficient for free electrons is evaluated by the fully kinetic simulation coupled with collisional reactions in order to maintain consistency between the fluid and particle models. The diffusion coefficient of ion D_i is assumed as 0 because the time scale of plasma propagation is much shorter than that of the ion movement. The time step for the diffusion equation is set as one period of the incident microwave because the time scale for plasma evolution is much longer than the wave propagation [11–14], and time integration is implicitly conducted using the Crank-Nicolson method. The electron current density is estimated as $\mathbf{J} = -en_e\mathbf{v_e}$ in the fluid model, where $\mathbf{v_e}$ is the electron velocity. The electron velocity is calculated by integrating an equation of motion with the electric force, Lorentz force, elastic collision, and electron-pressure gradient;

$$\frac{\partial \mathbf{v_e}}{\partial t} = -\frac{e(\mathbf{E} + \mathbf{v}_e \times \mathbf{B})}{m_e} - \nu_m \mathbf{v_e} - \frac{k_B}{m_e n_e} \nabla(n_e T_e), \tag{4}$$

where m_e is the electron mass, μ_m is the elastic collision frequency, k_B is the Boltzmann constant, and T_e is the electron temperature. The time step for the equation of motion is the same as for the FDTD simulation, and the Euler explicit method is employed to integrate the equation of motion for electrons. The current density is fed back to the FDTD simulation to reproduce interactions between the electromagnetic wave propagation and the plasma.

2.3. Fully kinetic simulation to evaluate transport coefficients under a magnetic field

It is necessary to assess the transport coefficients (ionization frequency, elastic collision frequency, diffusion coefficient, and electron temperature) to integrate the diffusion equation and equation of motion of electrons in the fluid description of a plasma [16]. Data sets of transport coefficients are evaluated against the root-mean-square (RMS) field value using the 1D particle-in-cell (PIC) with the Monte Carlo collision (MCC) model. The local RMS field is calculated by solving Maxwell's equation in the plasma fluid and FDTD coupling simulation. Using the local RMS field, the transport coefficients in the fluid model are interpolated from data sets evaluated by the PIC-MCC simulation. Charged particles have cyclotron motion under the external magnetic field, and these must be considered by coupling the 1D PIC-MCC model with the FDTD simulation when transport coefficients are estimated.

In the 1D PIC module with three velocity components, the equations of motion for electrons and ions are numerically integrated to trace the motion of the charged particles under electromagnetic fields calculated from the FDTD simulation. The Buneman-Boris method [17] is utilized to integrate the equation of motion, which guarantees that no work is induced by the magnetic field. Current density is not fed back to the FDTD module for the assessment of transport coefficients using the particle model of the plasma. To compare plasma patterns between the fluid and particle models, the current density is fed back to the FDTD simulation when the 1D breakdown structure is reproduced using the fully kinetic model. The current density at the computational grids is evaluated by a linear weighting method in the fully kinetic model. The MCC module is coupled with the PIC module to combine collisional reactions with the motion of the charged particles. The Nanbu method [18] is utilized to determine the reaction type and whether a collision exits for pure nitrogen gas. Phelps's cross section is used to introduce eighteen reactions (e.g., elastic collision, vibrational excitation, rotational excitation, electron-impact excitation, and ionization) for nitrogen gas [19]. The MCC module is coupled with the PIC module using the Birdsall method [20].

To evaluate the transport coefficients [16], the ionization frequency ν_i is deduced in the fully kinetic model by:

$$\nu_i = \frac{\ln[N_e(t)] - \ln[N_e(t_0)]}{t - t_0}, \tag{5}$$

where N_e is the total particle numbers of electrons and t_0 is the initial time. The electron-diffusion coefficient is assessed by:

$$D_e = \frac{\langle [x_e(t) - \langle x_e(t_0)\rangle]^2\rangle - \langle [x_e(t) - \langle x_e(t_0)\rangle]^2\rangle}{2(t - t_0)}, \tag{6}$$

where x_e is the electron position, and $\langle \cdots \rangle$ indicates the average value of all particles [16]. The electron temperature T_e is calculated as:

$$T_e = \frac{2}{3k_B}\langle \frac{1}{2} m_e(\mathbf{v}_e - \bar{\mathbf{v}}_e)^2\rangle, \tag{7}$$

where $\langle \cdots \rangle$ indicates the average over all electrons in the simulation domain. The elastic collision frequency is estimated by counting the number of collisions in the particle code.

3. Simulation condition

The 1D breakdown simulation is conducted in nitrogen gas to compare the fluid model with the particle model when an external magnetic field is applied to the breakdown volume, which is induced by microwave irradiation. The maximum intensity of the incident microwave E_y is 5 MV/m with a frequency of 110 GHz in an ambient pressure of 0.1 atm. The external magnetic field B_z is selected as 3.93 T to obtain electron cyclotron resonance (ECR) heating via coupling of electron cyclotron motion with field oscillations from the incident microwaves. The electrons and ions diffuse in the x-direction, the electric field oscillates in the y-direction, and the magnetic field is applied in the z-direction. The 1D simulation domain has 2.25λ ($\lambda = 0.27$ cm, which is the microwave wavelength), and the number of grid points is 1,800 for the fluid model. As a comparison, 40,000 grid points are used in the fully kinetic simulation to resolve the Debye length of the dense plasma. The initial plasma spot with a Gaussian distribution is set on the 1D simulation domain. The electrons initially have velocities corresponding to a temperature of 1 eV in the particle simulation. Neutral particles of 300 K are assumed as a background gas. A microwave is irradiated from the left to right boundaries to reproduce a breakdown pattern in nitrogen, and the plasma propagates toward the left boundary from the initial plasma spot.

4. Results

4.1. Evaluation of transport coefficients using 1D PIC-MCC model

To introduce the ionization frequency, elastic collision frequency, electron-diffusion coefficient, and the electron temperature into the fluid model, they are evaluated using the 1D PIC-MCC simulation for nitrogen gas at ambient pressure of 0.1 atm. Microwave beams of 0 to 15 MV/m are irradiated to assess the transport coefficients under an external magnetic field of 3.93 T. The current density is not fed back to the FDTD simulation for evaluation of the transport coefficients. The electron-diffusion coefficient of 3.93 T is smaller than 0 T because the external magnetic field disturbs the electron transport (Fig. 1), suppressing propagation of the ionization front inside the microwave-rocket nozzle. Confinement effects from the external magnetic field

IOP Conf. Series: Journal of Physics: Conf. Series **905** (2017) 012024 doi:10.1088/1742-6596/905/1/012024

are successfully considered in the estimation of transport coefficients by utilizing the fully kinetic model. The transport coefficients evaluated by the PIC-MCC model are locally interpolated in the FDTD and plasma fluid coupling simulation.

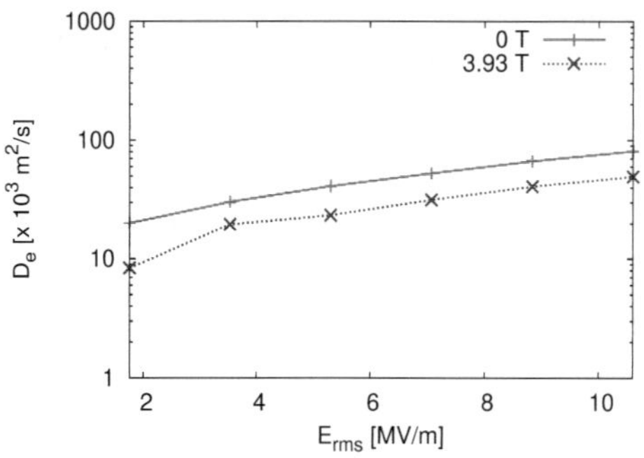

Figure 1. Electron-diffusion coefficients evaluated by the PIC-MCC simulation at an ambient pressure of 0.1 atm. The 110-GHz microwaves are irradiated to a Gaussian plasma spot under magnetic fields of 0 T and 3.93 T.

4.2. Comparison of 1D plasma structures between particle and fluid models under magnetic fields

The 1D fluid and particle simulations are conducted in nitrogen gas to compare plasma structures and propagation speeds obtained using both models at an ambient pressure of 0.1 atm. Microwaves of 5-MV intensity and a frequency of 110 GHz are injected from the left to the right boundaries under a magnetic field of 3.93 T. The current density is fed back to the FDTD simulation to reproduce an interaction between the dense plasma and the electromagnetic wave propagation in the fluid and particle models.

Overcritical plasma is formed due to the quick ionization at the initial plasma spot (Fig. 2(a)), which reflects the incident microwave. The ionization front propagates toward the left boundary because seed electrons diffuse from the bulk plasma. Sequential ionization occurs in front of the bulk plasma when the seed electrons are heated due to the strong electric field generated by an overlap of the incident and reflected microwaves. The maximum electron density of the fluid model is larger than that of the particle model. The density profile obtained by the fluid model agrees with that of the particle model because the transport coefficients in the fluid model are evaluated using the particle simulation. The time evolution of the ionization-front position from the initial plasma spot is assessed by tracking the electron densities exceeding 1×10^{13} cm^{-3} (Fig. 2(b)). The propagation speed of the ionization front is estimated as 680 km/s at 0.1 atm under the magnetic field of 3.93 T, and the propagation speeds of both fluid and particle models are the same. Under the external magnetic field, the multidimensional plasma pattern or the longer time-scale physics can be examined using the fluid model because consistency of the density profile and the propagation speed is maintained between the particle and fluid models.

5. Conclusion

Under an external magnetic field, the breakdown pattern of a microwave plasma was reproduced by coupling the fluid or particle descriptions of the plasma transports with an electromagnetic wave propagation. Confinement effects from the magnetic field were introduced in the fluid model from the transport coefficients utilized in the fluid model. These coefficients were evaluated by

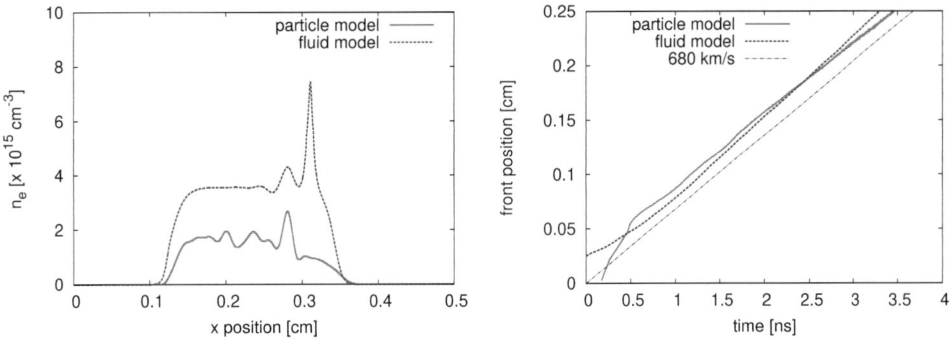

(a) Electron number densities (b) Front position of plasma

Figure 2. Electron number densities and front position of plasma obtained by the fluid and particle models. At 0.1 atm, 110-GHz microwave is irradiated to breakdown volume under external magnetic field of 3.93 T.

the particle code, which reproduces cyclotron motions of charged particles under a magnetic field. The particle and fluid simulations were conducted with current feedback to the FDTD simulation to compare the microwave-plasma patterns obtained by each model. The electron-density profile and the propagation speed of the ionization front obtained by the fluid model agree with those found in the particle model because consistency of electron transports was maintained between the particle and fluid models. It is possible to conduce the multidimensional or longer time-scale simulations can be performed using the fluid model when a magnetic field is applied to the breakdown volume induced by microwave irradiation.

Acknowledgements
The development of numerical models was supported by Dr. Jean-Pierre Boeuf at the University of Toulouse. The computations in this work were performed on a FUJITSU PRIMEHPC FX100 at the Japan Aerospace Exploration Agency. This work was supported by JSPS KAKENHI Grant Number JP16J09910.

References
[1] T. Nakagawa, Y. Mihara, and K. Komurasaki, J. Spacecraft and Rockets **41** (1), 151–153 (2004).
[2] Y. Oda and K. Komurasaki, J. Appl. Phys. **100**, 113307 (2006).
[3] Y. Oda, Doctor Thesis (The University of Tokyo, Tokyo, 2008).
[4] M. Fukunari, R. Komatsu, A. Arnault, T. Yamaguchi, K. Komurasaki, and Y. Arakawa, Vacuum **88**, 155–159 (2013).
[5] M. Fukunari, A. Arnault, T. Yamaguchi, and K. Komurasaki, Appl. Optics **53** (31), 16–22 (2014).
[6] M. Takahashi and N. Ohnishi, Appl. Phys. Lett. **105**, 223504 (2014).
[7] M. Takahashi and N. Ohnishi, Trans. Japan Soc. Aero. Space Sci., Aero. Tech. Japan (accepted).
[8] M. Takahashi and N. Ohnishi, J. Appl. Phys (accepted).
[9] K. S. Nam and P. J. Verboncoeur, Phys. Rev. Lett. **103**, 055004 (2009).
[10] Q. Zhou and Z. Dong, Appl. Phys. Lett. **98**, 161504 (2011).
[11] J. P. Boeuf, B. Chaudhury, and G. Q. Zhu, Phys. Rev. Lett. **104**, 015002 (2010).
[12] B. Chaudhury, J. P. Boeuf, and G. Q. Zhu, Phys. Plasmas **17**, 123505 (2010).
[13] K. Kourtzanidis, J. P. Boeuf, and F. Rogier, Phys. Plasmas **21**, 123513 (2014).
[14] M. Takahashi and N. Ohnishi, Jpn. J. Appl. Phys. **55**, 07LD02 (2016).
[15] G. Mur, IEEE Trans. EMC **EMC-23** (4), 377–382 (1981).
[16] C. Li, W. J. M. Brok, U. Ebert, and J. J. A. M. van der Mullen, J. Appl. Phys. **101**, 123305 (2007).
[17] O. Buneman, J. Comput. Phys. **1**, 517–535 (1967).
[18] K. Nanbu, Jap. J. Appl. Phys. **33**, 4752–4753 (1994).

IOP Conf. Series: Journal of Physics: Conf. Series **905** (2017) 012024 doi:10.1088/1742-6596/905/1/012024

[19] A. V. Phelps, Compilation of Electron Cross Sections (University of Colorado, 2008) [http://jila.colorado.edu/~avp/collision_data/electronneutral/ELECTRON.TXT].
[20] C. K. Birdsall, IEEE Trans. on Plasma Sci. **27**, 65–85 (1991).

IOP Conf. Series: Journal of Physics: Conf. Series **905** (2017) 012025 doi:10.1088/1742-6596/905/1/012025

Random Field Ising Models: Fractal Interfaces and their Implications

A Bupathy[1], M Kumar[2], V Banerjee[1] and S Puri[2]

[1] Department of Physics, Indian Institute of Technology, New Delhi – 110016, India.

[2] School of Physical Sciences, Jawaharlal Nehru University, New Delhi – 110067, India.

E-mail: varsha@physics.iitd.ac.in

Abstract. We use a computationally efficient graph-cut (GC) method to obtain exact ground-states of the $d = 3$ random field Ising model (RFIM) on simple cubic (SC), body-centered cubic (BCC) and face-centered cubic (FCC) lattices with Gaussian, Uniform and Bimodal distributions for the disorder Δ. At small-r, the correlation function $C(r, \Delta)$ shows a cusp singularity characterised by a non-integer roughness exponent α signifying rough fractal interfaces with dimension $d_f = d - \alpha$. In the paramagnetic phase ($\Delta > \Delta_c$), $\alpha \simeq 0.5$ for all lattice and disorder types. In the ferromagnetic phase ($\Delta < \Delta_c$), $\alpha \simeq 0.66$ with minor variations for the different lattice types. Our predictions are confirmed by scattering data from diluted antiferromagnets (DAFFs). Fractal interfaces have important implications on growth and relaxation.

1. Introduction

The random field Ising model (RFIM) is one of the simplest models that provides a framework to study the effects of quenched disorder and randomness on static and dynamic properties. It is realized by many physical systems such as dilute antiferromagnets in a field (DAFFs) [1], diluted dipolar magnet $LiHo_xY_{1-x}F_4$ [2], relaxor ferroelectrics [3], colloid-polymer mixtures [4], non-equilibrium phenomena like Barkhausen noise in magnetic hysteresis [5], etc. The RFIM comprises of N Ising spins on a d-dimensional lattice with the Hamiltonian given by [6]:

$$\mathcal{H}(\{\sigma_i\}) = -J\sum_{\langle ij\rangle}\sigma_i\sigma_j - \sum_i h_i\sigma_i, \quad \sigma_i = \pm 1 \tag{1}$$

Here, $J > 0$ is the strength of the nearest neighbour coupling and promotes ferromagnetic order and h_i's are the disordering random fields usually drawn from one of the following distributions:

$$\text{Gaussian}: P_1(h_i) = \frac{1}{2\pi\Delta^2}e^{-h_i^2/(2\Delta^2)}, \tag{2}$$

$$\text{Uniform}: P_2(h_i) = \begin{cases} \left[\sqrt{12}\Delta\right]^{-1} & : |h_i| \leq \sqrt{3}\Delta, \\ 0 & : |h_i| > \sqrt{3}\Delta, \end{cases} \tag{3}$$

$$\text{Bimodal}: P_3(h_i) = \frac{1}{2}\left[\delta(h_i - \Delta) + \delta(h_i + \Delta)\right]. \tag{4}$$

The standard deviation Δ is a measure of the strength of disorder. All the three distributions have mean zero and variance Δ^2. As the disorder strength is increased, the system undergoes a transition from the ferro to para phase at a critical disorder strength Δ_c.

The presence of competing interactions gives rise to a complex free energy landscape having several local minima separated by energy barriers that grow exponentially with N. The evolving system therefore gets trapped in a local minimum which can be arbitrarily far from the global minimum. As a result there is a multitude of time-scales which lead to slow growth and anomalous relaxation. Such energy functions are very difficult to minimize. Traditional Monte Carlo (MC) procedures such as the Metropolis algorithm or parallel tempering involve only $O(1)$ spin flips at a time and are unable to approach the ground state (GS). In recent years, several sophisticated techniques based on max-flow/min-cut or graph-cut (GC) methods have been employed to study the RFIM [7]. These methods attempt to find a local minimum by a simultaneous re-labeling of a large number of spin variables. An exponentially large portion of the phase-space can be sampled in one move, thereby facilitating a quick search for the global minimum. Further, the RFIM Hamiltonian satisfies a *regularity condition* which guarantees the GS via the GC.

The random fields cause pinning and roughening of interfaces. They are then characterized by a roughness exponent α which is related to the fractal dimension $d_f = d - \alpha$. When is the interfacial texture consequential? Consider a thermal quench from a high temperature paramagnetic phase to a low temperature ferromagnetic phase. The evolution of the system is via the motion of interfaces due to surface tension, but is impeded by the presence of energy barriers due to the disordering fields. The domain growth therefore is due to thermally activated barrier hopping. According to Villain [8], the barrier energy in the RFIM is $E_B \sim R^m$ where R is the characteristic length scale in the system and the barrier exponent $m = 2 - \alpha$. Consequently, this class of disordered systems exhibits slow *logarithmic growth* [9]. On a related note, de-pinning of interfaces on application of an external field has generated technological interest [10, 11]. The creep motion that preceeds the de-pinning transition is greatly affected by the interfacial roughness.

In this paper, we obtain GS morphologies ($T = 0$) of the RFIM on: (i) simple cubic (SC), body-centered cubic (BCC) and face-centered cubic (FCC) lattices with Gaussian disorder and (ii) SC lattice with Gaussian (G), Uniform (U) and Bimodal (B) distributions, by a computationally efficient GC protocol introduced by Boykov and Kolmogorov (BK) [12]. It's polynomial time complexity of $O(N)$ enabled us to access ground states of substantially larger system sizes than in previous studies which used procedures with complexity of $O(N^3)$ and more. We analyze the textures of GS morphologies using the correlation function $C(r, \Delta)$ and the structure factor $S(k, \Delta)$. The main results of our paper are as follows:

(a) The GS morphologies comprise of correlated domains of "up" spins and "down" spins separated by rough interfaces. These correlated regions grow in size as the disorder strength $\Delta \to \Delta_c^+$.

(b) By evaluating the Binder cummulant, our estimates of the critical disorder Δ_c are (a) 2.278 ± 0.002 (SC, G), (b) 3.316 ± 0.002 (BCC, G), (c) 5.16 ± 0.002 (FCC, G), (d) 2.234 ± 0.002 (SC, U) and (e) 2.20 ± 0.002 (SC, B).

(c) At small-r, the correlation function $C(r, \Delta)$ shows a cusp singularity with distinct roughness exponents α for the ferromagnetic phase ($\Delta < \Delta_c$) and paramagnetic phase ($\Delta > \Delta_c$). The structure factor $S(k, \Delta)$ correspondingly shows a non-Porod tail signifying scattering off fractal interfaces.

(d) In the paramagnetic phase ($\Delta > \Delta_c$), the roughness exponent $\alpha = 0.5 \pm 0.01$ for all three lattice types and disorder distributions, showing that the interfacial properties are universal.

(e) In the ferromagnetic phase ($\Delta < \Delta_c$), the roughness exponent $\alpha \simeq 0.66$ with minor variations for the different lattice types. Though tiny, these variations have important implications.

(f) We explain the non-Porod behavior of the structure factor obtained in small-angle neutron-scattering experiments on dilute antiferromagnets.

This paper is organized as follows. In Sec. 2 we present the GC procedure used to obtain the ground states and the tools for characterizing these morphologies. In Sec. 3, we analyze the RFIM morphologies obtained for the SC, BCC and FCC lattices. We also verify our predictions of fractal interfaces in a large number of dilute antiferromagnets by studying the small-angle neutron scattering data. Finally, we conclude this paper with a summary and discussion in Sec. 4.

2. Methodology

2.1. Graph Cuts for Energy minimization

In the GC method, the spin lattice is represented as a graph $\mathcal{G}$ with vertices $\mathcal{V}$ and edges $\mathcal{E}$ connecting them. Each vertex i is assigned a label $s_i \in \mathcal{L}$. The energy function associated with such a labelling is given as

$$E(\{s_i\}) = \sum_{\{ij\}\in\mathcal{N}} V_{ij}(s_i, s_j) + \sum_i D_i(s_i). \tag{5}$$

Here, V_{ij} is the *weight* of edge ij and represents the penalty of assigning labels s_i and s_j to vertices i and j respectively, and D_i is the cost of assigning the label s_i to vertex i. For simplicity, we only discuss the case of binary labels, i.e., $\mathcal{L} = \{0, 1\}$. A partitioning of the vertices $\mathcal{V}$ into two sets $\mathcal{S}$ and $\mathcal{T}$, then corresponds to labelling the vertices and is called a *cut* $\mathcal{C}$. Any edge $ij \in \mathcal{E}$ with $i \in \mathcal{S}$ and $j \in \mathcal{T}$ (or vice versa) is a cut edge. The cost of the cut is the sum of the weights of the cut edges. The problem is then to find the cut with the smallest cost or *min-cut*.

The RFIM Hamiltonian belongs to a special class of energy functions which satisfies a *regularity condition* [13]. The latter assures convergence to *exact* ground states if energy minimization is via graph-cuts. In recent papers [14, 15], we have described in detail our application of the Boykov-Kolmogorov GC method (BK-GCM) [12] to the RFIM. For the sake of brevity, we will not reiterate the algorithmic details here. The $O(N)$ polynomial complexity of the BK-GCM method allowed us to perform simulations on large systems and obtain smooth data.

2.2. Characterization of Interfacial Texture

Surface roughness has usually been described using the *solid-on-solid* model [16, 17]. In this model, the surface is described by a single-valued function $h(\vec{r})$ which represents the height of the surface at a position vector $\vec{r}$ on a $(d-1)$-dimensional substrate. The *interface width* is the r.m.s. fluctuation in the height:

$$w^2 = \frac{1}{A}\int d\vec{r}\,\left[h(\vec{r}) - \bar{h}\right]^2 = \frac{1}{A}\int d\vec{r}\;\delta h(\vec{r})^2, \tag{6}$$

where A is the substrate area and $\bar{h}$ is the average height. For a translationally invariant system, we can also write

$$w^2 = \left\langle h(\vec{r})^2\right\rangle - \left\langle h(\vec{r})\right\rangle^2, \tag{7}$$

where the angular brackets denote an ensemble average: $\langle h(\vec{r})\rangle = \bar{h}$. We expect $w^2 \sim L^{2\alpha}$, where α is the *roughness exponent* of the interface [16]. For a self-affine surface, $\alpha = d - d_f$, where d_f is the fractal dimension of the surface.

A more detailed tool for quantifying domain morphologies is the correlation function:

$$C(\vec{r}, \Delta) = \langle\sigma_i\sigma_j\rangle - \langle\sigma_i\rangle\langle\sigma_j\rangle, \tag{8}$$

where $\vec{r} = \vec{r_j} - \vec{r_i}$ and the angular brackets denote an ensemble average. In the isotropic case, $C(\vec{r}, \Delta)$ depends only on the vector magnitude $r = |\vec{r}|$. The correlation length ξ is usually defined as the distance over which $C(r, \Delta)$ decays to (say) 0.2 $\times$ maximum value. The typical GS morphology of the RFIM comprises of compact domains of "up" ($\sigma_i = +1$) and "down" ($\sigma_i = -1$) spins of size $\sim \xi$ separated by rough interfaces. For such a morphology, the small-r behaviour of the correlation function exhibits a *cusp regime* [18]:

$$\bar{C}\left(r\right) \equiv 1 - C\left(r\right) \simeq A\left(r/\xi\right)^{\alpha}, \tag{9}$$

where α is the *cusp* or *roughness* exponent. For rough fractal interfaces $0 \leq \alpha < 1$ and the corresponding fractal dimension is $d_f = d - \alpha$. For smooth interfaces, $\alpha = 1$ and is termed the *Porod law* [19].

Scattering experiments measure the structure factor, which is the Fourier transform of the correlation function:

$$S(\vec{k}, \Delta) = \int \mathrm{d}\vec{r}\; e^{i\vec{k}.\vec{r}}\; C(\vec{r}, \Delta), \tag{10}$$

where $\vec{k}$ is the wave-vector of the scattered beam. In the isotropic case $S(\vec{k}, \Delta)$ depends only on the vector magnitude $k = |\vec{k}|$. The short-distance cusp singularity has important implications for the structure factor $S(k)$. It decays with an asymptotic power-law form [20, 21, 22]

$$S(k) \sim \tilde{A}\left(\xi k\right)^{-(d+\alpha)}. \tag{11}$$

The dominant large-k behaviour in Eq. (11) is the *Non-Porod* regime $S(k) \sim (\xi k)^{-(d+\alpha)}$ implying scattering off rough fractal interfaces. In the case of smooth interfaces, $S(k) \sim (\xi k)^{-(d+1)}$ yielding the well known *Porod law*. We emphasize that the correlation-function data yields a more accurate measure of α, as it is obtained by averaging over all domains and interfaces in the system.

3. Numerical Results

The ground states of the RFIM were obtained using the BK-GCM on lattices of size L^3 ($L = 128$). Unless specified otherwise, we employ periodic boundary conditions in all directions. The BK-GCM yields states with 99% overlap with the GS in the first iteration itself, and we observe a mild critical slowing down as $\Delta \rightarrow \Delta_c$. All the results have been averaged over 100 realizations of $\{h_i\}$ for a given value of Δ. For the BCC and FCC lattices, we introduce extra sites with $\sigma_i = 0$, converting them to SC lattices with spacing $a/2$. With this procedure, we could use the standard fast Fourier transform routines for evaluating $S(k)$.

We have evaluated the critical disorder strength Δ_c using the fourth order Binder cummulant for the three lattices. These values are: (a) 2.278$\pm$0.002 for SC lattice, (b) 3.316$\pm$0.002 for BCC lattice and (c) 5.160$\pm$0.002 for FCC lattice [14, 23]. It is convenient to define $\delta = (\Delta - \Delta_c)/\Delta_c$, as the critical disorder strength Δ_c is different for the different lattice types. Fig. 1 shows the typical GS morphologies of the RFIM on a SC lattice in (a) the paramagnetic regime ($\delta = 0.08$) and (b) ferromagnetic regime ($\delta = -0.05$). In the paramagnetic regime, the GS has many domains of up and down spins separated by rough interfaces. In the ferromagnetic regime, however, there is a single large cluster of up (down) spins with small impurity clusters of down (up) spins. As a result, it is difficult to extract the interfacial properties in this phase. Therefore, at $\Delta < \Delta_c$, we create interfaces by the following procedure. We apply periodic boundary conditions in the x- and y-directions and fix the spin values at $z = 1, L$ as

$$\begin{aligned} \sigma(x, y, z = 1) &= -1, \\ \sigma(x, y, z = L) &= +1. \end{aligned} \tag{12}$$

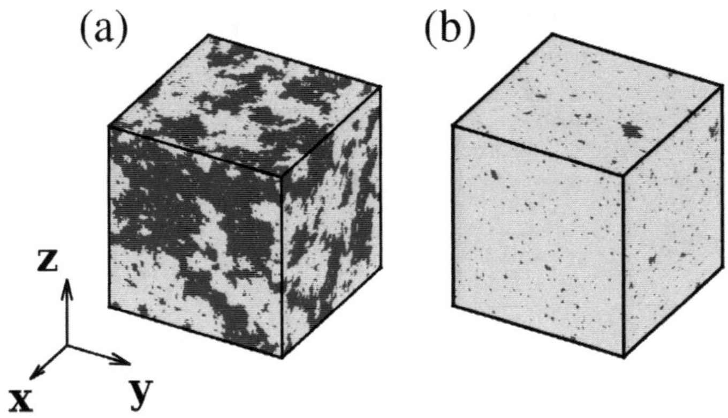

Figure 1. Typical ground state morphologies of the RFIM on a SC lattice of size 128^3 in: (a) the paramagnetic phase ($\delta = 0.08$) and (b) the ferromagnetic phase ($\delta = -0.05$). Green (gray) and blue (black) regions represent up and down spins, respectively.

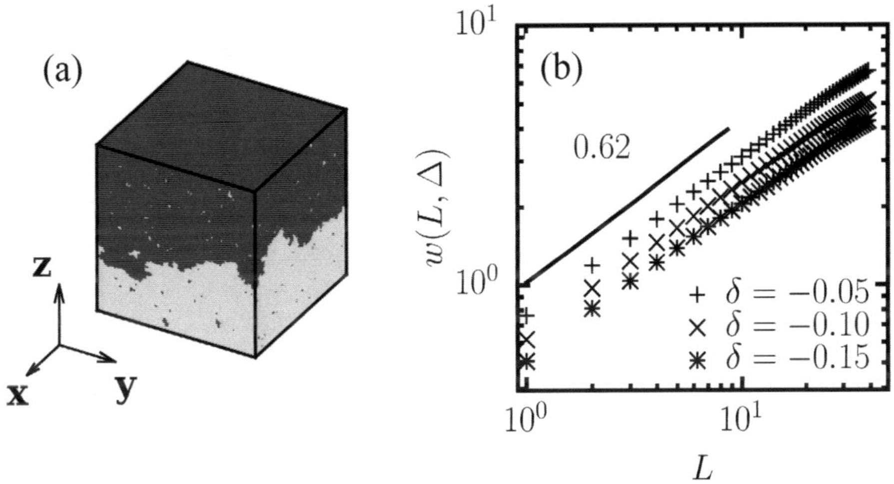

Figure 2. (a) Ground state morphology of the RFIM on a SC lattice of size 128^3 with anti-parallel boundary conditions in the z-direction in the ferromagnetic regime ($\delta = -0.05$). Green (gray) and blue (black) regions represent up and down spins, respectively. (b) Log-log plot of $w(L, \Delta)$ vs. L for the GS of the RFIM on a SC lattice at $\delta = -0.05, -0.1, -0.15$. The slope of the best-fit line to the small-r regime gives $\alpha = 0.62 \pm 0.025$.

Fig. 2(a) shows the typical GS morphology of the RFIM on SC lattice at $\delta = -0.05$ subject to these boundary conditions. There is a well-defined rough interface between up and down domains. In Fig. 2(b), we show the log-log plot of the rms width $w(L)$ vs. L of such interfaces on a SC lattice for $\delta = -0.05, -0.1$ and -0.15. From the slope of the best linear fit, we obtain the roughness exponent $\alpha \simeq 0.62 \pm 0.025$.

We now study the effect of lattice structure on the GS morphologies. In Fig. 3(a), we plot $1 - C(r, \Delta)$ vs. r/ξ on a log-log scale for SC, BCC and FCC lattices in the ferromagnetic regime ($\delta = -0.01$). Our system is not translationally invariant in the z-direction. Therefore,

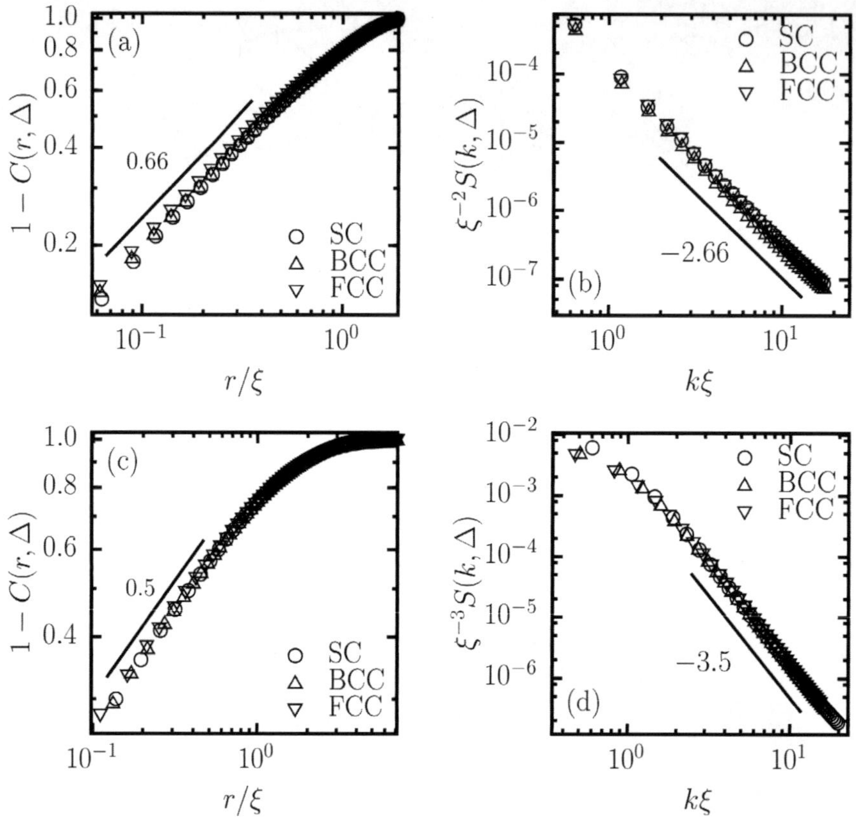

Figure 3. (a) Log-log plot of $1 - C(r, \Delta)$ vs. r/ξ for the RFIM on SC, BCC and FCC lattices with Gaussian disorder in the ferromagnetic regime ($\delta = -0.01$). The corresponding log-log plot of the scaled structure factor $\xi^{-2}S(k, \Delta)$ vs. $k\xi$ is shown in (b). The structure factor shows a non-Porod tail: $S(k, \Delta) \sim k^{-(d+\alpha)}$. Best-fits to the data yield $\alpha = 0.66 \pm 0.02$. (c) Log-log plot of $1 - C(r, \Delta)$ vs. r/ξ for the RFIM on SC, BCC and FCC lattices with Gaussian disorder in the paramagnetic regime ($\delta = 0.08$). The corresponding log-log plot of the scaled structure factor $\xi^{-2}S(k, \Delta)$ vs. $k\xi$ is shown in (d). Best-fits to the data yield $\alpha = 0.5 \pm 0.01$.

we compute the spin-spin correlation function $C(r)$ in the (xy)-plane at $z = \bar{h}$, where $\bar{h}$ is the average height of the interface from the $z = 0$ plane. The linear behaviour of the small-r region yields slightly different values of α for the three lattices in the range $\approx 0.64 - 0.68$. This is consistent with the roughness exponent measured from $w(L)$ in Fig. 2(b). We stress that the correlation function data yields a more accurate measure of α, as it is obtained by averaging over all interfaces in the system. In Fig. 3(b), we plot the scaled structure factor $\xi^{-2}S(k, \Delta)$ vs. $k\xi$ for the correlation functions in Fig. 3(a). It shows a distinct non-Porod behaviour: $S(k) \sim k^{-(d+\alpha)}$. In Figs. 3(c) and (d), we show the corresponding $1 - C(r, \Delta)$ and $S(k, \Delta)$ plots for the SC, BCC and FCC lattices in the paramagnetic regime ($\delta = 0.08$). Best fits to the data yields $\alpha = 0.5 \pm 0.01$, distinct from the ferromagnetic case. Note that the scaling is excellent, indicating that the interfacial textures are unaffected by the lattice structure in the paramagnetic regime.

In Fig. 4(a) we show $1 - C(r, \Delta)$ vs r/ξ on a log-log scale for G, U and B disorder on SC lattice in the paramagnetic regime ($\delta = 0.1$). The corresponding scaled structure factor $\xi^{-3}S(k, \Delta)$ is plotted in Fig. 4(b) on a log-log scale. We find that all the three data sets show excellent

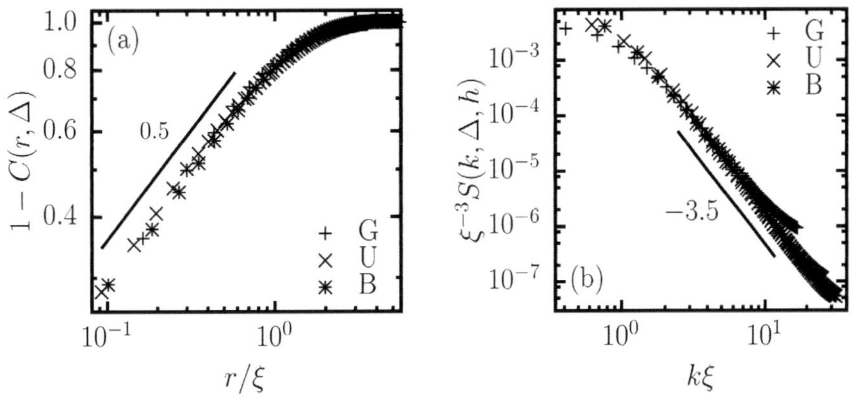

Figure 4. (a) Log-log plot of $1 - C(r, \Delta)$ vs. r/ξ for the RFIM on SC with Gaussian, Uniform and Bimodal disorder in the paramagnetic regime ($\delta = -0.01$). The corresponding log-log plot of the scaled structure factor $\xi^{-2}S(k, \Delta)$ vs. $k\xi$ is shown in (b). The slope of the best-fit line to the data yields $\alpha = 0.5 \pm 0.01$.

collapse. From the best fits to the data we obtain the roughness exponent $\alpha = 0.5 \pm 0.01$. Thus in the paramagnetic phase, the interfacial properties in the RFIM are unaffected by the form of the underlying lattice or disorder.

Finally, we compare our numerical results with the neutron scattering data of randomly diluted antiferromagnets (DAFFs) in a uniform field. They have been some of the most well studied realizations of the RFIM ever since Fishman and Aharony established their equivalence [1, 24, 25, 26]. In Fig. 5 we plot $S(k)$ vs k for the following data sets on a log-log scale. The critical temperature, the experimental temperature and the applied magnetic field strength are provided in the parantheses:
(1) $Fe_{0.46}Zn_{0.64}F_2$ (32.11 K, 32.34 K, 3.0 T) [24];
(2) $Co_{0.35}Zn_{0.65}F_2$ (13.25 K, 2 K, 5.0 T) [25];
(3) $Fe_{0.6}Zn_{0.4}F_2$ (46.13 K, 46.3 K, 2.0 T) [26];
The solid line with slope -4 corresponds to the Porod law, which arises from scattering off smooth interfaces. Clearly, all data sets are non-Porod with slopes greater than -4 indicating that the interfaces separating the phases are fractal. We do not expect to detect the small variations in the slopes due to different lattice structures as the accuracy of the scattering data is limited by the resolution function of the detector.

4. Summary and Discussion

The random field Ising model (RFIM) is simplest example of a system with quenched disorder and provides a framework to study the effects of randomness and disorder on interfacial properties. In this paper we have analysed the influence of (i) different lattice types, viz., simple cubic (SC), body-centered cubic (BCC) and face-centered cubic (FCC) and (ii) different disorder distributions, viz., Gaussian (G), Uniform (U) and Bimodal (B) on the ground state (GS) morphologies of the RFIM. We used a computationally efficient graph-cut procedure to obtain the exact ground states at $T = 0$. We summarize our results as follows:
(a) The GS morphologies of the RFIM comprise of compact domains separated by rough fractal interfaces. We have analyzed these morphologies using spin-spin correlation function $C(r)$ and structure factor $S(k)$.
(b) The correlation function $C(r, \Delta)$ shows a cusp singularity at small-r characterized by the

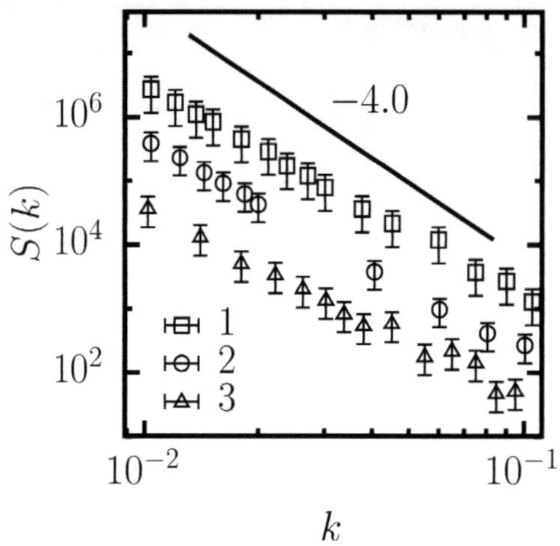

Figure 5. Log-log plot of neutron scattering data of DAFFs: (1) $Fe_{0.46}Zn_{0.64}F_2$ ($T_c = 32.11$ K, $T = 32.34$ K, $H = 3.0$ T) [24]; (2) $Co_{0.35}Zn_{0.65}F_2$ ($T_c = 13.25$ K, $T = 2$ K, $H = 5.0$ T) [25]; (3) $Fe_{0.6}Zn_{0.4}F_2$ ($T_c = 46.13$ K, $T = 46.3$ K, $H = 2.0$ T) [26]; The data sets have been shifted vertically for clarity. The solid line of slope -4 corresponds to Porod scattering from smooth interfaces. All data sets are non-Porod with slope greater than -4, implying scattering off fractal interfaces.

cusp or roughness exponent α. The structure factor shows a corresponding non-Porod behaviour $S(k,\Delta) \sim k^{-(d+\alpha)}$ implying scattering off fractal interfaces.
(c) In the ferromagnetic phase, the roughness exponent $\alpha \simeq 0.66$ for the SC lattice, with minor variations for BCC and FCC lattices. They may be attributed to the different nn environments in the three lattices. The corresponding fractal dimension of the interfaces $d_f = d - \alpha = 2.34 \pm 0.02$.
(d) In the paramagnetic phase, the roughness exponent $\alpha = 0.5 \pm 0.01$, yielding $d_f = d - \alpha = 2.5 \pm 0.01$. The interfacial properties are unaffected by the choice of lattice type or disorder distribition, suggesting that α is a universal exponent in the paramagnetic regime.
(e) Our predictions of fractal interfaces are confirmed in many experimental realizations of the RFIM by analyzing the tail of the structure factor obtained from neutron scattering experiments.

Interfaces in disordered systems have generated a lot of research interest in the past few decades. To understand their behaviour and to manipulate them may be the key to realizing interesting technological applications such as race track memories [10], magnetic logic gates [11], etc. The energy required by a domain of size R to grow or the barrier energy $E_B \sim R^{2-\alpha}$ [8, 9, 14, 23]. The interfacial texture therefore plays a crucial role in growth and relaxation. We believe that our methodologies and analysis may be beneficial from both theoretical and experimental points of view.

Acknowledgments

AB, MK and VB would like to acknowledge partial financial support from Department of Science and Technology (DST), India and Indo-French Centre for Promotion of Advanced Scientific Research (IFCPAR), India.

References

[1] Fishman S and Aharony A 1979 *J. Phys. C: Solid State Phys.* **12** L729.
[2] Schechter M 2008 *Phys. Rev. B* **77** 020401.

[3] Kleemann W, Dec J, Lehnen P, Blinc R, Zalar B and Pankrath R 2002 *Europhys. Lett.* **57** (1) 14.
[4] Vink R L C, Binder K and Löwen H 2006 *Phys. Rev. Lett.* **97** 230603.
[5] Sethna J P, Dahmen K A and Myers C R 2001 *Nature* **410** 242.
[6] Nattermann T and Villain J 1988 *Phase Transitions* **11** 5.
[7] Angls d'Auriac J-C and Sourlas N 1997 *N Europhys. Lett.* **39** 473.
[8] Villain J 1984 *Phys. Rev. Lett.* **52** 1543.
[9] Lai Z W, Mazenko G F and Valls O T 1988 *Phys. Rev. B* **37** 9481.
[10] Thomas L, Moriya R, Rettner C and Parkin S S P 2010 *Science* **330** 1810.
[11] Parkin S S P, Hayashi M and Thomas L 2008 *Science* **320** 190.
[12] Boykov Y and Kolmogorov V 2004 *IEEE Transactions on PAMI* **26** 1124.
[13] Kolmogorov V and Zabih R 2004 *IEEE Transactions on PAMI* **26** 147.
[14] Shrivastav G P, Kumar K, Banerjee V and Puri S 2014 *Phys. Rev. E* **90** 032140.
[15] Shrivastav G P, Krishnamoorthy S, Banerjee V and Puri S 2011 *Europhys. Lett.* **96** 36003.
[16] Barabasi A L and Stanley H E 1995 *Fractal Concepts in Surface Growth* (Cambridge: Cambridge University Press).
[17] Meakin P, Ramanlal P, Sander L M and Ball R C 1986 *Phys. Rev. A* **34** 5091.
[18] Wong P-Z 1985 *Phys. Rev. B* **32** 7417.
[19] Porod G 1982 *Small-Angle X-ray Scattering* Glatter O and Kratky (New York: Academic Press).
[20] Sorensen C M 2001 *Aerosol Sc. Tech.* **35** 648.
[21] Wong P-Z and Bray A J 1988 *Phys. Rev. Lett.* **60** 1344.
[22] Wong P-Z and Bray A J 1988 *Phys. Rev. B* **37** 7751.
[23] Bupathy A, Banerjee V and Puri S 2016 *Phys. Rev. E* **93** 012104.
[24] Belanger D P, Jaccarino V, King A R and Nicklow R M 1987 *Phys. Rev. Lett.* **59** 930.
[25] Hagen M, Cowley R A, Satija S K, Yoshizawa H, Shirane G, Birgeneau R J and Guggenheim H J 1983 *Phys. Rev. B* **28** 2602.
[26] Belanger D P, King A R and Jaccarino V 1985 *Phys. Rev. B* **31** 4538.

IOP Conf. Series: Journal of Physics: Conf. Series **905** (2017) 012026 doi:10.1088/1742-6596/905/1/012026

MORTICIA, a statistical analysis software package for determining optical surveillance system effectiveness.

A Ramkilowan[1, 2] **and D J Griffith**[1]
[1] Optronic Sensor Systems (OSS), Council for Scientific and Industrial Research (CSIR), P O Box 395, Pretoria 0001, South Africa
[2] Corresponding Author, aramkilowan@csir.co.za

Abstract. Surveillance modelling in terms of the standard Detect, Recognise and Identify (DRI) thresholds remains a key requirement for determining the effectiveness of surveillance sensors. With readily available computational resources it has become feasible to perform statistically representative evaluations of the effectiveness of these sensors. A new capability for performing this Monte-Carlo type analysis is demonstrated in the MORTICIA (Monte-Carlo Optical Rendering for Theatre Investigations of Capability under the Influence of the Atmosphere) software package developed at the Council for Scientific and Industrial Research (CSIR). This first generation, python-based open-source integrated software package, currently in the alpha stage of development aims to provide all the functionality required to perform statistical investigations of the effectiveness of optical surveillance systems in specific or generic deployment theatres. This includes modelling of the mathematical and physical processes that govern amongst other components of a surveillance system; a sensor's detector and optical components, a target and its background as well as the intervening atmospheric influences. In this paper we discuss integral aspects of the bespoke framework that are critical to the longevity of all subsequent modelling efforts. Additionally, some preliminary results are presented.

1. Introduction

The expected performance of an electro-optical sensor system is often uncertain due to the large number of factors that influence it. Range prediction for these surveillance sensors involves many dimensions of variability, both those under design control as well as environmental factors which can often be modelled with varying degrees of physical fidelity, but which lie outside engineering control. The MORTICIA (Monte-Carlo Optical Rendering for Theatre Investigations of Capability under the Influence of the Atmosphere) software framework has been developed to produce a statistical representation of the range capabilities of surveillance sensors in both generic and specific deployment theatres. This platform independent tool currently in the alpha stage of development, when complete, would be of particular use in the hands of commanders, providing knowledge of what the current surveillance capability is, both for own and opposing forces. The operational tool could show the current DRI range capability based at least partly on local, in situ, measurements of the prevailing conditions. However, the in situ measurements would likely have to be supplemented with relevant parameters that have been predicted using models or measured by remote sensing e.g. weather satellite observations. The greatest value would be achieved through dynamic integration of in situ measurements, model outputs and real-time or near real-time satellite observations.

In the sections to come, the image quality metric used for DRI predictions will be explored followed by a description of a typical Monte Carlo simulation run. Initial results and concluding remarks round off the paper.

2. Johnson's Criteria and Targeting Task Performance Metric

The probability of DRI for a particular target set can be quantified in terms of a threshold image quality metric. One of the oldest sets of such metrics is known as "Johnson's criteria" [1], which simply states that the surveillance system must resolve a certain number of spatial cycles over a certain critical target spatial dimension in order to execute the task with 50% probability of success. While useful as a ballpark estimate, there are many deficiencies in the original method and many improvements to surveillance tasking probability prediction have arisen. One of the latest image quality metrics for DRI that has been published is known as the Targeting Task Performance (TTP) metric [2]. The TTP metric is similar to Johnson's criteria in that thresholds for task performance, denoted V^{50} for a particular, target set are determined, preferably by experiment. These V^{50} thresholds, which essentially express the task difficulty, can then be used to predict performance in other conditions and with other sensors operating in the same spectral band. The equivalent thresholds for Johnson's criteria are denoted N^{50}.

A number of published studies have been consulted [2], [3] and [4] in order to establish reasonable V^{50} task cycle criteria.

3. Statistical Modelling

The surveillance problem is usefully conceived as a chain of elements from light source (chiefly the sun in the shortwave spectrum) via atmosphere and target, then back through the atmosphere to the surveillance sensor and onwards ultimately to the human observer. Each element in the surveillance chain generates spectral, spatial and temporal modulations in the optical signal received by the observer. For sake of brevity only some of the considerations for each of these elements are discussed below.

3.1. Target

The high fidelity implementation of target generation in MORTICIA utilises 3D synthetic scene generation using hyperspectral or band Radiance Environment Maps (REM) and raytracing of a target model. This is achieved primarily through the use of libRadtran [5], a collection of tools for atmospheric radiative transfer computation.

The default synthetic target used for modelling efforts is a lambertian sphere or ellipsoid with latitude and longitude spokes, creating what is essentially a 3D Siemens star, referred to in prose as a "beachball" target. An example of which is shown below in Figure 1.

Figure 1. Rendered Image of Beachball Target

The "beachball" target has been selected as a generic target that is fast and easy to render and whose only definitive properties in pure lambertian form are the spectral reflectance values for the light and dark sector spokes and the number of light/dark spoke pairs. For realistic simulations, the modelled diameter of the sphere as well as the mean and modulation of spectral reflectance must be representative of the target or set of targets for which it is to be a generic surrogate.

3.2. Background

Currently MORTICIA only caters for vegetative backgrounds. This model assumes a uniform background, but future efforts plan to use a more comprehensive approach such as the synthesising of backgrounds using statistical methods, further even, would be to use aerial photography of the deployment theatre. Since detection depends quite strongly on background spatial structure, the detection thresholds alluded to in the section on performance metrics are regarded as the least reliable of the DRI metrics. The field of view of the sensors utilised for these applications is assumed sufficiently narrow that the background has no spatial variation carried over from the REM either.

3.3. Environment

The emphasis here is on geophysical environmental variables that have most impact on the radiant environment i.e. the amount of light traveling in different directions at different wavelengths (spectral radiance).

3.4. Atmospheric Profile

It is necessary to specify the atmospheric profile with respect to height above ground as an input to the RTC. The libRadtran suite provides a number of standard profiles as well the option to customise a profile using site specific aerosol data.

3.5. Clouds

The latest version of MORTICIA makes use of only a single water cloud model and vertical distribution. This is potentially a significant shortcoming, but the real impact on DRI has not been assessed. Only the Cloud Optical Depth (COD) and Cloud Cover Fraction (CCF) can be manipulated in the simulation, while the vertical distribution and effective droplet remain fixed for all MC sample points.

3.6. Aerosols

Cumulative frequency distributions for total Aerosol Optical Depth (AOD) and aerosol phase function data representative of the theatre of interest are required for modelling the effects of aerosols on overall image quality. Through narrow-angle scatter, aerosols introduce blur into the image. The other effects of aerosols are to scatter and absorb desired photons from the target and to scatter undesirable photons into the sensor Field of View (FOV), referred to as path radiance. For longer ranges and higher scattering optical depth, exacerbated by high absorption optical depth, all of which tend to go together, aerosol Modulation Transfer Function (MTF) can become a significant or even dominant contributor to the overall atmospheric MTF [7]. The model used in this work does not currently have an aerosol MTF component and this is possibly a significant shortcoming.

3.7. Atmospheric Turbulence

With respect to optical imaging along sightlines of substantial length passing through the atmospheric surface layer, atmospheric turbulence is quite often the limiting factor on DRI range performance. Atmospheric turbulence is simulated based on a peak C_n^2 value at a user-defined reference height above ground, where C_n^2 is a measure of the variation in refractive index between two points in space. The time dependent C_n^2 at the reference height is computed based on the empirically observed variation between the times of sunrise and sunset. The variation of C_n^2 with height is based on a simple power law decay. The coherence diameter and corresponding Fried parameter for each sightline zenith angle at a specific time of day is then computed by performing a simple integration of C_n^2 over the height from target to sensor. With growing interest in elevated surveillance platforms, such as aerostats and Unmanned Aerial Vehicles (UAV), the reliable knowledge of vertical distribution of turbulence and aerosols becomes a vital consideration. This is elaborated on in work by Sprung et al.[8].

3.8. Sensor and Platform

Once the at-target radiance is known as well as the atmospheric path (sightline between sensor and target) transmittance, radiance and Fried coherence length, the at-sensor radiance and path MTF can be computed. What remains is to compute the image data from the sensor based on knowledge of the optics, detector and platform induced vibration and smear. These tasks are performed by the python translated Basic Optical Sensor Model (BOSM), code based on a Swedish Defense Research Agency (FOI) report by Kopeika [7] and Winzell [9] amongst others.

4. Implementing the simulation

The implementation described here has adopted the statistical approach in which data sources for a number of key theatre environment variables have been exploited. The surveillance theatre is defined by a single quadrilateral for which adequate sampling of climatological conditions is required. The overall strategy is to sample the climatology at a random day/time point and compute a REM for every date/time sample inside the simulation loop. The MC loop process runs through each of the engineering options and computes DRI metrics for samples of the environment covering the broadest possible range of system operating conditions. The loop continues until the task probability is 50%. At this point the ground range for the viewing zenith angle is computed and logged together with all MC sample data and relevant metadata.

5. Applications

MORTICIA is primarily concerned with determining the effectiveness of a specific surveillance system within a particular environment. However, once effectiveness modelling is in place, it becomes possible to do simple trade-off studies i.e. deciding which surveillance system from a limited choice of possibilities will be more effective in a particular application. Taking this one step further opens the possibility of performing optimization studies.

6. Results

The main results of the Monte Carlo simulation are written to a table containing all or most parameters on which results may be aggregated for graphical display or further statistical analysis. One of the most important result summaries is the polar DRI range plot. This polar graph shows the threshold (50% probability of success) ground range for the targets and tasks defined for the simulation as a function of Solar relative View Azimuth Angle (SVAZ). Results can also be also be aggregated over true azimuth (relative to north) or with respect to any other input or output model parameter. This allows for exploration of possible relationships and mining for insights contained within the model results.

Figure 2 below provides typical range plots for a task in which six separate sensors were required to detect a human target from a platform height of 500m above ground level. Relative daytime is used where 0 is sunrise and 1 is sunset. Local noon is at relative daytime of 0.5. The plots are all relative to solar azimuth, where looking into the sun is azimuth 90° and looking away from the sun is azimuth 270°. Threshold range is seen to be reduced when viewing into the solar azimuth, particularly early and late in the day when the sun is low, and atmospheric turbulence is reduced. This is due to a reduction in apparent target contrast because of the scattering of sunlight into the sensor FOV.

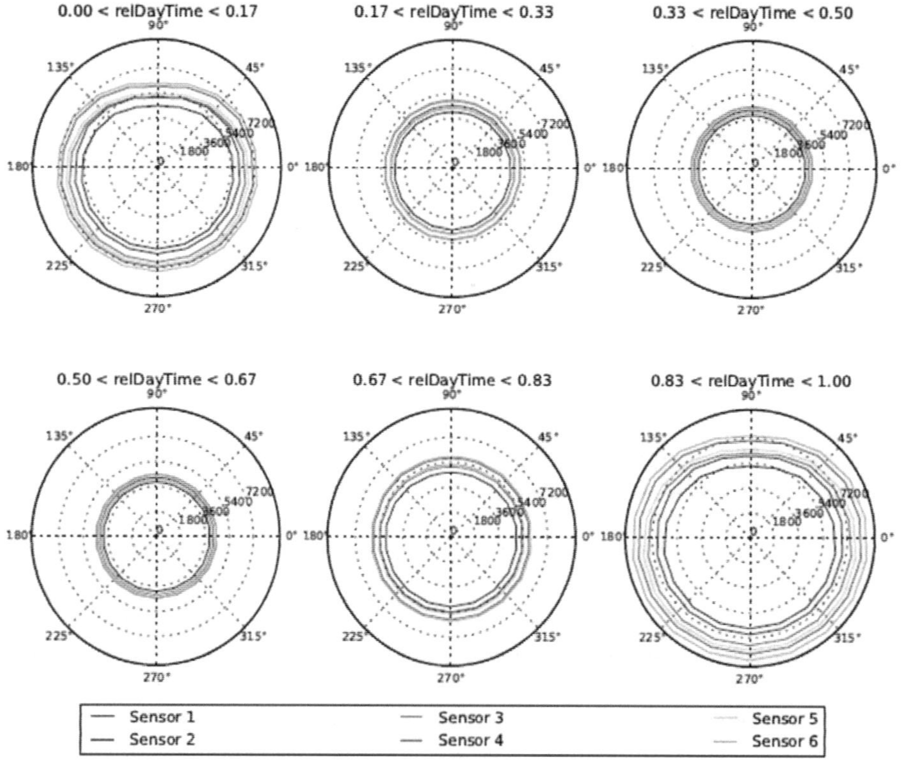

Figure 2 Ground Range Threshold for Target Task at Platform Height of 500m

IOP Conf. Series: Journal of Physics: Conf. Series **905** (2017) 012026 doi:10.1088/1742-6596/905/1/012026

7. Conclusion

While surveillance modelling in general has a rich body of literature, there is little coverage of the statistical approach attempted with MORTICIA. The proof of concept of statistical, theatre-orientated surveillance modelling has been successfully demonstrated in this study. A number of refinements and optimisations to the code base are underway. In addition, validation exercises are planned for the immediate future.

References

[1] Johnson J 1985 Selected papers on infrared design part I and II *Proc. SPIE. Conf. on Analysis of Image Forming Systems (San Diego)* vol 513, ed Johnson R B and Wolfe W L p 761

[2] Vollmerhausen R H and Jacobs E 2004 The targeting task performance (TTP) metric, a new model for predicting target acquisition performance *Modeling and Simulation Division, Night Vision and Electronic Sensors Directorate, United States Army Laboratory Report* p 50-2

[3] Hixson J G, Jacobs E L and Vollmerhausen R H 2004 Target detection cycle criteria when using the targeting task performance metric *Proc. SPIE, Electro-Optical and Infrared Systems: Technology and Applications (London)* vol 5612, ed Driggers R G and Huckridge D A p 275–83

[4] Teaney B P, Du Bosq T W, Reynolds J P, Thompson R, Aghera S, Moyer S K, Flug E, Espinola R and Hixson J 2012, Human target acquisition performance *Proc. SPIE. Infrared Imaging Systems: Design, Analysis, Modeling, and Testing (Baltimore)* vol 8355, ed Holst G C and Krapels K A p 510–14

[5] O'Connor J D, O'Shea P, Palmer J E, and Deaver D M 2006 Standard target sets for field sensor performance measurements *Proc. SPIE. Infrared Imaging Systems: Design, Analysis, Modeling, and Testing (Kissimmee)* vol 6207, ed Holst G C p 0U–7

[6] Mayer B and Kylling 2005 A Technical note: The libRadtran software package for radiative transfer calculations - description and examples of use *Atmos. Chem. Phys.* **5** p 1855-77

[7] Kopeika N S 1998 *A System Engineering Approach to Imaging (Press Monographs, SPIE Optical Engineering Press)* chapter 15 p 487-514

[8] Sprung D, Sucher E, Ramkilowan A and Griffith D J 2014 Investigation of optical turbulence in the atmospheric surface layer using scintillometer measurements along a slant path and comparison to ultrasonic anemometer measurements *Proc. SPIE. Remote Sensing of Clouds and the Atmosphere (Amsterdam)* vol 9242, ed Comeron A et al p 1I–12

[9] Winzell T 2006 *Basic Optical Sensor Model (Swedish Defence Research Agency)* p 1-38

IOP Conf. Series: Journal of Physics: Conf. Series **905** (2017) 012027 doi:10.1088/1742-6596/905/1/012027

A fast - Monte Carlo toolkit on GPU for treatment plan dose recalculation in proton therapy

M Senzacqua[1], A Schiavi[1], V Patera[1], S Pioli[2], G Battistoni[3], M Ciocca[4], A Mairani[4], G Magro[4] and S Molinelli[4].

[1] Sapienza - Università di Roma, Department of Basic and Applied Sciences for Engineering (SBAI), Via A. Scarpa 14, 00161 Roma, Italy
[2] INFN - Laboratori Nazionali di Frascati, Via Enrico Fermi 40, 00044 Frascati (Roma), Italy
[3] INFN - Sezione di Milano, Via Celoria, 16, 20133 Milano, Italy
[4] Fondazione CNAO, Strada Campeggi, 53 - 27100 Pavia, Italy

E-mail: martina.senzacqua@uniroma1.it

Abstract. In the context of the particle therapy a crucial role is played by Treatment Planning Systems (TPSs), tools aimed to compute and optimize the tratment plan. Nowadays one of the major issues related to the TPS in particle therapy is the large CPU time needed. We developed a software toolkit (FRED) for reducing dose recalculation time by exploiting Graphics Processing Units (GPU) hardware. Thanks to their high parallelization capability, GPUs significantly reduce the computation time, up to factor 100 respect to a standard CPU running software. The transport of proton beams in the patient is accurately described through Monte Carlo methods. Physical processes reproduced are: Multiple Coulomb Scattering, energy straggling and nuclear interactions of protons with the main nuclei composing the biological tissues. FRED toolkit does not rely on the water equivalent translation of tissues, but exploits the Computed Tomography anatomical information by reconstructing and simulating the atomic composition of each crossed tissue. FRED can be used as an efficient tool for dose recalculation, on the day of the treatment. In fact it can provide in about one minute on standard hardware the dose map obtained combining the treatment plan, earlier computed by the TPS, and the current patient anatomic arrangement.

1. Introduction

Charged Particle Therapy (CPT) is a radiation tumor treatment technique that uses protons or light ions. It aims to deliver a high precision treatment of solid tumors by exploiting the characteristic shape of the Bragg curve of charged hadrons: the dose deposition as a function of depth of traversed matter exhibits a sharp peak (the Bragg peak) at the end of the particle range. Thanks to its dose release profile accurate and efficient irradiation of the tumor can be obtained reducing the dose to the surrounding healthy tissue, thus achieving less post-irradiation complication probability with respect to the standard X-ray radiotherapy. According to recent statistics [1], at the end of 2014 more than 137,000 patients have been now treated worldwide with charged hadrons (about 86% with protons). The number of clinical centers dedicated to CPT currently in operation is 63 but by the end of 2018, 17 new proton therapy facilities will start to treat patients. The high spatial selectivity of CPT asks for stringent requirements on the accuracy that has to be achieved. The uncertainty on the position of the dose release in treatments can be due to different factors, such as uncertainties in particle range, the calibration of the Computed Tomography (CT) images and also patient mis-positioning and organ motion during the treatment itself. Furthermore, it has to be considered that morphological changes may occurr between the CT and each of the several irradiation sessions of a CPT treatment, operated in different days. All the effects mentioned can contribute to a total uncertainty of the order of few millimeters. In order to preserve the intrinsic advantages of hadrontherapy, fast and accurate dose calculation tools are necessary to check, verify, and eventually correct, initial treatment planning. Softwares dedicated to the dose optimization and calculation are called Treatment Planning System (TPS) and different calculation algorithms are used and still in progress of study to improve performances both in terms of accuracy and time computing. The two main approaches of computation methods are Monte Carlo (MC) techinque and analythical pencil beam algorithm. The first method has high accuracy in dose calculation but the simulation time needed to obtain adequate statistics is still too long to permit use in clinical routine. The analythical TPS are quicker. The aim of the project FRED is to build an innovative TPS tool using Monte Carlo methods for simulation of proton energy deposition in tissues and exploiting the mosta advanced GPU devices. The implementation of Multiple Coulomb Scattering (MCS), energy fluctuation and nuclear interaction is described. In particular, we present the implementation of approximated algorithms that allow accurate and fast evaluation of theoretical models, whose exact resolution is a time consuming task. Results about linear dose profiles are shown in comparison with full MC code FLUKA [2]. Finally, performances in terms of time computing using different hardware architecture and software models are presented.

2. Transport methods

The transport of protons in tissues is simulated by a condensed-history Monte Carlo method. The patient region of interest is imaged by a Computed Tomography (CT) diagnostic test. The volume is divided into small parallelepipeds- or *voxels*- considered uniform in atomic composition and density. In crossing materials, each primary proton undergoes elastic and inelastic collisions with nuclei and electrons, as well as nuclear interactions that make it lose energy and change direction.

The particle is tracked from the source to the end of its range. The track is obtained as the sum of steps whose length is limited by different criteria:

- geometric limitator: each step can not cross the border of two adjacent voxels so the maximum length walkable is the distance from the particle position to the voxel border;
- energy limitator: the energy loss for each step can not exceed the 1% of the particle energy, in order to allow the approximation of constant mean energy loss along a single step;
- discrete event limitator: if along the step a nuclear inelastic interaction occurs, the end of the step is set to the interaction point.

The mean energy loss is computed following the method of *Fippel and Soukup* [3]. The ratio between the stopping power in the tissue S and in water S_w is expressed as a function of the proton kinetic energy T_p and the material mass density ρ. S_w is obtained from stopping power tables PSTAR[4] and it depends on T_p and the step length δz.

Energy straggling due to statistical fluctuations of number of collisons suffered by protons and of the energy transferred in each collision is reproduced using two matching approximations. In thick absorber regime the number of ionization events is high, and the distribution of energy loss can be considered gaussian. The standard deviation is computed through the following formula[5]:

$$\Omega^2 = 0.0855\rho\delta z\frac{T_{\max}}{\beta^2}\left(1-\frac{1}{2}\beta^2\right)\,\text{MeV} \tag{1}$$

IOP Conf. Series: Journal of Physics: Conf. Series **905** (2017) 012027 doi:10.1088/1742-6596/905/1/012027

Where $T_{\max} = \frac{2m_e\beta^2\gamma^2}{1+2\gamma m_e/m_p+(m_e/m_p)^2}$ is the maximum energy transferable in a collision to a single electron, where m_e and m_p are the mass of the electron and of the primary respectively. In case of very thin absorber, the distribution the distribution of energy loss is not gaussian anymore. The theory in thin absorber regime has been developed by Landau first[6] and then Vavilov[7]. The computation of analytic Vavilov solution is extremly time-consuming, for this reason we implemented our original algorithm to approximate the distribution. The method consists of fitting the Landau function with a logarithmic normal distribution L_N [1]. The parameters σ, θ and m are expressed as a function of the energy of the particle and the material properties ρ and δz. The logarithmic-normal approximation for thin absorbers shows a remarkable efficiency, in fact the accuracy in energy loss distribution reproduction is good and the sampling time is comparable to the gaussian thick aborbers.
For what concerns the Multiple Coulomb Scattering (MCS) the assumption of small angle approximation is done. This means that the angle θ between the direction of the proton before and after the step is considered small and can be computed through the quadratic sum of the projected angles θ_x and θ_y ($\theta = \sqrt{\theta_x^2 + \theta_y^2}$). The distribution of the projected angles is fitted with three different functions that can be set by the user according to his needs of accuracy and time costs. The three functions are, in increasing order of both accuracy and calculation time:

(i) Single gaussian approximation. The standard deviation of the distribution is computed using the Highland forumla[8]. This model well reproduces the core of the distribution, but it misses the tails due to large angle scattering;

(ii) Double gaussian approximation. A second gaussian is superimposed to the core one to cover partially the tails of the distribution. The standard deviation of the second broader gaussian is obtained by fitting histograms generated with FLUKA, and then indexed as a function of the particle energy and of the step length.

(iii) Gauss-Rutherford-like approximation. The name of the third model is due to the expression of the distribution superimposed to the core gaussian that is similar to the Rutherford hyperbola. The function is:

$$f_{GR}(\theta_{x,y}) = \frac{1-w}{\sqrt{2\pi}\sigma_1} \exp\left[-\frac{1}{2}\left(\frac{\theta_{x,y}}{\sigma_1}\right)^2\right] + w\frac{a}{(\theta_{x,y}+b)^c} \quad , \quad c \sim 2.0$$

where, $0 \le w \le 1$ is the weight of the Rutherford-like component, σ_1 is the Highland standard deviation and a, b and c are the parameters of the tail function. This model allows to reach an accuracy of the order of 0.001%, within 4 standard deviations of the scattering angle θ distribution.

The nuclear interactions are simulated using the cross sections provided by ICRU Report 63[9]. The atomic composition is recreated by using the conversion table provided by *Schneider et al.*[10] that indicates the atomic composition of biological tissues as a function of the Hounsfield Number. The nuclear elastic and inelastic cross sections are computed for each step. When a nuclear reaction occurs, the type, the multiplicity and the energy of secondary particles produced are extracted. Secondaries are treated differently according to their type:

- secondary protons and deuterons are tracked;
- neutrons are neglected and their energy is considered lost outside the region of interest;
- heavier particles such as alpha and light ions are stopped immediatly and their energy deposited in the voxel where the production occurs;

[1] $L_n(\lambda_v) = \frac{1}{(\lambda_v-\theta)\sqrt{2\pi}\sigma} \exp -\frac{(\ln(\frac{\lambda_v-\theta}{m}))}{2\sigma^2}$, with λ_v the Vavilov variable (see [7]) and σ, θ and m are the parameters of the function

3. Results

In figure 1 we show histograms of linear dose profiles for a pencil beam of 150 MeV protons in water, with an initial FWHM set to 0. FRED results (red line) are compared with Full Monte-Carlo FLUKA simulations. Particular attention has to be paid to the lateral dose profiles on the right. They refer to the dose distribution at a depth 90% of the BP position, that means around 14 cm in the water phantom. At that point, the beam lateral distribution is strongly changed with respect to the initial configuration and conditioned by all the process suffered by the particles. As it can be observed, even at this critical position there is a good agreement between FLUKA and FRED histograms. Also the tails are well reproduced, up to 3 orders of magnitude.

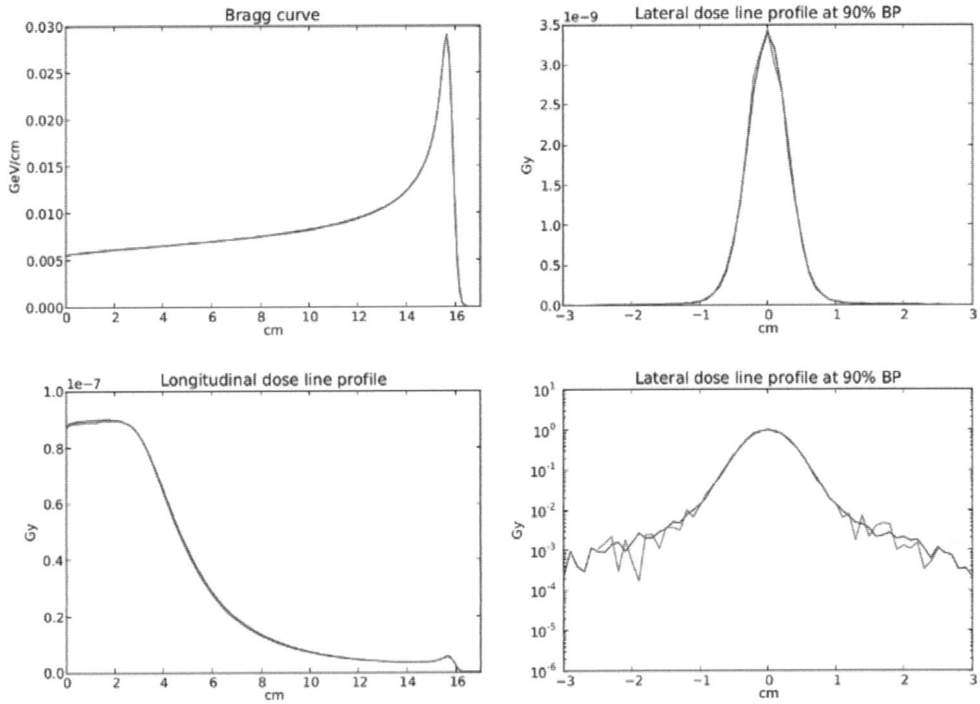

Figure 1. Linear dose profiles of a 150 MeV point section (FWHM=0.0) proton beam in water. In blue full-Monte Carlo data, in red FRED histograms. Upper left: longitudinal dose profile, also called *Bragg Peak*. On the bottom left: dose release inside the central voxels in the longitudinal direction. The decreasing trend of the longitudinal dose line profile testifies the emptying of the central voxels, due to scattering events that spread the beam. At the end of the path is visible a small peak representing the Bragg Peak. On the right: lateral dose profile about 1.5 cm before the Bragg Peak, both in linear (upper) and logarithmic (bottom) scale.

4. Time performances

In table 4 time performances for different architectures are displayed and compared with the time consumption of a simulation performed with the full Monte-Carlo tool FLUKA. As it can be observed, in the same running modality, FRED is 20 times faster than FLUKA on the same CPU hardware. This is due to the simplification of physics models implemented, and that's where the denomination *fast Monte-Carlo* comes from. The non-linear scaling of time with GPU number is due to the different types of GPU employed. As shown, running on GPU even

		Threads	primary/s	$\mu s/primary$
CPU	FLUKA	1	0.75K	1340
	FRED	1	15 K	68
	FRED	16	48 K	21
	FRED	32	80 K	12.5
GPU	FRED	1 GPU[1]	800 K	1.35
	FRED	2 GPU[2]	3500 K	0.3
	FRED	4 GPU[3]	20000 K	0.05

Table 1. Computing times for different hardware architectures.
[1] LAPTOP: MacBookPro(AMD Radeon R9 M370X).
[2] DESKTOP: Mac Pro (AMD FirePro D300).
[3] LINUX WorkStation with 4 NVIDIA GTX 980 GPUs

with a standard laptop the gain in terms of time is about three orders of magnitude with respect the full Monte Carlo and about 50 with respect to FRED single CPU modality.

5. Conclusions

The fast-Monte Carlo FRED showed a good agreement with full-Monte Carlo simulations in the computation of the dose distribution originated by proton beams passing through biological tissues. At the same time, the employement of GPU devices allows to significantly reduce the computation time. Thanks to its performances in terms of accuracy and calculation time, FRED represents an efficient tool that can be used in clinical routine for the recalculation of the treatment plan needed during the validation protocol. Furthermore, the short times implied in computation could allow the use of FRED in online dose monitoring techniques. By detecting the secondary products, in facts, the distal point of a treatment beam could be inferred. By taking into account the interactions undergone by the proton in the exit path using the information from a charged particle tracker detector and patient anatomic information available from a Computed Tomography scan, FRED can significantly improve the accuracy of the online dose monitoring device.

References

[1] http://www.ptcog.ch
[2] Böhlen T T, Cerutti F, Chin M P W, Fass A, Ferrari A, Ortega P G, Mairani A, Sala P R, Smirnov G and Vlachoudis V 2014 *Nuclear Data Sheets* **120** 211-14
[3] Fippel M and Soukup M 2004 *Med. Phys.* **31**(8) 2263-73
[4] Berger M J 1993 MD *Techn. Rep. NBSIR 4999 Nat. Inst. of Stand. and Techn.*
[5] Fippel M 2006 *New Techn. in Rad. Onc.* 197-206
[6] Landau L 1944 *J. Phys. (USRR)* **8** 201
[7] Vavilov P V 1957 *Sov. Phys. JETP* **5** 749
[8] Highland V L 1075 *Nucl. Instr. and Meth.* **129** 497-99
[9] 2000 *ICRU Report-63. Nuclear data for neutron and proton radiotherapy and radiation protection.*
[10] Schneider U, Pedroni E and Lomax A 1996 *Phys. Med. Biol* **41** 111-124

IOP Conf. Series: Journal of Physics: Conf. Series **905** (2017) 012028 doi:10.1088/1742-6596/905/1/012028

Monte Carlo computation of the effective Sherman function

M Drągowski[1], M Włodarczyk[1], J Ciborowski[1], G Weber[2], J Enders[3], Y Fritzsche[3] and A Poliszczuk[1]

[1] University of Warsaw, Faculty of Physics, Pasteura 5, 02093 Warsaw, Poland
[2] Helmholtz Institut Jena, Fröbelstieg 3, 07743 Jena, Germany
[3] Technische Universität Darmstadt, Institut für Kernphysik, Schlossgartenstraße 9, 64289 Darmstadt, Germany

E-mail: mdragowski@fuw.edu.pl

Abstract. The PEBSI Monte Carlo code was recently upgraded towards usefulness to compute the effective Sherman function, which is used in electron Mott polarimetry. The results of the simulation were compared to experimental data for a 100 keV polarized electron beam. The properties of the effective Sherman function, as well as the problem of experiment optimization were analyzed using the data from the simulation.

1. Introduction

The method of Mott polarimetry is used to determine the spin–polarization of electron beams, taking advantage of the dependence of Mott scattering cross section on spin direction due to the spin–orbit interaction. It is used in a wide range of electron energies (from keV up to several MeV), both in basic research, as well as numerous applications (e.g., spin–polarized scanning electron microscopy and spin–resolved photoemission spectroscopy).

The method is based on measuring the azimuthal asymmetry[1] in Mott scattering of a polarized electron beam on thin targets made of heavy elements (e.g., Au). The degree of the observed asymmetry is related to the Sherman function, dependent on scattering angle and beam energy.

However, the pure Sherman function, which may be calculated theoretically, corresponds to scattering on a single atom. On the other hand, the measurement is performed on targets of finite thicknesses. In this case the effective Sherman function has to be used, in which multiple interactions of the electron with several target atoms are accounted for. These effects cannot be calculated analytically and simulation tools must be used for modeling the passage of polarized electrons through matter. Other background processes changing the polarization state of the electrons also take place (in particular Møller scattering and bremsstrahlung emission), leading to the decrease of the analyzing power of the polarimeter.

Furthermore, experimental data covering a wide range of scattering angles exist for a few energies only[2]. Therefore, simulation seems to be a natural choice for a systematic study of the

[1] i.e., with detectors placed symmetrically about the beam axis at a given angle with respect to the beam direction
[2] the construction of a typical Mott polarimeter does not allow for convenient change of the angle

IOP Conf. Series: Journal of Physics: Conf. Series **905** (2017) 012028 doi:10.1088/1742-6596/905/1/012028

properties of the Sherman function (especially in energy ranges where few or no experimental data exist) and is beneficial because of the fact that it allows to investigate additional effects besides the value of Sherman function and angular distribution of scattered electrons (e.g., sources of background, optimal detector cuts etc.).

The angular and energy distributions of scattered electrons were studied, both for all particles and based on interaction types which took place in the target. Electrons originating from Møller scattering were found to have important impact on the effective Sherman function if low energy particles are taken into account, thus leading to significant decrease of the analyzing power of the polarimeter. Furthermore, simple dependence on target thickness is not observed anymore in this case. This effect can be avoided by imposing proper detector cuts on electron energy.

Optimizing the measurement comes down to choosing the optimal scattering angle and target thickness for electrons of a given energy. Therefore, the angular dependence of the Sherman function was analyzed from this perspective. Simulation allows to exclude regions less suitable for measurement (e.g., where the Sherman function undergoes rapid changes with scattering angle[3]). The dependence of the Sherman function on target thickness (for a given scattering angle) was also analyzed and compared to commonly used parametrizations.

2. Interactions of electrons with matter

2.1. Mott scattering

The cross section for Mott scattering (i.e., the elastic scattering of electrons on atomic nuclei) depends on electron polarization according to the following relation [1]:

$$\frac{\mathrm{d}\sigma_{\mathrm{Mott}}}{\mathrm{d}\Omega} = \left(\frac{\mathrm{d}\sigma_{\mathrm{Mott}}}{\mathrm{d}\Omega}\right)_0 (1 + S(\theta)\vec{P}\cdot\vec{n}),$$

where $(\mathrm{d}\sigma_{\mathrm{Mott}}/\mathrm{d}\Omega)_0$ is the cross section corresponding to scattering of unpolarized electrons, $S(\theta)$ is the Sherman function, θ is the scattering angle, $\vec{P}$ is the polarization vector of the incident electron and $\vec{n}$ is the unit vector normal to the scattering plane.

The Mott scattering cross section is calculated from the Dirac equation, taking into account the spin–orbit interaction (contrary to the Rutherford cross section which neglects this effect). The numerical values depend on the atomic potential assumed in the calculations; at low energies and low scattering angles the screening of the nuclear Coulomb potential by the electron cloud cannot be neglected (as in the Rutherford formula assuming pure Coulomb field of the nucleus).

Experiments are carried out using detectors placed symmetrically about the beam axis at a given angle θ with respect to the beam direction. The measurement is typically performed in backscattering, as the largest effect (S) is obtained for large scattering angles. The quantity of interest is the azimuthal (left–right) asymmetry in the number of electrons detected in the two detectors, A_{LR}, which allows to calculate the polarization of the electron beam according to the following formula:

$$\vec{P}\vec{n} = \frac{A_{LR}}{S}.$$

In practice the theoretical value of S is replaced with its effective value S_{eff} in the above equation. It is worth stressing that this method is sensitive only to the spin projection on the direction perpendicular to the scattering plane.

Both the unpolarized cross section and the Sherman function are defined by two scattering amplitudes, $f(\theta)$ – spin–conserving amplitude and $g(\theta)$ – spin–flip amplitude, according to the following equations [2]:

$$\left(\frac{\mathrm{d}\sigma_{\mathrm{Mott}}}{\mathrm{d}\Omega}\right)_0 = |f(\theta)|^2 + |g(\theta)|^2,$$

[3] this could not be taken into account in experiments in which the Sherman function was determined by a single measurement

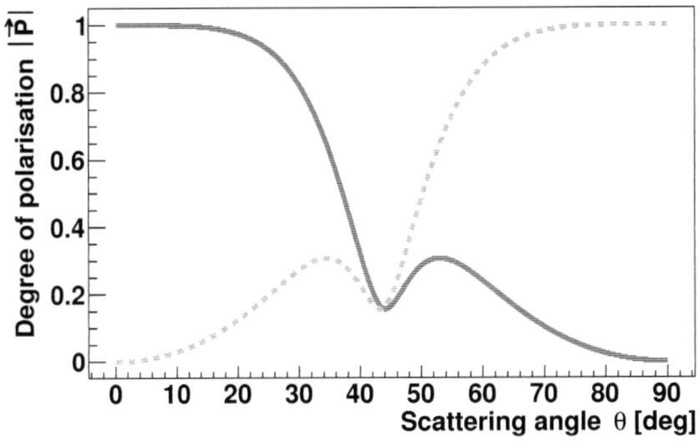

Figure 1. Transfer of polarization from a 100 keV, 100% polarized beam electron to the secondary Møller electrons; dependence of the length of the polarization vector on the scattering angle θ for two electrons in the final state (θ is measured for the electron whose polarization is plotted with the solid line).

$$S(\theta) = i\frac{fg^* - f^*g}{|f(\theta)|^2 + |g(\theta)|^2}.$$

Furthermore, the polarization vector of the electron after the scattering, $\vec{P'}$, is expressed by the initial polarization, $\vec{P}$, Sherman function $S(\theta)$ and two additional functions, $T(\theta)$ and $U(\theta)$ [2]:

$$\vec{P'} = \frac{(\vec{P}\cdot\vec{n} + S(\theta))\vec{n} + T(\theta)\vec{n}\times(\vec{n}\times\vec{P}) + U(\theta)(\vec{n}\times\vec{P})}{1 + S(\theta)\vec{P}\cdot\vec{n}},$$

which are also defined by the scattering amplitudes:

$$T(\theta) = \frac{|f(\theta)|^2 - |g(\theta)|^2}{|f(\theta)|^2 + |g(\theta)|^2},$$

$$U(\theta) = \frac{fg^* + f^*g}{|f(\theta)|^2 + |g(\theta)|^2}.$$

The S, T and U functions have the following interpretation: (i) S describes the change of the polarization vector component perpendicular to the scattering plane, (ii) T describes the change of the polarization vector component parallel to the scattering plane, and (iii) U describes the polarization appearing in the scattering plane, but perpendicular to the initial polarization vector.

2.2. Møller scattering

Møller scattering (i.e., the elastic scattering of two electrons) is the second most important process if the cross sections are considered. Its significance follows also from the fact, that it results in a relatively large change of both the momentum and the polarization of the interacting electron.

Interestingly, the final spin states of individual Møller electrons after the scattering are not unambiguously determined, as the particles may be entangled [3]. It is, however, possible to calculate the mean polarization vectors from the spin density matrix of the final state.

In a typical experimental case when the beam electron is highly polarized and the target electron is unpolarized, the two electrons in the final state share the initial polarization (i.e., they are only partly polarized). This effect is illustrated quantitatively in Fig. 1, in which the dependence of the degree of polarization (the length of the polarization vector) is shown for both electrons in the final state as a function of the scattering angle. It can be seen that in case of symmetric scattering both electrons have the same polarization (due to their indistinguishability), while in general one of the electrons inherits majority of the initial polarization. The electron with higher energy is always more polarized.

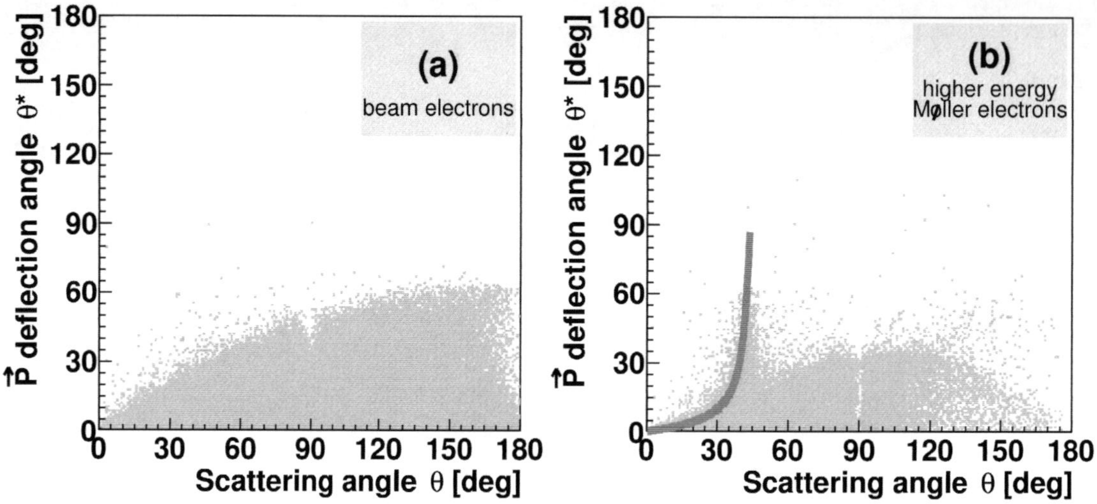

Figure 2. Distribution of simulated events as a function of the angles θ^* and θ (between the initial and final polarization and the initial and final momentum vectors, respectively) for a 10 nm target; (a) only Mott–scattered electrons (no energy cut), (b) higher energy Møller electrons (above 50 keV). The curve demonstrates the theoretical prediction for Møller electrons.

2.3. Implementation in PEBSI

The PEBSI Monte Carlo simulation was originally written to simulate bremsstrahlung accompanying passage of polarized electrons through thin target foils [4]. It was recently upgraded towards usefulness to compute the effective Sherman function. The description of Mott scattering was improved and polarization transfer in Møller scattering was included in the code.

The model of Mott scattering is based on the theoretical formalism presented above. However, calculation of the scattering amplitudes requires the use of numerical methods due to the mathematical complexity of the problem; the method was originally proposed by Sherman [5]. In the PEBSI package their values are calculated using the ELSEPA code, in which they are derived from the Dirac equation assuming the Thomas–Fermi–Dirac atomic potential [6].

In order to simulate Møller scattering, analytical formulae derived according to [3] are used both for the cross section and the polarization transfer. Møller scattering regards, however, scattering on (quasi) free electrons only. Therefore, in case when the binding energy of the target electron cannot be neglected compared to the kinetic energy (which is typically the case for electron shells other than the outermost one) a combination of the formula proposed by Bote et al. [7] and the MRBEB model by Guerra et al. [8] was used to simulate electron impact ionization.

Bremsstrahlung is the least significant process to be considered, as the emission of a bremsstrahlung photon is unlikely to significantly change the polarization state of the electron. Therefore, this effect was neglected in the present work.

3. Results

3.1. Depolarization

Accounting for the polarization change caused by Møller scattering in the PEBSI code allowed us to observe interesting effects related to this interaction. The dependence between the change of polarization and momentum directions is illustrated in the scatter plots in Fig. 2; θ^* denotes

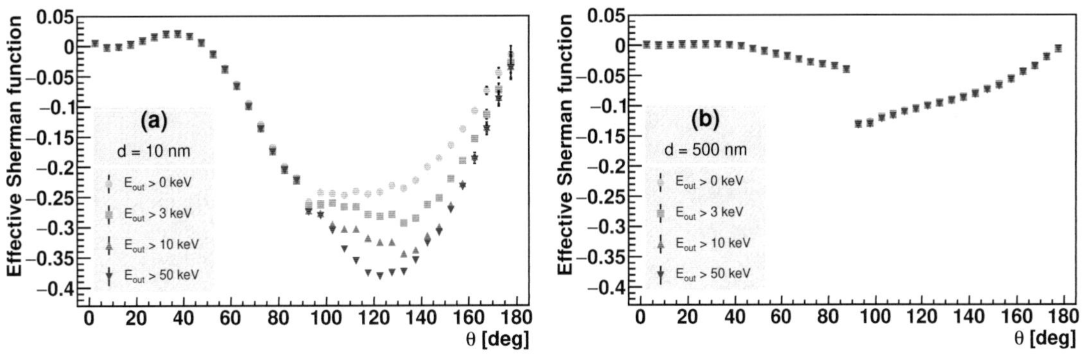

Figure 3. Simulated effective Sherman function; effect of the low energy cuts on electron energy for 100 keV electrons impinging on 10 and 500 nm Au targets.

the angle between the initial and final polarization vectors, θ – between the initial and final momentum vectors (which is equivalent to the scattering angle[4]). The plots were made separately for electrons that did not undergo Møller scattering and for more energetic Møller electrons (above 50 keV in case of 100 keV incident beam). It can be seen that these two classes of events differ from each other, as for Møller electrons one may expect greater change in the polarization direction, even in case of a small change in momentum direction. This result is in agreement with the theoretical model describing Møller scattering; the theoretical prediction was marked in the figure with a solid line.

3.2. Simulated effective Sherman function

The impact of Møller scattering on the effective Sherman function can be analyzed by computing its value with different low energy cuts on the energy of the electrons leaving the target (which corresponds to the detector cuts determining the energy of the electrons recorded in the experiment). The resulting dependences of the effective Sherman function on the scattering angle, plotted for several values of the energy cuts, are shown in Fig. 3 for two different target thicknesses (10 and 500 nm). It can be noticed that in case of the thin target the low energy electrons significantly affect the value of the effective Sherman function, leading to a large decrease of the analyzing power of the polarimeter. This effect decreases with increasing target thickness and is no longer observed in case of the thickest target. Nevertheless, this phenomenon is strictly related to Møller scattering, as by analyzing the energy distributions of the electrons leaving the target we found that electrons with that low values of energy always originate from this interaction.

The reliability of the simulation was proved by comparison with available experimental data for a 100 keV polarized electron beam incident on 10 — 500 nm Au targets, published by Kohashi et al. [9]. The result for a 500 nm target is shown in Fig. 4. In this case the cut on electron energy was chosen as 50 keV, which corresponds to the conditions of the experiment. The comparison was made in a wide scattering angle range (from 20 to 160 degrees) and the results were found in good agreement. Although there is a systematic difference, the agreement can be considered satisfactory given the simulation is based on a purely theoretical model, with no empirical corrections.

[4] here we refer to the effective scattering angle that may be observed in an experiment, as opposed to the theoretical scattering angle in a single scattering event

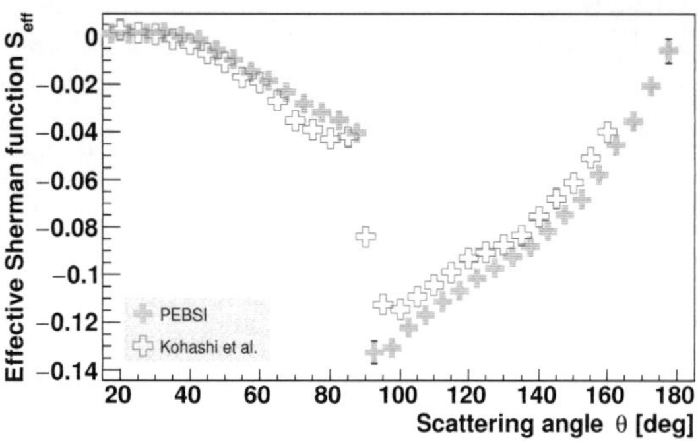

Figure 4. Simulated effective Sherman function; comparison with measurement for 100 keV electrons impinging on a 500 nm Au target.

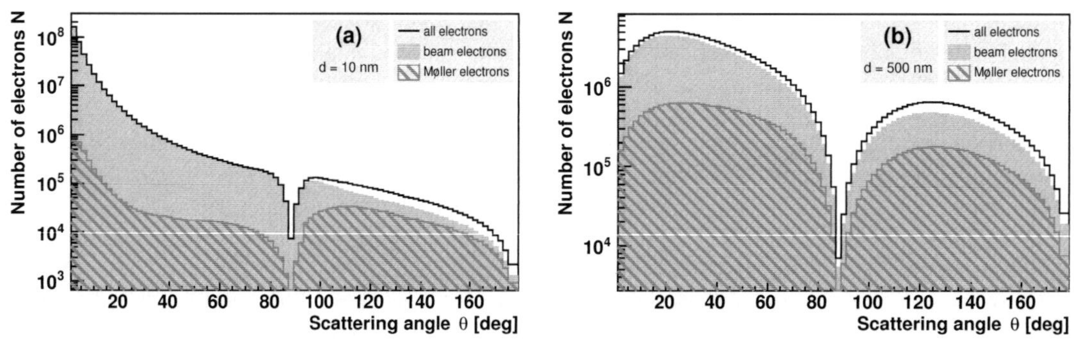

Figure 5. Scattering angle distribution for 100 keV electrons impinging on 10 and 500 nm Au targets.

3.3. Parametrizations

The scope of our study was not limited to the computation of the effective Sherman function for certain conditions. In particular, thanks to the possibility to generate an amount of data corresponding to different target thicknesses, we shortly addressed the problem of experiment optimization and extrapolation procedures for the Sherman function.

Optimal experimental conditions correspond to the maximum of the figure–of–merit, FoM, defined as [9, 10]:

$$\mathrm{FoM}(E, \theta, d) \propto S^2_{\mathrm{eff}}(E, \theta, d)\, N(E, \theta, d),$$

where $N(E, \theta, d)$ is the total number of electrons recorded in the detectors placed at a given angle θ. The form of this function is intended to balance the resolution of the measurement (which is proportional to S_{eff}) and the statistical uncertainty (which is proportional to $\sqrt{N}$). The optimal target thickness corresponds to the maximum of this function for a given energy and scattering angle.

The distribution of the scattering angle changes significantly with the target thickness, which results in a complicated $N(d)$ dependence. This is illustrated in Fig. 5 where the scattering angle distribution is shown for two different target thicknesses (10 and 500 nm). It can be seen that in case of thicker targets electrons are more likely to be scattered at large angles. However, the value of the effective Sherman function decreases with increasing target thickness which makes it necessary to take both effects into account.

The effective Sherman function is often parametrized in terms of target thickness in order to provide a functional form of the figure–of–merit and to interpolate the value of the effective Sherman function between the measured or simulated data points. There are two

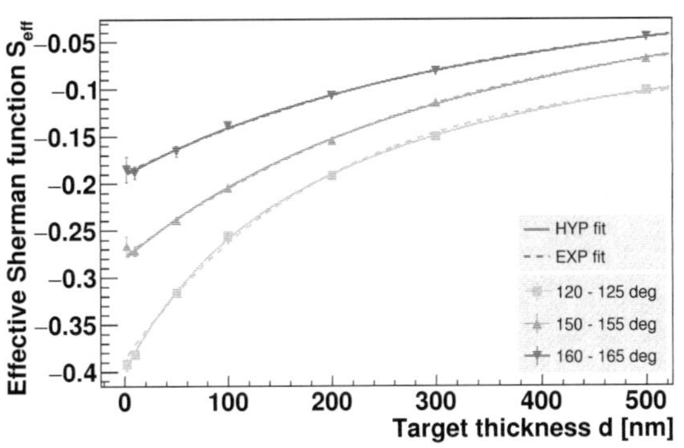

Figure 6. Simulated effective Sherman function; dependence on target thickness for 100 keV electrons impinging on 2 — 500 nm Au targets.

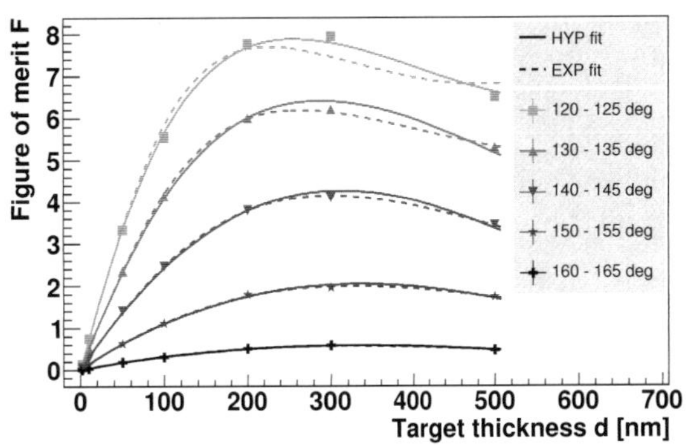

Figure 7. Figure–of–Merit FoM $= S_{eff}^2 N$ for 100 keV electrons impinging on 2 — 500 nm Au targets.

parametrizations commonly used in the literature [9, 10, 11]:

$$\text{hyperbolic: } S_{\text{eff}}^{\text{HYP}}(d) = S_c + \frac{S_0 - S_c}{1 + \alpha_H\, d},$$

$$\text{exponential: } S_{\text{eff}}^{\text{EXP}}(d) = S_c + (S_0 - S_c)\exp(-\alpha_E\, d).$$

The result of the comparison to the data from our simulation is shown in Fig. 6. It can be seen that both parametrizations describe the simulated data reasonably well. The difference between the two fits is more pronounced in case of lower scattering angles.

Having the results of the analyses of both the number of electrons scattered to the detectors and the dependence of the effective Sherman function on target thickness allowed us to calculate the figure–of–merit. The dependence of FoM on target thickness is shown in Fig. 7, for several values of the scattering angle. It can be noticed that in this case there is a significant difference between the two parametrizations of $S_{\text{eff}}(d)$; one of the two functions is found more appropriate depending on the scattering angle. Nevertheless, applying the procedure described above to the simulated data allows one to determine the optimal target thickness for Mott polarimetry measurements, which is of high practical significance.

4. Summary

The values of the simulated effective Sherman function were compared to the experimental data for 100 keV electrons and the results of these first comparisons were found highly encouraging. Further analyses of the behavior of the Sherman function and of the general properties of studied phenomena revealed many interesting effects that did not get enough attention in the past.

These first results suggest many prospective applications for our code, including, but not limited to: (*i*) computation of the effective Sherman function, (*ii*) experiment optimization, (*iii*) depolarization of electron beams passing through matter. Firstly, PEBSI allows to compute the Sherman function for arbitrary experimental conditions (i.e., also when no experimental data exist). Secondly, a comprehensive analysis and optimization of an experiment requires collecting a large amount of data, which may be difficult to perform by a direct measurement. Thirdly, its applications are definitely not limited to the computation of the effective Sherman function only. The PEBSI code allows to investigate any effect related to the change of the polarization state of an electron beam passing through matter. Such studies may be essential for many other applications, which require the precise knowledge of the polarization of an electron taking part in an interaction.

In order to prove the applicability of our simulation in other energy ranges in which Mott polarimetry is used it would be necessary to compare to other experimental data, especially for higher energies. This computation is, however, time consuming due to the low cross sections for studied processes. Nevertheless, further validation is currently ongoing.

Acknowledgments

We thank Prof. J. Rembieliński and Prof. P. Caban for discussions regarding polarization effects in Møller scattering; Prof. A. Jabłoński for consultations regarding the ELSEPA package. This work was supported from the funds of the National Science Centre: (1) as a part of the research project, contract/decision DEC-2012/06/M/ST2/00430; (2) as a postdoc project (M.W.) contract/decision DEC-2013/08/S/ST2/00551.

References

[1] Mott N F 1929 *Proc. R. Soc.* A **124** 425
[2] Kessler J 1985 *Polarized Electrons* 2nd ed. (Berlin: Springer)
[3] Caban P, Rembieliński J and Włodarczyk M 2013 *Phys. Rev.* A **88** 032116
[4] Weber G et al. 2012 *Nucl. Instr. Meth. Phys. Res.* B **279** 155
[5] Sherman N 1956 *Phys. Rev.* **103** 1601
[6] Salvat F, Jablonski A and Powell C J 2005 *Comput. Phys. Commun.* **165** 157
[7] Bote D, Salvat F, Jablonski A and Powell C J 2009 *Atomic Data and Nuclear Data Tables* **95** 871
[8] Guerra M, Amaro P, Machado J and Santos J P 2015 *J. Phys.* B **48** 18
[9] Kohashi T, Konoto M and Koike K 2006 *Jpn. J. Appl. Phys.* **45** 6468
[10] Gay T J and Dunning F B 1992 *Rev. Sci. Instrum.* **63** 1635
[11] Gay T J et al. 1992 *Rev. Sci. Instrum.* **63** 114

Computational modelling parity nonconservation and electroweak interaction effects in heavy atomic systems within the nuclear-relativistic many-body perturbation theory

O Yu Khetselius, A V Glushkov, M Yu Gurskaya, A A Kuznetsova, Yu V Dubrovskaya, I N Serga and L A Vitavetskaya

Odessa State Environmental University, L'vovskaya str.15, Odessa-16, 65016, Ukraine

E-mail: dirac13@mail.ru

Abstract. Computing the hyperfine coupling constants and parity non-conservation (PNC) effect parameters in a few heavy atomic systems has been performed and based on the combined relativistic nuclear mean-field theory and relativistic many-body perturbation theory (PT) formalism with accounting for the interelectron correlation and dominant QED corrections. Results of estimating hyperfine structure constants and PNC parameters for different heavy atoms (caesium, ion of barium, thallium, ytterbium) are presented and compared with other theoretical and experimental data. The spin-dependent contributions to the PNC amplitude for the caesium are presented too.

1. Introduction

Although the spectacular experimental achievements of particle physics in the last decade have strengthened the Standard Model (SM) as an adequate description of nature, they have also revealed that the SM matter represents a mere 5% or so of the energy density of the Universe, which clearly points to some physics beyond the SM despite the desperate lack of direct experimental evidence [1-6]. The sector responsible for the spontaneous breaking of the SM electroweak symmetry is likely to be the first to provide experimental hints for this new physics. The detailed review of these topics can be found in Refs. [1-6], in particular, speech is about brief introducing the SM physics and the conventional Higgs mechanism and a survey of recent ideas on how breaking electroweak symmetry dynamics can be explained. Further one could remind that the observation of a static electric dipole moment of a many-electron atom which violates parity, P , and time reversal, T , symmetry, represents a great fundamental interest in searching for a new physics beyond the SM of elementary particles [1-23]. The interaction mixes a parity of atomic states and also induces a static electric dipole moment of the atom. At the present time, different methods have been used in calculation of the hyperfine structure parameters, PNC and similar effects characteristics.

The most popular methods for computing the electron structure, hyperfine coupling constants, parity and time reversal symmetry violations parameters in heavy many-electron atoms are the multiconfiguration Dirac-Fock (MCDF) and relativistic cluster-coupled (RCC) methods. Besides there have been developed and intensively applied the different versions of a many-body relativistic and QED perturbation theory (PT), namely, the PT with the relativistic Hartree-Fock (RHF), the Dirac-Fock (DF), the Dirac-Kohn-Sham (DKS) zeroth approximations [6-23]. In order to provide a consistent studying the hyperfine coupling constants, PNC effect parameters in the heavy atomic systems, these methods are combined with the different approaches to treat a nuclear structure and the corresponding nuclear effects such as the Bohr-Weisskopf and Breit-Rosenthal-Crawford-Schawlow effects, the skin neutron correction etc. Here the most popular approach is the relativistic mean-filed (RMF) model (see [24-27] and references there in) in different realizations. In Refs.[28-33] we have presented our version of the relativistic PT with an optimized DKS zeroth approximation and correct accounting for the interelectron correlations and dominant QED corrections, combined with the RMF nuclear model (N-QED). In present paper this approach is used for computing the hyperfine structure constants, electroweak interaction and PNC effect parameters in the heavy atomic systems. The N-QED approach allows to fulfil studying the spectra of the multi-electron atomic systems with the correct account of the relativistic, interelectron correlation, nuclear, radiative effects [33-45]. All interelectron correlation corrections of the second order and dominated classes of the higher orders diagrams have been taken into account. Results of estimating these constants for different heavy atoms (caesium, ion of barium, thallium, ytterbium) are presented and compared with other theoretical and experimental data. The spin-dependent contributions to the parity PNC amplitude are presented too.

2. Nuclear-relativistic perturbation theory approach to hyperfine, electroweak effects in heavy atoms

The main elements of the relativistic PT formalism with the optimized DKS zeroth approximation and correct accounting for the interelectron correlations and dominant QED corrections, combined with the RMF nuclear model, are in details presented in Refs. [24-40]. Here we are limited only by the key elements. The relativistic PT zeroth approximation electron wave functions are found from the Dirac equation solution with a self-consistent potential, which includes the core ab initio DKS potential, electric, polarization potentials of a nucleus. All correlation corrections of the second and high orders of PT (electrons screening, particle-hole interaction etc.) are accounted for [24-28].

The nuclear model is based on the relativistic mean-field (RMF) model, which is usually designed as a renormalizable meson-field theory for nuclear matter and finite nuclei (see details in [23-28]). Its relatively successful application in nuclear calculations is usually explained by the fact that the RMF model can incorporate certain ground-state correlations and yields a ground-state description beyond the literal mean-field picture [25]. In our computing there is used NL3-NLC scheme, which is among the most successful parameterizations available. As it is indicated in Ref.[25], the strong attractive scalar (*S:* -400 MeV) and repulsive vector (*V:* +350 MeV) fields provide both the binding mechanism (*S* + *V:* -50 MeV) and the strong spin-orbit force (*S* – *V:* -750 MeV) of both right sign and magnitude.

In order to compute radiation transition amplitudes we have used an advanced gauge-invariant version of the generalized relativistic energy approach (see details in Refs. [11,32,36]). Simultaneously, this theory provides powerful ab initio method of construction of the optimized PT basis and consistent treating gauge-dependent multielectron contributions $\mathrm{Im}\Delta E_{\mathrm{ninv}}$ of the lowest relativistic PT corrections to the atomic level radiation width. This value is considered to be the typical representative of the multielectron exchange-correlation effects contributions. The method used allows to estimate an effectiveness of accounting for the multielectron exchange-correlation effects and provides the practical way of the refinement of the atomic characteristics calculations, based on the "first principles".

Let us consider the key elements of the scheme of calculating the PNC transition amplitude. The dominative contribution to the PNC amplitude is provided by the spin-independent part of the operator of the weak interaction, which should be added to the standard atomic Hamiltonian [2,6]:

$$H = H_{at} + \mu\sum_j H_W(j), \tag{1}$$

$$H_W^1 = \frac{G}{2\sqrt{2}} Q_W \gamma_5 \rho(r), \tag{2}$$

where G_F is the Fermi constant of the weak interaction, γ_5 is the Dirac matrice, $\rho(r)$ is a density of the charge distribution in a nucleus and Q_W is a weak charge of a nucleus, linked with number of neutrons N and protons Z and the Weinberg angle θ_W in the Standard model (c.f. [1-3]) as:

$$Q_W = Z(1-4\sin^2\theta_W) - N \tag{3}$$

With account for the radiative corrections, equation (3) can be rewritten as [2]:

$$\begin{aligned} Q_W &= \{Z(1-[4.012 \pm 0.010]\sin^2\theta_W) - N\}\cdot(0.9857 \pm 0.0004)(1+0.0078\,T) \\ \sin^2\theta_W &= 0.2323 + 0.00365\,S - 0.00261\,T) \end{aligned} \tag{4}$$

The parameters S,T parameterize the looped corrections in the terms of conservation (S) and violation (T) of an isospin. As it is known, the spin-dependent contribution to the PNC amplitude has three distinct sources: a nuclear anapole moment (it is usually considered as an electromagnetic characteristics of system, where the PNC takes a place; generally speaking, speech is about an emerged spin structure and the magnetic field distribution is similar to the solenoid field), the *Z*-boson exchange interaction from nucleon axial-vector currents (A_nV_e), and the combined action of the hyperfine interaction and spin-independent *Z*-boson exchange from nucleon vector (V_nA_e) currents [2,6,7]. The anapole moment contribution strongly dominates. The above-mentioned interactions can be presented by the following Hamiltonian

$$H_W^i = \frac{G}{\sqrt{2}} k_i (\alpha \cdot I)\rho(r) \tag{5}$$

where $k(i=a)$ is an anapole contribution, $k(i=2)=k_{Z0}$ is axial-vector contribution, $k(i=\mathrm{kh})=k_{Qw}$ is a contribution due to the combined action of the hyperfine interaction and spin-independent Z exchange one. Calculation of the corresponding matrix elements is reduced to the computing the integrals as follows [6,15]:

$$< i \,|\, H_W^1 \,|\, j >= i\frac{G}{2\sqrt{2}} Q_W \delta_{k_i - k_j} \delta_{m_i m_j} \int_0^\infty dr [F_i(r)G_j(r) - G_i(r)F_j(r)]\rho(r) \tag{6}$$

The reduced matrix element can be written as:

$$< i \,\|\, H_W^1 \,\|\, j >= i\frac{G}{2\sqrt{2}} Q_W \int_0^\infty dr [F_i(r)G_j(r) - G_i(r)F_j(r)]\rho(r) \tag{7}$$

Further the general expression for the corresponding PNC amplitude for the *a-b* transition is written as follows:

$$< a \,|\, PNC \,|\, b >= -\sum_n [\frac{< b \,|\, e\alpha_\nu A^\nu \,|\, n >< n \,|\, H_W^{(1)} \,|\, a >}{\varepsilon_a - \varepsilon_n} + \frac{< b \,|\, H_W^{(1)} \,|\, n >< n \,|\, e\alpha_\nu A^\nu \,|\, a >}{\varepsilon_b - \varepsilon_n} \tag{8}$$

The corresponding spin-dependent PNC contribution is :

$$< a \,|\, PNC \,|\, b >^{sd} = k_a < a \,|\, PNC \,|\, b >^{(a)} + k_2 < a \,|\, PNC \,|\, b >^{(2)} + k_{hf} < a \,|\, PNC \,|\, b >^{(hf)} \tag{9}$$

where

$$<a|PNC|b>^{(hf)}=$$

$$=\sum_{\substack{m\neq a\\ n\neq a}}\frac{<a|H_W^{(1)}|n><n|H_W^{(hf)}|m><m|e\alpha_\nu A^\nu|b>}{(\varepsilon_a-\varepsilon_m)(\varepsilon_a-\varepsilon_n)}+\sum_{\substack{m\neq a\\ n\neq a}}\frac{<a|H_W^{(hf)}|n><n|H_W^{(1)}|m><m|e\alpha_\nu A^\nu|b>}{(\varepsilon_a-\varepsilon_m)(\varepsilon_a-\varepsilon_n)}+$$

$$+\sum_{\substack{m\neq a\\ n\neq b}}\frac{<a|H_W^{(1)}|m><m|e\alpha_\nu A^\nu|n><n|H_W^{(hf)}|m>}{(\varepsilon_a-\varepsilon_m)(\varepsilon_b-\varepsilon_n)}+\sum_{\substack{m\neq a\\ n\neq b}}\frac{<a|H_W^{(hf)}|m><m|e\alpha_\nu A^\nu|n><n|H_W^{(hf)}|b>}{(\varepsilon_a-\varepsilon_m)(\varepsilon_b-\varepsilon_n)}+$$

$$+\sum_{\substack{m\neq b\\ n\neq b}}\frac{<a|e\alpha_\nu A^\nu|n><n|H_W^{(1)}|m><m|H_W^{(hf)}|b>}{(\varepsilon_b-\varepsilon_m)(\varepsilon_b-\varepsilon_n)}+\sum_{\substack{m\neq b\\ n\neq b}}\frac{<a|e\alpha_\nu A^\nu|n><n|H_W^{(hf)}|m><m|H_W^{(1)}|b>}{(\varepsilon_b-\varepsilon_m)(\varepsilon_b-\varepsilon_n)}-$$

$$-<a|H_W^{(hf)}|a>\sum_{m\neq a}\frac{<a|H_W^{(1)}|m><m|e\alpha_\nu A^\nu|b>}{(\varepsilon_a-\varepsilon_m)^2}-\sum_{n\neq b}\frac{<a|e\alpha_\nu A^\nu|n><n|H_W^{(1)}|b>}{(\varepsilon_b-\varepsilon_n)^2}<b|H_W^{(hf)}|b>. \tag{10}$$

Here the following notations are used: $|a>=|aIF_F M_F>, |b>=|bIF_I M_I>$, I – spin of a nucleus, $F_{I,F}$-is a total momentum of an atom and M – its z component (I,F are the initial and final states). It should be noted the expressions for the matrix elements $<a|PNC|b>^{(a)}$, $<a|PNC|b>^{(2)}$are similar to equation (6). The detailed description of the corresponding matrix elements, procedures of their computing and other aspects of the general method and PC code is presented in Refs.[28-45].

3. Results and conclusions

Here we present the results of computing the hyperfine structure parameters for the neutral caesium and Cs-like ion of barium Ba^+. In Table 1 the experimental (A^{Exp}) and our ($A^{N\text{-}Qed}$) data for magnetic dipole constant A (MHz) for the valent states of ^{133}Cs ($I=7/2$, $g_i=0.7377208$) are presented. The calculation results, obtained within the standard (A^{RHF}) RHF and RHF with account of the second and higher PT corrections, the MCDF approximation and QED formalism [22,23] are listed too. In a whole there is a physically reasonable agreement of our data with experimental ones that is provided by using the gauge-invariant relativistic orbital basis and the correct accounting for the exchange-correlation, nuclear, QED effects.

Table 1. The values (MHZ) of the hyperfine structure constant A for valent states of ^{133}Cs: A^{Exp} - experiment; A^{RHF}, dA^{RHF} - RHF calculation plus the second and higher PT orders contribution; A^{QED} – data [17,18]; $A^{N\text{-}Qed}$ – the present paper;

State	A^{MCDF}	A^{RHF}	$A^{RHF}+dA$	A^{Qed}	$A^{N\text{-}Qed}$	A^{Exp}
$6s_{1/2}$	1736.9	1426.81	2291.00	2294.45	2296.85	2298.16
$6p_{1/2}$	209.6	161.09	292.67	292.102	291.97	291.90(13)

In Table 2 we present the results of calculating the hyperfine structure constant A for different states of the Cs-like ion of barium: $[5p^6]6s_{1/2},6p_{1/2}$. The following notations are used: A^{RCC} – results of calculation on the basis of the RCC method; A^{DF} – DF method; A^{RHF} - RHF method and A^{QED}- the QED calculation (from Refs. [13-23]); $A^{N\text{-}Qed}$ – our results.

Table 2. Theoretical and experimental data for the hyperfine structure constant A for the states of the Cs-like ion Ba (see text)

State	A^{Exp}	A^{RCC}	A^{DF}	A^{RHF}	A^{QED}	$A^{\text{N-Qed}}$
$[5p^6]6s_{1/2}$	4018	4072.83	4193.02	4208.2	4014.52	4016.76
$[5p^6]6p_{1/2}$	742.04	736.98	783.335	---	742.96	742.54

Further in Table 3 we list the nuclear spin dependent corrections to the PNC ^{133}Cs 6s-7s amplitude E_{PNC}, calculated by different theoretical methods (in units of the $k_{a,2,hf}$ coefficient): many-body PT (MBPT), DF-PT, the shell model, N-QED PT (present paper) (from Refs. [7-20]). In Table 4 there are presented the theoretical values of the weak charge Q_W for three heavy atoms, predicted in the Standard Model and different approaches (from Refs. [6-9]).

Table 3. The nuclear spin-dependent corrections to the PNC ^{133}Cs: 6s-7s amplitude E_{PNC}, calculated by different methods (in units of $k_{a,2,hf}$ coeff.): MBPT, DF-PT, shell model, N-QED PT (see text).

Correction	MBPT	Shell model	DF	This work
K (sum)	0.1169	0.1118	0.112	0.1159
k_2- the Z-boson exchange interaction from nucleon axial-vector currents (A_nV_e)	0.0140	0.0140	0.0111 0.0084	0.0138
k_{hf} - the combined action of the hyperfine interaction and spin-independent Z exchange	0.0049	0.0078	0.0071 0.0078	0.0067
k_a –anapole moment	0.0980	0.090	0.0920	0.0954

Table 4. The estimated values of the weak charge Q_W for heavy atoms, predicted in different approaches (see text)

Atom:	Standard Model	RHF+Breit	MCDF	MBPT	This work
^{133}Cs	-73.19(3)	-72.66 -71.70	-69.78 -71.09	-72.69 -72.18	-72.62
^{173}Yb	-96.85(6)	-	-	-	-95.41
^{205}Tl	-116.81(4)	-116.20	-	-	-116.15

The key requirements to the precise computing PNC amplitudes are connected with the correct accounting for the inter electron correlations, relativistic, Breit and radiative and nuclear corrections (including the Bohr-Weisskopf and Breit-Rosenthal-Crawford-Schawlow effects, i.e. a magnetic moment distribution in a nucleus, the skin neutron and other corrections). The analysis of results shows that in principle a majority of theoretical approaches provides physically reasonable agreement with the SM data, but the important question is how much exact this agreement. The theoretical data on the constants directly indicate the necessity of new, more precise experiments. From the other side, it is of a great interest to study the PNC effect in other heavy atoms such lanthanides and actinides atoms. The rare-earth elements are especially interesting as they have very complicated spectra of autoionization resonances (with very unusual their behavior in a weak electric and laser fields from physical viewpoint [35]). The elementary analysis shows that the perspectives of the PNC experiments with Stark pumping of the individual states in the rare-earth and other heavy atoms (and probably more effective multicharged ions of these elements) and simultaneously polarized laser field dressing (with a cold-atom fountain or interferometer) could provide comfortable conditions for precise observation of the electroweak effects.

Acknowledgements
The authors would like to thank Prof. Nithaya Chetty, Prof. Marius Potgieter and colleagues for invitation to make contributions to CCP-2016 (Gauteng, South Africa). The useful comments of the anonymous referees are very much acknowledged too.

References

[1] Grojean C 2007 *Phys.–Uspekhi* **50** 3
[2] Khriplovich I B 1991 *Parity Nonconservation in Atomic Phenomena* (Philadelphia: Gordon and Breach)
[3] Tsigutkin K, Dounas-Frazer D, Family A, Stalnaker J E, Yashchuk V V and Budker D 2009 *Phys. Rev. Lett.* **103** 071601
[4] Bouchiat M A 2007 *Phys. Rev. Lett.* **98** 043003
[5] Auerbach N 2008 *J. Phys. G: Nucl. Part. Phys.* **35** 014040
[6] Johnson W R, Sapirstein J and Blundell S A 1993 *Phys. Scripta.* **46** 184.
[7] Safronova M S, Rupsi Pal, Jiang D, Kozlov M G, Johnson W R and Safronova U I 2009 *Nucl. Phys. A.* **827** 411c
[8] Johnson W R, Safronova M S and Safronova U I 2003 *Phys. Rev. A* **67** 062106
[9] Haxton W C, Liu C-P, Ramsey-Musolf M J, 2002 *Phys. Rev. C.* **65** 045502
[10] Tomaselli M, Schneider S M, Kankeleit E and Kuhl T 1995 *Phys. Rev. C.* **51** 2989
[11] Bieron J and Pyykko P 2005 *Phys. Rev. A.* **71** 032502
[12] Glushkov A V and Ivanov L N 1992 *Phys. Lett.* A **170** 33
[13] Dzuba V A, Flambaum V V and Sushkov O P 1997 *Phys. Rev. A.* **56** R4357
[14] Safronova M S, Johnson W R and Derevianko A 1999 *Phys. Rev. A.* **60** 044103
[15] Dzuba V A, Harabati C, Johnson W R and Safronova M S 2001 *Phys. Rev. A.* **63** 044103
[16] Kozlov M G, Porsev S G and Tupitsyn I I 2001 *Phys. Rev. Lett.* **86** 3260
[17] Shabaev V M, Tupitsyn I I, Pachucki K, Plunien G and Yerokhin V 2005 *Phys. Rev. A.* **72** 062105
[18] Khetselius O Yu 2009 *Phys. Scr.* **T135** 014023
[19] Khetselius O Yu 2009 *Int. J. Quant. Chem.* **109** 3330
[20] Safronova M S, Johnson W R, Safronova U I and Cowan T E 2006 *Phys. Rev. A* **74** 022504
[21] Dzuba V A, Flambaum V V and Safronova M S 2006 *Phys. Rev. A* **73** 022112
[22] Sahoo B K, Gopakumar G, Chaudhuri R K, Das B P, Merlitz H, Mahapatra U S and Makherjee D 2003 *Phys. Rev. A* **68** 040501
[23] Derevianko A and Porsev S G 2005 *Phys. Rev. A* **71** 032509
[24] Serot B D and Walecka J D 1986 *Adv. Nucl. Phys.* **16** 1
[25] Reinhard P G 1989 *Rep. Prog. Phys.* **52** 439
[26] Burvenich T J , Evers J and Keitel C H 2006 *Phys. Rev. Lett.* **74** 044601
[27] Furnstahl R J 2004 Next generation relativistic models *Extended Density Functionals in Nuclear StructurePhysics (*vol 641*)* ed G A Lalazissis, P Ring and D Vretenar (Berlin: Springer) chapter 1 pp 1-30
[28] Glushkov A V , Rusov V D , Ambrosov S V and Loboda A V 2003 Resonance states of compound superheavy nucleus and EPPP in heavy nucleus collisions *New Projects and New Lines of Research in Nuclear Physics (*vol 1*)* ed Fazio and F Hanappe (Singapore: World Scientific) pp 126-132
[29] Glushkov A V and Malinovskaya S V 2003 Cooperative laser nuclear processes: Border liones effects *New Projects and New Lines of Research in Nuclear Physics (*vol 1*)* ed Fazio and F Hanappe (Singapore: World Scientific) pp 242-250
[30] Glushkov A V, Khetselius O Yu, Loboda A V and Malinovskaya S V 2007 Discharge of metastable nuclei during ngative moun capture: Energy approach *Meson-Nuclear Physics and the Structure of the Nucleon (*vol 2*)* ed S Krewald and H Machner (Juelich: IKP) pp 139-143; SLAC eConf C070910 2007 (Menlo Park, CA) **2** 186
[31] Glushkov A V, Khetselius O Yu, Gurnitskaya E P, Loboda A V , Florko T A, Sukharev D E and Lovett L Gauge-invariant QED perturbation theory appraoch to calculating nuclear quadrupole moments, hyperfine structure constants for heavy atoms and ions *Frontiers in Quantum Systems in Chemistry and Physics (Progress in Theoretical Chemistry and Physics* vol 18*)* ed S Wilson, P J Grout, J Maruani, G Delgado-Barrio and P Piecuch (Berlin:

Springer) pp 505-522
[32] Glushkov A V, Khetselius O Yu and Lovett L 2009 Electron-β-nuclear spectroscopy of atoms and molecules and chemical environment effect on the β-decay parameters *Advances in Theory of Atomic and Molecular Systems Dynamics, Spectroscopy, Clusters, and Nanostructures (Progress in Theoretical Chemistry and Physics* vol 20*)* ed P Piecuch, J Maruani, G Delgado-Barrio and S Wilson (Dordrecht: Springer) chapter 7 pp 125-151
[33] Glushkov A V, Khetselius O Yu and Svinarenko A A 2012 Relativistic theory of cooperative muon-gamma-nuclear processes: Negative muon capture and metastable nucleus discharge *Advances in the Theory of Quantum Systems in Chemistry and Physics (Progress in Theoretical Chemistry and Physics* vol 22*)* ed P Hoggan, E Brandas, J Maruani, G Delgado-Barrio and P Piecuch (Dordrecht: Springer) chapter 3 pp 51-68
[34] Glushkov A V 2006 *Relativistic and Correlation Effects in Spectra of Atomic Systems* (Odessa: Astroprint)
[35] Glushkov A V, Khetselius O Yu, Svinarenko A A and Prepelitsa G P 2011 Energy approach to atoms in a laser field and quantum dynamics with laser pulses of different shape *Coherence and Ultrashort Pulsed Emission (*vol 1*)* ed F J Duarte (Rijeka: InTech) chapter 8 pp 159-186
[36] Glushkov A V, Ambrosov S V, Loboda A V, Chernyakova Y G, Khetselius O Yu and Svinarenko A A 2004 *Nucl. Phys. A*. **734**S 21
[37] Malinovskaya S V, Glushkov A V, Khetselius O Yu, Mischenko E V, Loboda A V and Svinarenko A A 2009 *Int. J. Quant. Chem.* **109** 3031
[38] Malinovskaya S V, Glushkov A V, Khetselius O Yu, Svinarenko A A, Mischenko E V and Florko T A 2009 *Int. J. Quant. Chem.* **109** 3325
[39] Malinovskaya S V, Glushkov A V, Khetselius O Yu, Loboda A V, Lopatkin Yu M, Nikola L V, Svinarenko A A and Perelygina T B 2011 *Int. J. Quant. Chem.* **111** 288
[40] Khetselius O Yu, Glushkov A V, Gurnitskaya E P, Loboda A V, Mischenko E V, Florko T A and Sukharev D E 2008 *Spectral Line Shapes* (AIP) **15** 231
[41] Florko T A, Ambrosov S V, Svinarenko A A and Tkach T B 2012 *J. Phys.: Conf. Ser.* **397** 012037
[42] Svinarenko A A 2014 *J. Phys.: Conf. Ser.* **548** 012039
[43] Svinarenko A A, Ignatenko A.V. Ternovsky V B, Nikola L V, Seredenko S S and Tkach T B 2014 *J. Phys.: Conf. Ser.* **548** 012047
[44] Svinarenko A A, Khetselius O Yu, Buyadzhi V V, Florko T A, Zaichko P A and Ponomarenko E L 2014 *J. Phys.: Conf. Ser.* **548** 012048
[45] Glushkov A V, Kondratenko P A, Buyadgi V V, Kvasikova A S, Sakun T N and Shakhman A N 2014 *J. Phys.: Conf. Ser.* **548** 012025

IOP Conf. Series: Journal of Physics: Conf. Series **905** (2017) 012030 doi:10.1088/1742-6596/905/1/012030

Using memory-efficient algorithm for large-scale time-domain modeling of surface plasmon polaritons propagation in organic light emitting diodes

Andrey Zakirov[1,2], Sergei Belousov[2,4], Ilya Valuev[2,3,6], Vadim Levchenko[1], Anastasia Perepelkina[1,5], Yasunari Zempo[5]

[1] Keldysh Institute of Applied Mathematics, Moscow, 125047, Russia;
[2] Kintech Lab Ltd., Moscow, 123298, Russia; [3] HiperCone Ltd., Moscow, 123298, Russia;
[4] National Research Center "Kurchatov Institute" 123182, Moscow, Russia; [5] Hosei University, Faculty of Computer and Information Sciences, Tokyo, 184-8584, Japan
[6] Joint Institute for High Temperatures, Moscow, 125412, Russia

E-mail: mogmi@narod.ru

Abstract. We demonstrate an efficient approach to numerical modeling of optical properties of large-scale structures with typical dimensions much greater than the wavelength of light. For this purpose, we use the finite-difference time-domain (FDTD) method enhanced with a memory efficient Locally Recursive non-Locally Asynchronous (LRnLA) algorithm called DiamondTorre and implemented for General Purpose Graphical Processing Units (GPGPU) architecture. We apply our approach to simulation of optical properties of organic light emitting diodes (OLEDs), which is an essential step in the process of designing OLEDs with improved efficiency. Specifically, we consider a problem of excitation and propagation of surface plasmon polaritons (SPPs) in a typical OLED, which is a challenging task given that SPP decay length can be about two orders of magnitude greater than the wavelength of excitation. We show that with our approach it is possible to extend the simulated volume size sufficiently so that SPP decay dynamics is accounted for. We further consider an OLED with periodically corrugated metallic cathode and show how the SPP decay length can be greatly reduced due to scattering off the corrugation. Ultimately, we compare the performance of our algorithm to the conventional FDTD and demonstrate that our approach can efficiently be used for large-scale FDTD simulations with the use of only a single GPGPU-powered workstation, which is not practically feasible with the conventional FDTD.

1. Introduction

Improvement of optical devices design by numerical simulation is generally less expensive than multiple iterative trial fabrication. However, the simulation methods are often limited by the computational power. Particularly, full-wave optical simulation requires fine space and time meshes. The treatment of the whole device in such conditions implies large memory and computing performance, which is on the threshold of the current computer technology, or even far beyond it. So there exists a range of problems that has not been solved numerically yet due to computer cost limitations, such as a whole-device full wave simulation of OLED [1], optical antennae, and others [2].

In this work we demonstrate the practical usage of a high performance FDTD [2] code

DTmaxwell [3]. The aim of its development was to fully utilize the computational performance of a GPGPU-powered computer (single node or cluster) so that by minimizing the computation time and reformulating memory requirements we may broaden the scope of simulated problems in the area of electromagnetic modeling. DTmaxwell was developed on the base of finite-difference time domain algorithm DiamondTorre [4, 5] for CUDA GPGPU. DiamondTorre belongs to the class of Locally-Recursive non-Locally Asynchronous (LRnLA) algorithms and will be explained briefly in the following sections. DTmaxwell code implements all necessary typical features for FDTD simulation:

- use of GPGPU accelerators;
- 3D geometry;
- 4th order approximation for finite differences in space; 2nd order approximation in time;
- Perfectly Matched Layer (PML) boundaries [6];
- Total Field/Scattered Field (TF/SF) source;
- arbitrary model for material equations, including Drude-Lorenz model for metals;
- performance of 1 billion Yee cells updates per second on one GPU;
- linear scalability on multi-GPU clusters.

Previously, we have studied the code performance and scalability [7]. High computation rate of more than 1 billion Yee cell updates per second on one device was achieved. The scaling for ~ 1000 devices is linear for weak scaling, and saturates at ~ 100 for strong scaling. The maximal achieved performance on a TSUBAME 2.5 supercomputer was $0.65 \cdot 10^{12}$ Yee Cells per second for a 3D domain with $0.3 \cdot 10^{12}$ Yee cells in total. This number of cells corresponds, for example, to 1 cubic millimeter domain for wave optics problems.

In this work, we apply DTmaxwell code for modeling of optical properties of OLEDs. We take our previous work as the base for the current study: [8] describes a numerical investigation of an OLED by the FDTD code EMTL [9]. Presently we use the problem described in the mentioned paper as a test case for assessing capabilities and performance of DTmaxwell code, and for comparing the results and performance with EMTL, which utilizes traditional FDTD time stepping algorithm.

2. Problem statement

OLEDs are multilayered electroluminescent devices [10]. Light emission is generated via the recombination of excitons in the organic layer in the close proximity to the metallic cathode. The generated emission excites SPPs at the surface of the metallic cathode in OLEDs thus creating a major loss channel leading to the low outcoupling efficiency seen in these devices [8]. One of the methods used to increase the light outcoupling efficiency of OLEDs consists in extracting the energy lost to SPPs via corrugation at the cathode surface. SPPs scatter off the corrugation as they propagate along the metal-organic interface thereby recovering the light outcoupling efficiency of OLEDs.

The improvement of the light outcoupling efficiency of OLEDs has been studied both theoretically [11–21] and experimentally [14, 17]. In order to accurately describe the interaction of emission in OLED with its multilayered and corrugated geometrical structure, numerical simulations are required, for which FDTD models are offering the best scalability and performance on parallel computers. However, large transverse computation domains are necessary so that SPPs propagation and decay is accurately captured. Thus, application of the FDTD for materials with long decay length in optical range is limited, since the computational domain size is restricted by the computer power and memory. Previous FDTD simulations of SPP decay in OLEDs were 2D [14, 16] or included only a fraction of decay length. Since aluminium has shorter decay length it has been simulated in 3D [21].

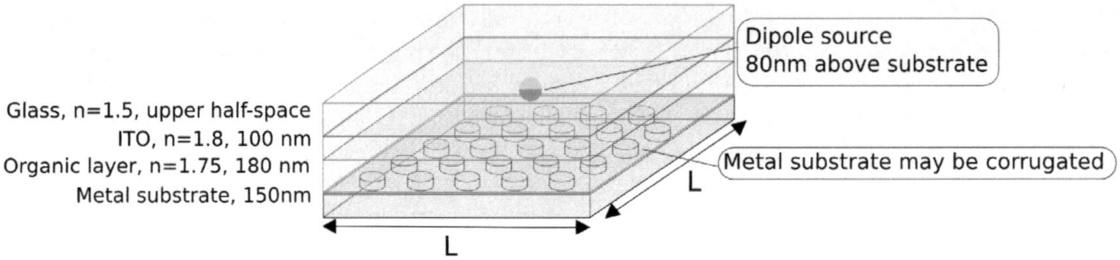

Figure 1. OLED structure in the numerical problem statement

In the current work, we use DTmaxwell code to demonstrate that it is possible to extend the simulation geometry sufficiently so that SPP decay dynamics is accounted for even for materials with long decay length such as silver. We further consider an OLED with periodically corrugated metallic cathode and show how the SPP decay length can be greatly reduced due to scattering off the corrugation. The basic structure of a simulated region is shown in Fig. (Figure 1).

The multi-layered OLED is placed inside a box with the lateral (X and Y) dimensions $L \times L$ (L is a variable numerical parameter). The Z-axis corresponds to the layers normal direction. The bottom layer is a metal substrate, its height is 150 nm. Its dispersion is simulated by the Drude model with the parameters corresponding to silver ($\omega_p = 1.3666 \cdot 10^{16}\mathrm{s}^{-1}$, $\gamma = 2.69 \cdot 10^{13}\mathrm{s}^{-1}$). The second layer is an organic layer with 180 nm height and its refractive index is equal to $n_{org} = 1.75$. The optical wave source is simulated as a point dipole inside this layer, 80 nm above the metal substrate, placed in the center of the computational domain ($x = y = L/2$). Through it a long pulse is injected with central wavelength $\lambda = 550$ nm. Next is ITO layer, its height is equal to 100 nm and $n_{ITO} = 1.8$. Upper halfspace, as well as space around the $L \times L$ box, is treated as glass, $n_g = 1.5$. Corrugation of the cathode is simulated by metal cylinders placed regularly on the cathode surface. This setup corresponds to the one studied in [8], however other types of corrugation (including the random one) may be used.

In order to probe the SPP propagation, the numerical field detectors are placed on the interface between the cathode and the organic layer at regular intervals, along the straight line ($y = L/2$) passing through the c enter of the computational domain ($x = L/2$).

3. Computational considerations

The problem under consideration is essentially multiscale. The mesh step should resolve the smallest scale (cathode skin length is estimated as $\delta_{SPP} \sim 20$ nm under the specified conditions). On the other hand, the transverse size L should be large enough in order to accurately describe the SPP propagation and decay. The SPP decay length is directly related to the SPP dispersion relation at the metal-organic interface $k_{SPP}(\omega)$: $L_{decay} \sim 1/\mathtt{Im}k_{SPP}$, where $k^0_{SPP} = \omega/c\sqrt{(\varepsilon_{org}\varepsilon_m(\omega))/(\varepsilon_{org} + \varepsilon_m(\omega))}$, with ε_m and $\varepsilon_{org} = n^2_{org}$ being the dielectric constants of the metallic and organic layers. For the given material parameters $L_{decay} \sim 50$ mkm.

For the 3D simulation, the necessary numerical parameters are estimated as follows. The mesh step is chosen to be $\Delta x = 5$ nm for the second order scheme in [8]. It is determined by the distance from the dipole to the cathode and the cathode skin length. The time step is limited by the Courant condition and is chosen as $\Delta t \sim 6.7 \cdot 10^{-18}$ s (light velocity is equal to one in the chosen system of units, units for distance are mkm; in this system $\Delta t = 0.002$). The required number of time steps for the light to travel 1 mkm is ~ 500.

Taking all the mentioned limitations into account, the total number of processed Yee cells

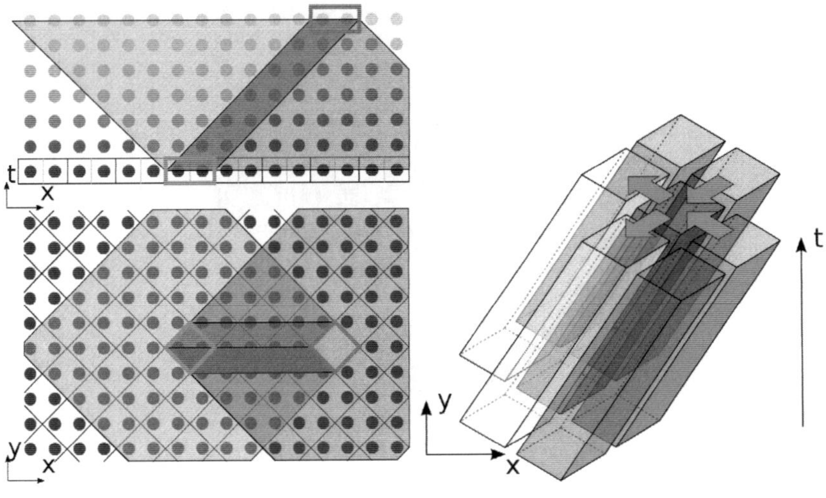

Figure 2. DiamondTorre algorithm. The intersection of domain of influence and domain of dependency of two 2D diamond-shaped bases in 3D time-space (left). Data dependencies between the adjacent DiamondTorres (right). They are directed from green prisms into purple one; from purple prism to yellow prisms. Red and purple prisms calculations are completely independent from each other and may be processed asynchronously.

for the problem (including time steps) should be of the order of 10^{11}. The two following issues stand in the way of efficiently performing such simulations with the traditional FDTD algorithms, especially on GPGPUs. The FDTD simulation is memory-bound. Some popular FDTD codes show sufficient (up to 10^9 Yee cells per second) performance on small sizes, but their performance drops significantly with the increase of data size [22]. This way the accessible problem size is limited by the cumulative GPU device memory, and the CPU RAM (which has much larger size) can not be efficiently utilized. The second issue is that with the use of many-node clusters inter-node data transfers become the limitation of computation time. As a consequence, big domains for SPP problems may not be simulated with the state-of-art solutions, even with the use of the modern supercomputing power.

The algorithm DiamondTorre, which is used in DTmaxwell, provides a solution for the aforementioned issues. The "calculation window" approach allows the use CPU memory and even on-disk swap-file, so the limitation on problem size is lifted by an order of $10 \div 100$. Due to the use of time-space approach, concurrency is implemented in such way that data transfers are completely concealed by computation. These concepts are explained below.

3.1. Time-space approach

DiamondTorre follows the line of LRnLA algorithms: decomposing the discrete time-space computational domains, occupied by Yee cells to be updated, into time-space shapes. These shapes are then processed in the the order satisfying the data dependencies determined by the local stencil used in the discretization of Maxwell's equations. Time-space decomposition allows one to implement processing functions for the shapes that conduct computation on multiple time layers [5, 7]. The prism in Figure 2 represents a decomposition shape for DiamondTorre algorithm — the computation proceeds with the cells, that fall inside it. These cells are on several time layers, but the construction of the algorithm ensures the correct processing order and final computation result.

A row (along the Y axis) of such prisms may be computed asynchronously (red and purple in

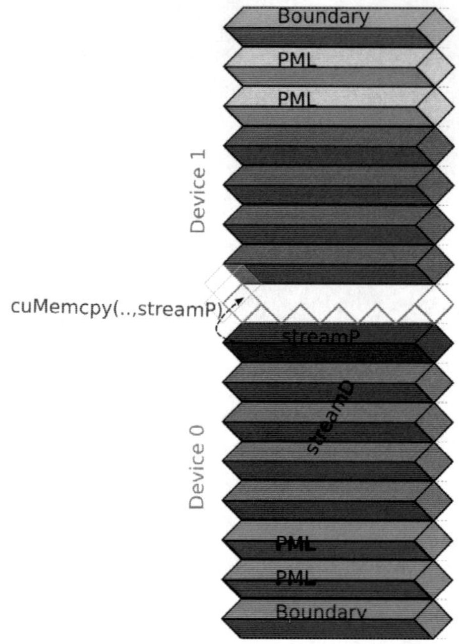

Figure 3. Asynchronous DiamondTorres. The different colors correspond to different CUDA kernels, which are possibly executed on different devices

Figure 2). Several stages of such asynchronous computations propel all cells of the computation mesh forward by a fixed amount of time steps (prism height).

3.2. Calculation window

To perform the DiamondTorre computation only the data that falls into its projection are needed in the device memory. A "calculation window" is implemented, which moves from right to left (according to Figure 2) of the X axis, and encloses an active DiamondTorre at each frame. Data load and save to/from global RAM are performed asynchronously with the computations. Performance does not decline in case the computation time of DiamondTorre is longer than the time, necessary for memory copy to/from the device. Since computation time increases linearly with its height, and the copy time is constant, with the right parameter choice host-device transfers are completely concealed.

3.3. Concurrency on Y axis

Since all DiamondTorre's standing side-by-side along Y axis are asynchronous, they may be processed by different devices inside one node, as well as by different nodes. (fig. 3) In case the number of DiamondTorres per device is high enough, the computation time conceals the data transfer.

As a result, the data transfers may be concealed completely. This alleviates the memory-bound restriction of the numerical scheme. It opens a way for the solution of a new wide range of industrial problems, for which computing power was not enough previously. Algorithm parameters (such as the prism height and problem size) allow not only qualitative but also quantitative estimates of the performance and parallel scaling.

4. Performance evaluation

The performance evaluation in [7] has been carried out for a model problem with no physical significance. Thus, the numerical parameters were chosen to show the largest performance possible.

Here we make the performance test for the target problem, which is characterized by

- wide and flat geometry ($Nx = Ny \gg Nz$);
- inclusion of the Drude model dispersion to the material equations
- on-the-fly diagnostics of the physical parameters

The performance measured is $\sim 0.54 \cdot 10^9$ Yee cell updates per second for one K20x GPGPU. One cell data occupies 68 B of memory, so the computation time for different problem sizes may easily be estimated assuming the complete conceal of device to CPU memory copies.

- $10 \times 10 \mu m^2$ domain — 70 GB, 9 hours;
- $20 \times 20 \mu m^2$ domain — 280 GB, 70 hours;
- $40 \times 40 \mu m^2$ domain — 1100 GB, 580 hours .

With the use of multiple GPUs, 100% efficiency parallel scaling was confirmed in the previous study, so, with the use of several devices, the simulation time is divided by their number.

The performance comparison with EMTL [9] code based on traditional algorithms is shown in Figure 4. The performance is measured for the same problem statement. The difference is in the numerical scheme (4th order spatial finite difference is used in DTmaxwell, 2nd order in EMTL). DTmaxwell has been run on one NVIDIA K20x GPGPU at a single workstation, EMTL has been run on 2 nodes (Xeon E5-2690) of the supercomputer MVS-10p of the Joint Supercomputer Center of Russian Academy of Sciences. Formally these two systems have comparable peak floating point performance of approximately of 100 GFlop/s. It can be seen from the comparison, that the performance of DTmaxwell is more than twice higher than that of EMTL. We also stress here that DTmaxwell has been run on a single workstation, which is only possible due to the memory efficient implementation of LRnLA algorithm.

5. Simulation results

We performed two simulations to demonstrate the importance of modelling of OLED's large geometric sizes and to show the effect of regular corrugation in the cathode. The maximum area size that fits into the CPU memory of one computational node (128 GB) is $18 \times 18 \times 0.64$ mkm. The maximal whole OLED size in our simulations is near to 40×40 mkm: 7680×7680×128 Yee cells with space step Δr=5ñm. We performed all simulations with high-order FDTD numerical scheme with accuracy of 4th order in space and 2nd order in time on hybrid cluster computer K100 (Keldysh Institute of Applied Mathematics, RAS) and Tsubame-2.5 supercomputer using several compute nodes.

In all simulations the source is a vertical dipole (E_z component) placed at the center of OLED. The distance between the dipole and cathode is 80 nm. The dipole generates signal with narrow-band spectra of 500 nm wavelength. Sensors on the cathode detect time-varying E_x component of electromagnetic field at different distances r from the center of simulated region. This field component corresponds to the surface plasmon polariton mode (SPP) of the emmited light. Decay of this mode depends on the properties of the cathode. Except for geometrical divergence of SPP wave with the distance r (proportional to $1/r$), this mode is also exponentially absorbed via interaction with cathode. The length of SPP decay can be very large (about 20-40 micromenters) and corrugation of the cathode significantly decreases this length.

The EM-field visualization for the planar OLED is shown in Figure 5. The propagation of SPP may be easily traced on the metal surface. Figure 6 shows the difference between the

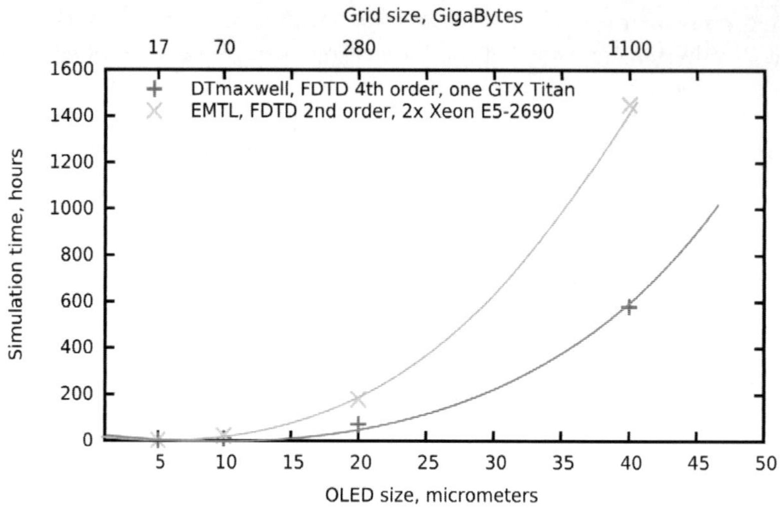

Figure 4. Performance of EMTL (parallel FDTD CPU code based on traditional algorithm) and DTmaxwell (GPGPU code based on LRnLA algorithm) on compute systems of comparable peak performance

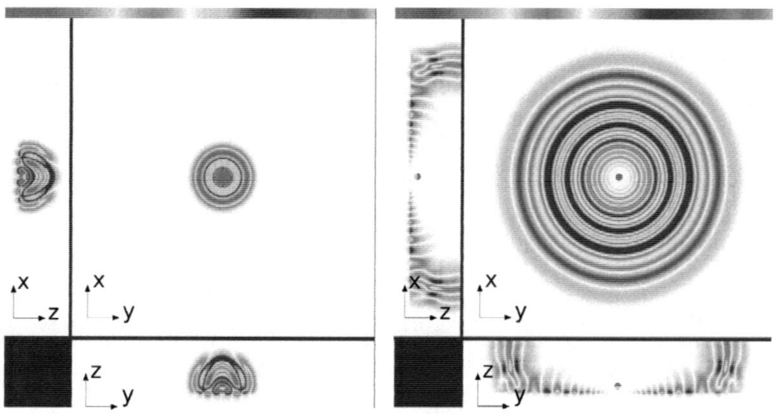

Figure 5. 2D slices of the 3D distribution of the EM field at the two time instants

corrugated and planar OLED. Estimated plasmon decay length is equal to 24 micrometers for the planar structure and only 1.7 micrometers for the corrugated structure.

6. Conclusion

The simulation of OLED geometry for the large area ($40 \times 40 \times 0.64$ mkm) has been performed on clusters of multiple GPGPUs. This way we have confirmed the applicability of DTmaxwell code to large physical problems.

Taking into account the scaling efficiency of the code, we confirm that significantly larger sizes may be simulated with the use of larger GPU clusters. In the simulation, we have traced the SPP propagation and its decay. The result shows that it is necessary to include large domain size in the simulation to reproduce all the relevant SPP phenomena in OLED simulation. The role of the corrugated cathode structure in enhancing the efficiency of light emission from the

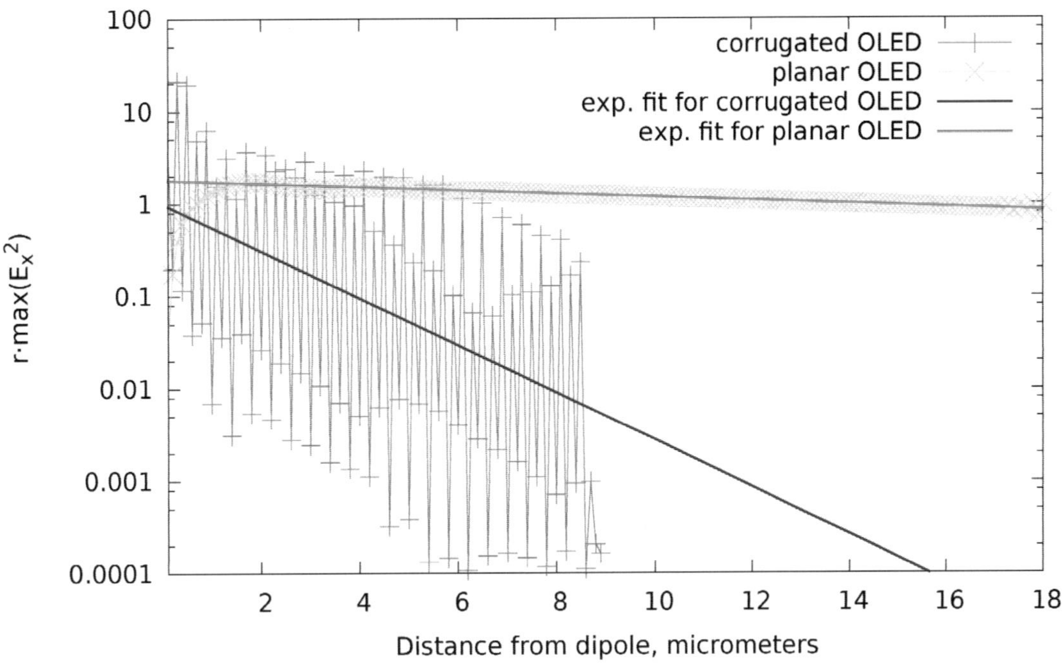

Figure 6. Maximum $|E_x|^2$ component of SPP as a function of the distance from source for planar and corrugated cathodes. The cathode is corrugated with cylindrical holes with depth of 50 nm and diameter of 200 nm. Corrugation period is 400 nm. Fitted lines are proportional to $exp(-r/1.7)$ for the corrugated cathode and $exp(-r/24)$ for the planar OLED (r is measured in micrometers).

OLED, which was studied earlier, is now confirmed by the simulation of much larger scale.

Abbreviations

LRnLA Locally Recursive non-Locally Asynchronous
FDTD Finite Difference Time Domain
PML Perfectly Matched Layer
EM Electro-Magnetic
TF/SF Total Field/Scattered Field
GPGPU General Purpose Graphical Processing Unit
GPU Graphical Processing Unit
SPP Surface Plasmon Polariton
OLED Organic Light Emitting Diode
ITO Indium Tin Oxide
CUDA Compute Unified Device Architecture

References

[1] Ozbay E 2006 *Science* **311** 189–193 ISSN 0036-8075 (*Preprint* `http://science.sciencemag.org/content/311/5758/189.full.pdf`) URL `http://science.sciencemag.org/content/311/5758/189`

[2] Taflove A and Hagness S C 2005 *Computational Electrodynamics: the Finite-Difference Time-Domain Method* 3rd ed (Norwood, MA: Artech House)

[3] Zakirov A 2016 DTmaxwell `https://github.com/zakirovandrey/DTmaxwell4`

[4] Levchenko V 2005 *J. of Inf. Tech. and Comp. Systems* 68

[5] Perepelkina A Y and Levchenko V D 2015 *Keldysh Institute Preprints* **18** 20

[6] Berenger J P 1994 *Journal of Comp. Phys.* **114** 185–200

[7] Zakirov A V, Levchenko V D, Perepelkina A Y and Zempo Y 2016 *Keldysh Institute Preprints* 44–22

[8] Belousov S, Bogdanova M and Teslyuk A 2016 *Journal of Physics D: Applied Physics* **49** 085102

[9] Bogdanova M, Belousov S, Valuev I, Zakirov A, Okun M, Shirabaykin D, Chorkov V, Tokar P, Knizhnik A, Potapkin B, Bagaturyants A, Komarova K, Strikhanov M, Tishchenko A, Nikitenko V, Sukharev V, Sannikova N and Morozov I 2014 *Procedia Computer Science* **29** 740 – 753 ISSN 1877-0509 URL `http://www.sciencedirect.com/science/article/pii/S1877050914002439`

[10] Brütting W, Frischeisen J, Schmidt T D, Scholz B J and Mayr C 2013 *physica status solidi (a)* **210** 44–65 ISSN 1862-6319 URL `http://dx.doi.org/10.1002/pssa.201228320`

[11] Adam J, Hauss J, Gleiss S, Riedel B, Geyer U, Lemmer U and Gerken M 2009 *ICTON Mediterranean Winter Conference, 2009. ICTON-MW 2009. 3rd* (IEEE) pp 1–6

[12] Chutinan A, Ishihara K, Asano T, Fujita M and Noda S 2005 *Organic electronics* **6** 3–9

[13] Kim J W, Jang J H, Oh M C, Shin J W, Cho D H, Moon J H and Lee J I 2014 *Optics express* **22** 498–507

[14] Zhang X L, Feng J, Song J F, Li X B and Sun H B 2011 *Optics letters* **36** 3915–3917

[15] Lee Y J, Kim S H, Kim G H, Lee Y H, Cho S H, Song Y W, Kim Y C and Do Y R 2005 *Optics express* **13** 5864–5870

[16] Kim J Y, Kim D, Kim D H and Choi K C 2014 *IEEE Photonics Technology Letters* **26** 896–899

[17] Nowy S, Krummacher B C, Frischeisen J, Reinke N A and Brütting W 2008 *Journal of Applied Physics* **104** 123109

[18] Qu Y, Slootsky M and Forrest S R 2015 *Nature Photonics* **9** 758–763

[19] Shi X, Wang J, Liu J, Huang S, Wu X, Chen C, Lu J, Su Y, Zheng Y, Kim W Y *et al.* 2014 *Organic Electronics* **15** 864–870

[20] Smith L H, Wasey J A, Samuel I D and Barnes W L 2005 *Advanced Functional Materials* **15** 1839–1844

[21] Sun Y and Forrest S R 2008 *Nature photonics* **2** 483–487

[22] Acceleware Ltd 2016 Acceleware FDTD solvers `http://www.acceleware.com/fdtd-solvers`

IOP Conf. Series: Journal of Physics: Conf. Series **905** (2017) 012031 doi:10.1088/1742-6596/905/1/012031

Temperature specification in atomistic molecular dynamics and its impact on simulation efficacy

R.O. Ocaya and J.J. Terblans

Department of Physics, University of the Free State, South Africa.

E-mail: ocayaro@ufs.ac.za

Abstract. Temperature is a vital thermodynamical function for physical systems. Knowledge of system temperature permits assessment of system ergodicity, entropy, system state and stability. Rapid theoretical and computational developments in the fields of condensed matter physics, chemistry, material science, molecular biology, nanotechnology and others necessitate clarity in the temperature specification. Temperature-based materials simulations, both standalone and distributed computing, are projected to grow in prominence over diverse research fields. In this article we discuss the apparent variability of temperature modeling formalisms used currently in atomistic molecular dynamics simulations, with respect to system energetics, dynamics and structural evolution. Commercial simulation programs, which by nature are heuristic, do not openly discuss this fundamental question. We address temperature specification in the context of atomistic molecular dynamics. We define a thermostat at 400K relative to a heat bath at 300K firstly using a modified *ab-initio* Newtonian method, and secondly using a Monte-Carlo method. The thermostatic vacancy formation and cohesion energies, equilibrium lattice constant for FCC copper is then calculated. Finally we compare and contrast the results.

1. Introduction

Temperature is a crucially important parameter in almost all systems and its specification is not necessarily straightforward, particularly for atomistic systems. Some simulations report temperature-based results that are within "tolerable" accuracy bounds, by relying implicitly on the temperature dependence of another parameter. For instance, some have investigated temperature effects by presuming its expansive effects on lattice parameters indirectly through empirically determined linear expansivity of the bulk material. However, motivated by the rapidly growing field of nanotechnology, it is understood that macroscopic, microscopic and nanoscopic properties are not necessarily commensurate. These emerging fields necessitate the correct thermodynamical modeling and specification of simulation temperature. The aim of this article is to compare and contrast the effects of two thermostat specifications on thermally sensitive parameters of a simple, face-centered cubic (FCC) arrangement of copper atoms. The simulations are standalone for a computationally cost-effective, small atomic array to aid comparison with respect to accuracy. Thermostats are used for a number of reasons in condensed matter simulations, such as in the search for high-temperature dynamics, annealing, formation energies, structural changes, and other effects. Additionally, limiting of steady state drifts caused by encroaching numerical errors in MD simulations, as well as provision for heat evacuation in non-equilibrium MD simulations can be done through thermostats. We examine two thermostat definitions based on

(i) Energy Equipartition theorem (EPT), and
(ii) the Monte-Carlo (MC) approach.

1.1. Thermostats in EPT methods

A time instant δt of simulation deterministically defines the evolution of length and energy parameters. This means that given a particular input, the application of the iterative algorithm will produce the same result at time δt. System energetics are conveyed by the Hamiltonian,

$$\mathcal{H}(\bar{x}, \bar{p}) = \mathcal{U}(\bar{x}) + \mathcal{K}(\bar{p}),$$

where $\mathcal{U}$ and $\mathcal{K}$ are total kinetic and potential energy respectively, $\bar{x}$ is displacement and $\bar{p}$ is momentum. Instantaneous temperature $T(t)$ is defined in terms of the instantaneous internal kinetic energy ε. Then, macroscopic system temperature equals averaged instantaneous temperature, i.e. T=$\langle T(t)\rangle$ [1]. In the thermostatic approach variance in the Hamiltonian, as might arise from computational error accumulation, is tolerable because the thermostat is referenced to a heat bath that tightly constrains temperature variations about the equilibrium value. At equilibrium, the standard deviation $\sigma(\mathcal{H})$ is related to isochoric heat capacity, c_V, by

$$\sigma^2(\mathcal{H}) = \langle \mathcal{H}^2 \rangle - \langle \mathcal{H} \rangle^2 = k_B T^2 c_V, \tag{1}$$

where k_B is Boltzmann constant. Internal particle motion within a system of particles is generally a complex combination of translations and rotations about a center of mass which contribute three and six degrees of freedom respectively. Computational simplifications can be achieved by subtraction of degrees of freedom when stochastic and friction are absent from the system because energy exchange between mechanically uncoupled degrees does not then occur [2, 3]. If the center of mass is at rest and the particle motions are irrotational, then

$$T = \frac{2}{3Nk_B}\varepsilon = \frac{2}{3Nk_B}\sum_{i=1}^{N}\frac{m_i}{2}\bar{v}_i^2. \tag{2}$$

Using the discrete notation $\bar{p}_i(t) := \bar{p}_i^n$, then in the absence of friction and stochastic forces the velocity Störmer-Verlet method [4, 5, 6] discretizes particle position and velocity as

$$\overline{x}_i^{n+1} = \overline{x}_i^n + \delta t \overline{v}_i^n + \frac{\overline{F}_i^n}{2m_i}\delta t^2, \quad \text{and} \quad \overline{v}_i^{n+1} = \overline{v}_i^n + \left(\frac{\overline{F}_i^n + \overline{F}_i^{n+1}}{2m_i}\right)\delta t. \tag{3}$$

Since instantaneous temperature $T(t)$ is directly related to atomic velocities, thermostats require a mechanism to control the rate of change of particle velocities. This can be achieved either by introducing friction or by direct velocity scaling. In the first method or Langevin approach [7, 8, 9], Newton's second law is modified by introducing a stochastic force $\bar{R}_i(t)$ and a friction coefficient $\gamma_i(t)$>0 into the acceleration of the i-th particle:

$$\dot{\bar{v}}_i(t) = \bar{F}_i(t)/m_i - \gamma_i(t)\bar{v}_i(t) + \bar{R}_i(t)/m_i \tag{4}$$

In the absence of stochasticity, the friction term loses its traditional meaning and only indicates heat flow direction relative to the infinite heat bath. In which case, $\gamma_i(t)$>0 implies decelerating particles and consequently loss of heat to the bath. Conversely, $\gamma_i(t)$<0 implies heat gain from the bath and accelerating particles. This can also be interpreted as isochoric system pressure fluctuations. The second method or velocity scaling starts at Eq. 2 to define a scaling factor β in terms of temperatures, kinetic energies and consequently velocities [10, 11, 12, 13] i.e.

$$\beta^2 := \frac{T'}{T} = \frac{\varepsilon'}{\varepsilon} = \left(\frac{\bar{v}_i'}{\bar{v}_i}\right)^2 \quad \Rightarrow \bar{v}' := \beta\bar{v}. \tag{5}$$

After every δt step, the velocity is therefore reassigned a value $\bar{v}'$ instead of $\bar{v}$ for the new temperature to be T'. Velocity scaling using constant β affects temperature distribution strongly, and a damping factor $\Gamma \in (0,1)$ can be introduced [11] s.t. $\beta = (1+\Gamma(T'/T-1))^2$. If β is adjusted at the end of every time step δt, then $dT/dt \sim (T - T')$, i.e. the temperature rate of change depends on the temperature difference. The method does not remove local correlations in particle motions. Setting $\gamma_i^n(t)=\xi^n m_i$, the non-stochastic net force is

$$m_i \frac{\bar{v}_i^{n+1} - \bar{v}_i^n}{\delta t} = \bar{F}_i^n - \xi^n m_i \bar{v}_i^n, \qquad \text{for } \delta t \to 0 \tag{6}$$

with new velocities

$$\bar{v}_i^{n+1} = \bar{F}_i^n \delta t / m_i + \beta \bar{v}_i^n, \tag{7}$$

where the term $\beta=(1 - \xi^n \delta t)$ describes the friction-time behavior. Using $\bar{F}_i^n$ and $\bar{F}_i^{n+1}$ in Eqs. 3 and 6 gives new position and velocity:

$$\overline{x}_i^{n+1} = \overline{x}_i^n + \beta \overline{v}_i^n \delta t + \frac{\overline{F}_i^n}{2m_i} \delta t^2, \qquad \overline{v}_i^{n+1} = \frac{1}{\varphi} \left[\beta \overline{v}_i^n + \left(\frac{\overline{F}_i^n + \overline{F}_i^{n+1}}{2m_i} \right) \delta t \right], \tag{8}$$

where $\varphi=(1 + \frac{\delta t}{2} \xi^{n+1})$. Therefore, friction amounts to scaling present velocity first by β, then overall by $1/\varphi$. For a thermostat kinetic energy is constant i.e. $d\varepsilon/dt$=0. For the i-th particle

$$m_i \dot{\bar{v}}_i = \bar{F} - \xi m_i \bar{v}_i, \qquad \frac{d\varepsilon}{dt} = \sum_{i=1}^{N} m_i \bar{v}_i \cdot \dot{\bar{v}}_i = \sum_{i=1}^{N} \bar{v}_i \cdot (\bar{F} - \xi m_i \bar{v}_i). \tag{9}$$

Writing $\bar{F} = -\bar{\nabla} V_{\bar{x}_i}$ leads to

$$\frac{d\varepsilon}{dt} = -\left(\frac{dV}{dt} + \xi \sum_{i=1}^{N} m_i \bar{v}_i^2 \right), \tag{10}$$

hence the constant temperature condition:

$$\xi = -\frac{dV/dt}{\sum_{i=1}^{N} m_i \bar{v}_i^2} = \frac{dV/dt}{2\varepsilon}. \tag{11}$$

Some thermostats introduce an extra degree of freedom in the form of heat bath temperature. This conveys the extent of system–bath coupling [12, 13]. Since ξ^n arises from friction, the rate of system kinetic energy loss is related to the friction-time behaviour:

$$\frac{d\xi}{dt} = \Big(\sum_{i}^{N} m_i \bar{v}_i^2 - 3Nk_B T^{\text{bath}} \Big) / M \qquad \text{for } M \in \mathbb{R}^+, \tag{12}$$

where $M >> 1$ implies less frictional coupling. To determine $\bar{v}_i^{n+1}$ in Eq.8, the term ξ^{n+1} must be known. Many nonlinear methods [14, 15, 16] arise from discretizations of Eq.12, e.g. midpoint time averaging:

$$\xi^{n+1} = \xi^n + \frac{\delta t}{2M} \left(\frac{d\xi}{dt}\Big|^n + \frac{d\xi}{dt}\Big|^{n+1} \right) \tag{13}$$

gives one possible thermostat when $d\xi/dt$ from Eq.12 is used. Simpler, more approximate discretizations are possible. In general, a Hamiltonian system may have stable integrators but the equations of motion may not be derivable. If the Hamiltonian is time-independent [17, 18] and can be written as

$$\mathcal{H}(\bar{x},\bar{p}) = \sum_{i=1}^{N} \frac{\bar{p}_i^2}{2m_i\gamma^2} + V(\bar{x}_1,\ldots,\bar{x}_N) + \frac{\bar{p}_\gamma^2}{2M} + 3Nk_BT^{\text{bath}}\ln\gamma, \tag{14}$$

then the equations of motion and phase space can be derived. The phase space may be defined in terms of intensive macroscopic variables, which are aggregates of microscopic contributions. Explicit reference to bath temperature generally avoids systematic energy drifts, implying ergodicity [19, 20, 21]. In such a system, particles cannot leave the ensemble. Therefore, assuming equal Maxwell-Boltzmann particle distribution probabilities and ergodicity *a priori* may result in a canonical ensemble (NVT) of microstates and momenta [17]. The probability distribution of microstates over radial and momentum space takes the form

$$\rho(\bar{x},\bar{p}) = e^{-\mathcal{H}(\bar{x},\bar{p})/k_BT} \Big/ \iint e^{-\mathcal{H}(\bar{x},\bar{p})/k_BT} d\bar{x}d\bar{p}. \tag{15}$$

Putting $\bar{p}_i = m_i\dot{\bar{x}}_i$ allows simplification in Cartesian coordinates. The particle velocities can be shown to also follow a Maxwell-Boltzmann distribution. There are other MD thermostats that are specified under various limiting conditions [17, 22, 23]. Caution is needed when comparing thermostats grouped holistically within the MD approach. This is because a particular MD thermostat implementation may facilitate determination of a particular parameter, but is not generally advantageous to all situations. For instance, Hoover-Evans [20, 24] and Haile-Gupta [23] thermostats are inherently time-reversible, whereas Andersen [10] and Berendsen [11] thermostats are not.

1.2. Thermostats using MC methods

A simpler thermostat can be defined non-deterministically by sampling relative particle positions using Monte-Carlo (MC) techniques [17, 25]. This involves testing the effort of a random displacement of a particle within the domain (i.e. box dimensions) against the change in potential energy $\Delta\mathcal{U}(\bar{x})$ without involving kinetic energy. The move is "acceptable" if a probability p can be found such that

$$p = \min\{e^{-\beta\Delta\mathcal{U}(\bar{x})}, 1\}, \quad \text{where } \beta = 1/k_BT. \tag{16}$$

The microstate probability distribution $\rho(\bar{x})$ is then

$$\rho(\bar{x}) = \frac{e^{-\beta\mathcal{U}(\bar{x})}}{\int e^{-\beta\mathcal{U}(\bar{x})}d\bar{x}}. \tag{17}$$

This form, where particle positions are sampled exponentially without geometrical constraints, encapsulates a thermostat in the canonical ensemble that obeys Maxwell-Boltzmann (MB) thermodynamics.

1.3. Timescale and macroscopics

Timescale can be thought of as times larger than the relaxation time in interatomic particle collisions but smaller than the shortest experimental observation time for the system. The deterministic approach, unlike the MC approach, is explicit in the timescale. This naturally leads to the fundamental question of whether time-correlation functions can be found in the MC approach to describe macroscopic time-based phenomena. Instantaneous particle positions depend on the time behavior and temperature spread over particles. It is therefore important to deal with temperature-time behavior in the MC method, at the very least qualitatively, in the absence of analytical expressions [10]. The average temperature per particle is taken as $\bar{T}$ for the system in contact with a bath at temperature $T^{\rm bath}$. Many time-based system parameters of interest such as diffusion coefficients are calculated as statistical averages in the Einstein formulation, or by time-correlation integrals in the Green-Kubo formulation. At constant volume V the change in average energy is $d\bar{E}{=}c_v d\bar{T}$. By Newton's law of cooling:

$$\frac{1}{c_v}\frac{d\bar{E}}{dt} = \frac{d\bar{T}}{dt} = \alpha(T^{\rm bath} - \bar{T}), \quad \text{where } \alpha = \frac{c_v}{\kappa V^{1/3}}. \tag{18}$$

The constant κ encapsulates temperature inhomogeneities and system shape and is estimated from the heat flow continuity and flux equations:

$$\begin{aligned} \bar{J}(\bar{x},t) &= -\kappa\bar{\nabla}T(\bar{x},t) \\ \frac{\partial T(\bar{x},t)}{\partial t} &= -\frac{V}{c_v}\bar{\nabla}\cdot\bar{J}(\bar{x},t). \end{aligned} \tag{19}$$

Solving Eq. (18) gives

$$\bar{T}(t) = T_0 + [\bar{T}(0) - T^{\rm bath}]e^{-\alpha t}, \tag{20}$$

on a timescale such that stochasticity in $\bar{T}(t)$ averages out. From purely theoretical considerations the choice of either a specific MD, or even the MC thermostat, is subjective. However, one expects correct dynamics with a thermostat that accommodates temperature fluctuations on the timescale.

2. Simulation

The present work investigates the effect of the foregoing thermostat definitions, deterministic and Monte Carlo, on the equilibrium lattice constant and formation energies on a isochoric, monoatomic array of n=1583 (plus one, if with a surface adatom) copper atoms in the FCC structure, and a heat bath at 300K where necessary. The system is decidedly simple for speedier computation and comparison. The deterministic approach employs the Sutton-Chen (SC) embedded atom model (EAM) form of the Finnis-Sinclair potential [26, 27, 28, 29]. The authors have previously reported their implementation of SC code [3], which was modified for the present work to accommodate velocity scaling-based thermostats using Störmer-Verlet integration [6], as well as the MC thermostat. EAM methods are not purely deterministic but their lower computational effort for more particles and better handling of larger timescales is attractive [30]. The EAM Hamiltonian can be written [28, 30, 31]

$$\mathcal{H} = \frac{1}{2}\sum_{ij}^{N} V(x_{ij}) + \sum_{i}^{N} F(\bar{\rho}_i), \tag{21}$$

where $V{=}V(\bar{x}_{ij})$ is the pairwise interaction potential energy between particles i and j separated a distance $\bar{x}_{ij}$, and $F{=}F(\bar{\rho}_i)$ is the embedding energy of atom i as a function of the host electron density. For FCC metals the translation and rotation invariant total potential energy [4, 26, 32, 33] can be written

$$V = \varepsilon\sum_{i=1}^{N}\left(\sum_{j=i+1}^{N}\left(\frac{\sigma}{x_{ij}}\right)^n - c\sqrt{S_i}\right), \tag{22}$$

where

$$S_i = \sum_{j=1,j\neq i}^{N}\left(\frac{\sigma}{x_{ij}}\right)^m, \tag{23}$$

and σ is a length parameter e.g. lattice constant, ε is an energy scaling factor. The fitting constants are c, m and n, where $m < n$. The force can be written as

$$\overline{F} = -\overline{\nabla} V(\bar{x}). \tag{24}$$

Hence, for the i-th particle, the force is

$$\overline{F}_i = \varepsilon \sum_{j=1,\ j\neq i}^{N} \left[n\left(\frac{\sigma}{x_{ij}}\right)^n - \frac{cm}{2}\left(\frac{1}{\sqrt{S_i}} + \frac{1}{\sqrt{S_j}}\right)\left(\frac{\sigma}{x_{ij}}\right)^m \right] \frac{\bar{x}_{ij}}{x_{ij}^2}, \tag{25}$$

where $\bar{x}_{ij} = (\bar{x}_j - \bar{x}_i)$ is the vector between particles i and j, expressed in terms of particle positions. The Hamiltonian ($\mathcal{H}$) is here accepted to within $O(\delta t^2)$ discretization error, but is ideally constant [34]. Other temperature dependent effects like diffusion can be investigated through the effect of the thermostat definition on existing literature models. For instance, diffusion [35] is probabilistically modeled by

$$D = D_0 \prod P_m P_v = D_0 e^{-E_m/k_B T} e^{-E_v/k_B T}, \tag{26}$$

where P_m and P_v are inter-lattice point migration and vacancy availability probabilities respectively, expressed in terms of migration energy E_m and vacancy formation energy E_v. These energies can be evaluated for each thermostat T, by considering the various vacancy formation and migratory mechanisms for the perfect crystal lattice under adatom perturbation [36]. We consider only for the {100} equivalent planes of the lattice-point to surface Schottky defect creation shown in Fig.1 and from-infinity to surface migration, shown in Fig.2. The

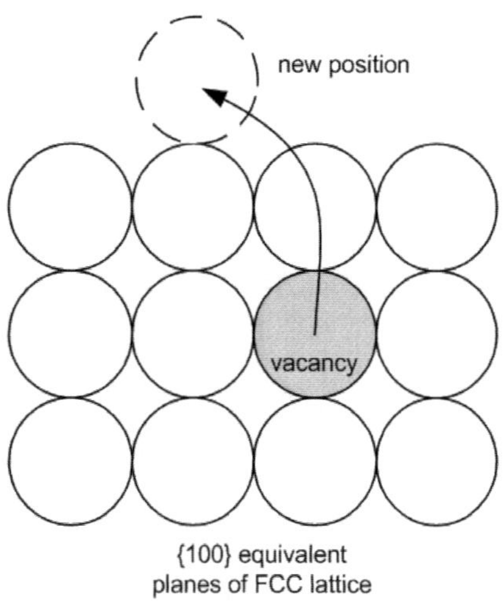

Figure 1. Schottky defect formation by surface migration.

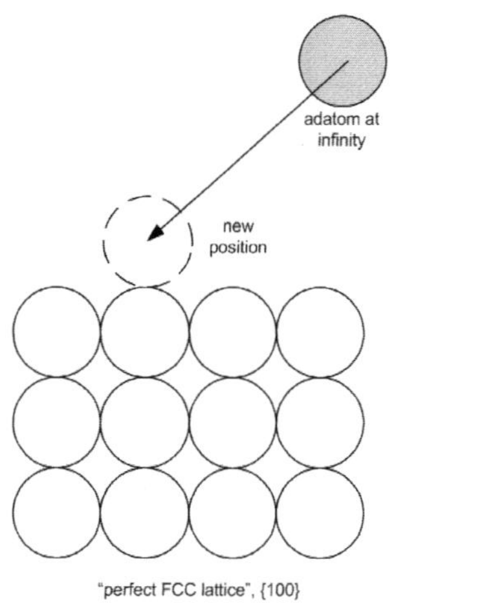

Figure 2. From-infinity to surface migration.

software reported in [3] calculates the average cohesive energy, E_{coh}, as the difference between the total potential energy of the perfect, undisturbed FCC lattice and when all N atoms are "blown apart" to infinite interatomic separation i.e.

$$E_{coh} = (U_{apart,tot} - U_{perfect})/N. \tag{27}$$

3. Results and discussion

The isotherm implementations using both the deterministic model and the Monte-Carlo method were taken to be at 400K relative to a bath temperature of 300K. The two thermostats were used to calculate cohesive (E_{ch}), vacancy formation (E_v), Schottky formation (E_{sch}) (1), infinity-to-surface atom capture (E_{inf}) and extraction energies (E_{ext}), and equilibrium lattice constant (a_{eq}) under constant volume. Table 1 lists the input parameters for the simulation of copper. All adatom effects were simulated under the assumption that they occurred relative to planes equivalent to the (100) plane. Table 2 shows the results of the two thermostats. The point at infinity is taken approximately 100 lattice parameters away from the bulk of the array, since the atoms are assumed to be in a closed box of fixed volume.

Table 1. SC simulation parameters for FCC copper.

ε (eV)	a ($\AA$)	c	n	m	N
0.012382	3.61	39.432	9	6	1583 (+1 adatom)

Table 2. Parametric comparison of deterministic (A) versus MC (B) thermostats on $Cu^{(100)}$ at 400K.

Method	a_{eq} ($\AA$)	E_v	E_{ext}	E_{coh} (eV)	E_{sch}	E_{inf}
A	3.63±0.01	1.31±0.02	4.39±0.01	1.28±0.01	4.41±0.01	2.96±0.01
B	3.74±0.02	1.66±0.30	5.84±0.62	2.17±0.60	5.98±0.60	3.50±0.50

The results shown in Table 2 were compared with literature values, where abundant empirical data exists in the literature at 400K for the deterministic approach in Method A [37, 38, 39, 40], than for the MC approach, Method B. The MC method requires aggregation over several steps per move to give a result obeying the canonical distribution, and the actual reported results in Table 2 are time-based averages of several computations per move, with a move acceptance probability of around 30%. The move perturbations are selected randomly through the Markov-chain approach, with a probability dependent on Δx^N, which represent all possible generated combinations due to perturbing inter-particle separation, x. The difficulty with the MC approach is therefore appreciable for larger ensembles owing to the total size of possible configurations. This accounts for the lower MC method accuracies depicted in the Table 2.

4. Conclusion

In this article we present the rationale for two broad thermostat definitions, the MD and MC thermostats. It then proceeds to describe their basic approaches and then compares and contrasts them. The choice of a thermostat for a given simulation is motivated by several factors, not least being the particular parameter of interest, rather than ease of implementation. In summary, the MD approach is generally deterministic, time-reversible, and canonical in the Hamiltonian and potential energy. The behavior of the system Hamiltonian is tolerant of drifts from the equilibrium, making the method more amenable iterative error accumulation. External degrees of freedom for the system, such as a bounding bath temperature, can readily be defined in terms of their ergodic impact on the system. The calculated dynamical trajectories can therefore be much smoother, even though method generally lacks the ability to constrain the total kinetic energy. The MC method has the main advantage of ease of thermostatic specification and description of external degrees of freedom. In this work, several difficulties were associated with the MC method, particularly that of determining time-related dynamics and the drastically increased combinations as a power of ensemble size. However, the article highlights the importance of correct temperature specification. Commercial packages are necessarily heuristic with respect to this specification and as systems evolve, particularly more towards nanoscale aggregates and ever shorter interaction timescales, the effect of temperature will become ever more important. This suggests clear directions for more future research.

Acknowledgments

The authors hereby acknowledge the National Research Foundation (NRF) for funding support through KIC 2016, Grant No. 104078.

References

[1] Chandler D *Introduction to Modern Statistical Mechanics* (New York: Oxford University Press)
[2] Graben HW and Ray JR 1991 *Phys. Rev. A* **43** 4100–4103
[3] Ocaya RO and Terblans JJ 2016 *SoftwareX* doi: 10.1016/j.softx.2016.05.005
[4] Griebel M, Knapek S, Zumbusch G, in Barth TJ et al.(Eds.) 2007 *Numerical Simulation in Molecular Dynamics, in: Texts in Computational Science and Engineering 5*, (Berlin: Springer)
[5] Hockney R 1970 *Methods Comp. Phys.* **9** 136–211
[6] Verlet L 1967 *Phys. Rev.* **159** 98–103
[7] Schneider T and Stoll E 1978 *Phys. Rev.* B **17** 1302–1322
[8] Wang W and Skeel RD 2003 *Mol. Phys.* **101** 2149–2156
[9] Berkowitz M, Morgan JD and McCammon JA 1983 *J. Chem. Phys.* **78** 3256–3261
[10] Andersen H 1980 *J. Chem. Phys.* **72** 2384–2393
[11] Berendsen H, Postma J, van Gunsteren W, Di Nola A and Haak J 1984 *J. Chem. Phys.* **72** 3684–3690
[12] Hoover W 1985 *Phys. Rev.* A **31** 1695–1697
[13] Nosé S 1984 *Mol. Phys.* **101** 255–268
[14] Frenkel D and Smit B 1996 *Understanding molecular simulations: from algorithms to applications* (New York: Academic Press)
[15] Gear C *Numerical initial value problems in ordinary differential equations* (New Jersey: Prentice-Hall)
[16] Bond S, Leimkuhler B and Laird B 1999 *J. Comput. Phys.* **151** 114–134
[17] Hünenberger PH 2005 *Adv. Polym. Sci.* **173** 105–149 doi:10.1007/b99427
[18] Allen MP and Tildesley DJ 1987 *Computer simulation of liquids* (New York: Oxford University Press)
[19] Rugh HH 1997 Phys. Rev. Lett. **78** 772–774
[20] Evans DJ and Sarman S 1993 Phys. Rev. E. **48** 65–70
[21] Butler BD, Ayton G, Jepps OG and Evans DJ 1998 J. Chem. Phys. **109** 6519–6522
[22] Woodcock LV 1971 *Chem. Phys. Lett.* **10** 257–261
[23] Haile JM and Gupta S 1983 *J. Chem. Phys.* **79** 3067–3076
[24] Hoover WG, Ladd AJC and Moran B 1982 *Phys. Rev. Lett.* **48** 1818–1820
[25] van Gunsteren WF, Nanzer AP, Torda AE 1995 *Molecular simulation methods for generating ensembles or trajectories consistent with experimental data*, in: Binder K, Ciccotti G (eds) *Monte Carlo and molecular dynamics of condensed matter systems*, Proceedings of the Euroconference, SIF, Bologna, Italy, **49** 777–788
[26] Finnis MW, Sinclair JE 1984 *Philos. Mag. A* **50** 45
[27] Daw MS, Foiles SM and Baskes MI 1993 *Mater. Sci. Rep.* **9** 251–310
[28] Daw MS and Baskes MI 1984 *Phys. Rev. B* **29** 6443–6453
[29] Todd BD and Lynden-Bell RM 1993 *Surface Science* **281** 191–206
[30] Chamati H, Papanicolaou NI, Mishin Y and Papaconstantopoulos DA 2006 *Surface Science* **600** 1793–1803 doi: 10.1016/j.susc.2006.02.010
[31] Mishin Y 2005 *in: Yip S (Ed.), Handbook of Materials Modeling* (The Netherlands: Springer) 459
[32] Doye JPK and Wales DJ 1998 *New J. Chem.* 733–744
[33] Swope W, Andersen H, Berens P and Wilson K 1982 *J. Chem. Phys.* **76** 637–649
[34] Landau L and Lifschitz E 1976 *Mechanics, Course of Theoretical Physics* **Vol. 1** (Oxford: Pergamon Press)
[35] Terblans JJ, Erasmus WJ and Viljoen EC 1999 *Surface Interface Anal.* **28** 70–72

[36] van der Walt C, Terblans JJ, Swart HC 2014 *Comput. Mater. Sci.* **83** 70–77
[37] Terblans JJ 2002 *Surf. Interface Anal.* **33** 767–770
[38] Kraftmakher Y 1998 *Phys. Rep.* **299** 79–188
[39] Fluss MJ, Smedskjaer LC, Siegel RW, Legnini DG and Chason MK *in: Hasiguti RR and Fujiwara K (Ed.)* Proceeding of the Fifth International Conference on Positron Annihilation, Lake Yamanaka, Japan, April 811, 1979, (Japan Institute of Metals) 97
[40] Tritshauser W and McGervey JD 1975 *Appl. Phys.* **6** 177–180

IOP Conf. Series: Journal of Physics: Conf. Series **905** (2017) 012032 doi:10.1088/1742-6596/905/1/012032

Ab initio studies of isolated boron substitutional defects in *graphane*

R E Mapasha[1], **N Chetty**[1,2]

[1] Department of Physics, University of Pretoria, Hatfield campus, Pretoria 0002, RSA.
[2] National Institute for Theoretical Physics, Johannesburg 2000, RSA.

E-mail: edwin.mapasha@up.ac.za

Abstract. We have systematically studied energetics, structural and electronic properties of different configurations of the B atoms substituting C-H pairs located on a single hexagonal ring in a *graphane* system using the first-principles density functional theory (DFT). A total number of 12 distinct B dopants configurations were identified and characterized. Based on the formation energy analysis, we found that relative stability of B dopants depends greatly on the defect configurations. Our results suggest that the B substitutions prefer to be distributed randomly but avoiding the formation of homo-elemental B-B bonds in a *graphane* system, at any concentration. Generally, the values of band gap decrease as the number of B dopants increases, but the low energy configurations have large band gaps compared to those that have homo-elemental bonds. As a result, the band gap of *graphane* can be fine tuned through the change in the structural arrangement of B atoms. The adequate control of the electronic structure of *graphane* through doping should be essential for technological device applications.

1. Introduction

A fully hydrogenated graphene monolayer (*graphane*) is a wide band gap semiconductor [1]. The previous density functional theory (DFT) studies reported a band gap of about 3.30 eV [2, 3, 4]. *Graphane* is likely to find its use in nanoelectronic technology owing to its relatively thin membrane [5]. On the other hand, *graphane* can be used as hydrogen storage template due to its extremely high hydrogen density [6].

The adequate control of the electronic properties of *graphane* through structural modification will be extremely important for technological device applications. For the fact that *graphane* has a low dimensionality, it can easily be modified at the atomic level. There are various techniques suggested [5] (for example, application of strain, creating vacancies etc.). For the latter, the hydrogen (H) vacancy and a pair of carbon and hydrogen (C-H pair) vacancy introduce a stable magnetism in a *graphane* system [5, 7, 8, 9, 10]. Foreign atom doping is another widely used method to alter electronic structure of a *graphane* system [5]. Substituting C atom with Boron (B) or Nitrogen (N) atom in a *graphane* monolayer surely causes the p-type or n-type system respectively and thereby causes a semiconductor-to-metal transition [11]. It has been reported that a *graphane* monolayer can be used as a base for creating new promising and nanotechnologically useful devices [12]. Experimentally, the B substitution in a *graphane* monolayer can occur through high energy ion implantation, as it was done in a free standing graphene [13]. In a *graphane* monolayer, this substitution or doping will affect the C atom as well as its bonded H atom (C-H pair). The creation of C-H pair vacancy leaves three C atom, each with dangling bond and therefore, the well know three valence electrons in a B atom should be able to saturate this vacancy induced dangling bonds through electronic pairing. Thus, understanding the effect of B atom on the electronic structure of *graphane* is important for its application in the nanotechnological devices.

In this work, we systematically investigated the energetics and electronic structure of B substituting C-H pairs (B_{CH}) on a single hexagonal ring in a *graphane* system using the first-principles DFT approach. To understand the role of B-B interactions in a *graphane* monolayer, the number of B dopants have been increased from 1 to 6, but restricted to a single hexagonal ring. At each number of B substitution dopants, different configurations where identified and examined. Firstly, our results show that the B substituting a C-H pair is more feasible than when substituting only a C atom in a *graphane* monolayer. The band gap of *graphane* can be fine tuned through the change in the structural arrangement as well as through the change in B atom concentrations. The adequate control of the electronic structure of *graphane* through doping should be essential for technological device applications.

2. Computational details

Our ground state electronic structure calculations were performed on the frame work of density functional theory implemented within the Vienna *ab initio* simulation package (VASP) code [14, 15, 16, 17]. The generalized gradient approximations (GGA) parameterization of Perdew, Burke and Ernzerhof (PBE) was used for the exchange-correlation interactions [18]. The electronic spin was included for the start of each calculation. The projector augmented wave (PAW) method was employed for the generation of the atomic pseudopotentials [19]. For the plane wavefunctions expansion, an energy cut-off was set to 500 eV. For accurate sampling of the Brillouin zone, a $10\times10\times1$ k-meshes generated using Monkhorst-Pack scheme [20] were used. The total energies were allowed to converge to 10^{-7} eV.

To accurately populate electronic states in our calculations, the well known Methfessel-Paxton (MP) scheme [21] was used. The structure is allowed to fully relax until the inter-atomic forces is less than 0.01 eVÅ^{-1}. To eliminate the unwanted interlayer spurious interactions in the periodic system, the 15 Å spacing was set between the *graphane* layers.

3. Results and discussion

3.1. Identification of B configurations

In this study, the investigation of B atom substituting a C-H pair (B_{CH}) on a single hexagonal ring in a 7×7 supercell of a *graphane* monolayer was considered. To understand the role of B-B interactions in a *graphane* monolayer, the number of B dopants were increased from 1 to 6, but restricted to a single hexagonal ring. At each number of B substitution dopants, different configurations where studied. This study was carried out on a chair-like-*graphane* system which is a thermodynamically stable conformer of *graphane* [2].

The labels 1 to 6 in Fig 1 indicate the C-H pair sites which should be occupied by the B atoms. The labeling help us to name and distinguish the identified B configurations. All the identified configurations are presented in Table 1. Configuration c_1 is called B_{CH_1} denoting a B dopant substituting C-H pair at site 1. For two B substitutions, we where able to identify three distinct configurations c_1, c_2 and c_3 namely, $B_{CH_{12}}$, $B_{CH_{13}}$ and $B_{CH_{14}}$ respectively. The configurations $B_{CH_{12}}$, $B_{CH_{13}}$ and $B_{CH_{14}}$ denote two B atom substituting C-H pairs at sites 1 and 2; sites 1 and 3, and at sites 1 and 4 respectively. In following this procedure, a total number of 12 unique B configurations on a single hexagonal ring of a *graphane* monolayer were obtained and shown in table 1. The energetics and electronic properties of each configuration where calculated and discussed.

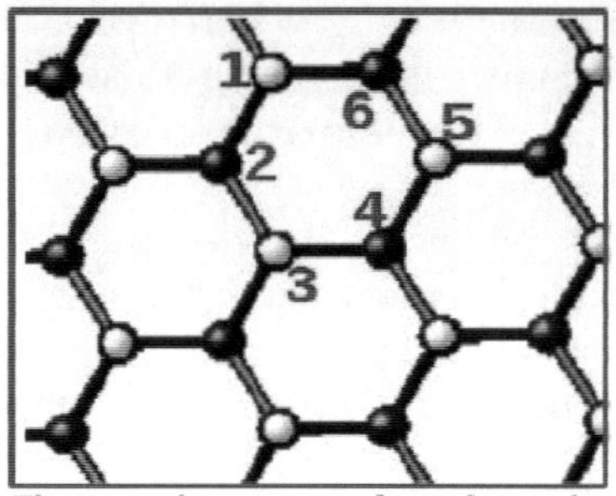

Figure 1. The B adsorption sites in an unrelaxed state of a *graphene* monolayer supercell obtained from Ref [22]. The C-H pairs labeled 1 to 6 indicate the C-H pair sites which are occupied by the B atoms.

Table 1. The B atom configurations identified from a 7×7 supercell of a *graphane* monolayer.

Configurations	B defects sites	No of B defects
c_1	B_{CH_1}	1
c_2	$B_{CH_{12}}$	2
c_3	$B_{CH_{13}}$	2
c_4	$B_{CH_{14}}$	2
c_5	$B_{CH_{123}}$	3
c_6	$B_{CH_{124}}$	3
c_7	$B_{CH_{135}}$	3
c_8	$B_{CH_{1234}}$	4
c_9	$B_{CH_{1235}}$	4
c_{10}	$B_{CH_{1245}}$	4
c_{11}	$B_{CH_{12345}}$	5
c_{12}	$B_{CH_{123456}}$	6

3.2. Formation energies and structural properties

To evaluate the relative stability and formation possibility of the identified B substitutional configurations presented in table 1, their formation energies $E_f(B_{CH})$ are calculated as,

$$E_f(B_{CH}) = E_{tot}(B_{CH}) - E_{tot}(G) - n_B\mu_B + n_H\mu_H + n_C\mu_C, \quad (1)$$

in which $E_{tot}(B_{CH})$, $E_{tot}(G)$, μ_B, μ_H and μ_C denote the total energy of the B doped *graphane* system, the corresponding pristine *graphane* monolayer, B in an orthorhombic structure, H molecule in a large box and C in a graphene monolayer respectively. The structure optimization was carried out with respect to atomic positions as well as cell parameters. The variables n_B, n_H and n_C denote the number of B dopants substituting the number of H and C atoms removed from a *graphane* monolayer.

The obtained $E_f(B_{CH})$ values of the identified B configurations are presented in Fig 2(a). The $E_f(B_{CH})$ value of a single B substituting a C-H pair B_{CH_1} (configuration c_1) is 0.99 eV. Although

IOP Conf. Series: Journal of Physics: Conf. Series **905** (2017) 012032 doi:10.1088/1742-6596/905/1/012032

the positive formation energy indicates the thermodynamic instability of a defect, this value is far less than that for a B dopant substituting only a C atom in a *graphane* system of 4.51 eV calculated at the same level of accuracy. Although the B_{CH} might not form spontaneously, can occur through ion implantation, with less energy compared to that of the B_C defect in a *graphane* system. We measured the bond distances (C-B and C-C) to examine the influence of the B dopants on the structural properties of a *graphane* system. The C-B bondlength shown in Fig 2(b) suggests that the bonding between a B defect and its surrounding C atoms is the sp^3-like bond just like in a pristine *graphane* system. The C-B bondlength is found to be 0.02 Å lower than the homogeneous C-C bondlengths in the same system, revealing that an isolated B dopant in a *graphane* system does not induces a significant structural distortions.

To study the effects of B-B substitution interactions in a *graphane* system, the number of B dopants were increased to six on a single hexagonal ring in a *graphane* monolayer. For each number of dopants, different unique configurations are identified and presented in table 1. Fig 2 shows that the value of the formation energies increases as the number of dopants increases, although not monotonically. For two dopants concentration, three distinct configurations i.e. $B_{CH_{12}}$ (c_2), $B_{CH_{13}}$ (c_3) and $B_{CH_{14}}$ (c_4) showing ortho, meta and para isomers are identified. For this concentration, the $E_f(B_{CH})$ follows an order of $c_3 < c_4 < c_2$. The $E_f(B_{CH})$ for c_3 is almost equal to that of c_1. Configuration c_2 to have high energy of 2.98 eV should be attributed to the formation of energetically unfavourable homo-elemental B-B bond of 1.66 Å in a *graphane* monolayer (shown in Fig 2(c)) as opposed to the formation of C-B bond. The formation of this relatively large B-B bond in (c_2) induces a significant structural distortion. As it is noted that C-B and C-C bond distances reduce to 1.51 Å and 1.49 Å respectively, as compared to those in c_1 (shown in Fig 2(b)) as well as in c_2 (shown in Fig 2(d)) and c_3 (shown in Fig 2(e)).

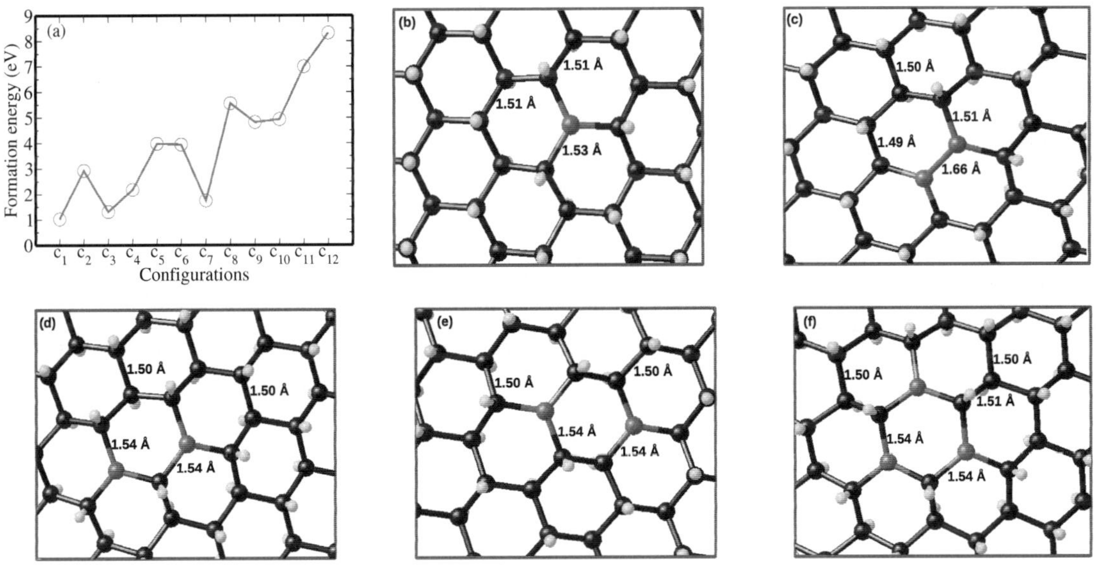

Figure 2. (a) The calculated formation energies of identified distinct configurations of B dopants in a *graphane* monolayer. Selected relaxed geometries of the identified configurations (b) c_1, (c) c_2, (d) c_3, (e) c_4 and (f) c_7 showing the significant variation in bondlengths due to the effect of B dopants.

In the case of three B dopants, three distinct configurations are noted i.e. c_5, c_6 and c_7. For this concentration, the $E_f(B_{CH})$ follows the order of $c_5 = c_6 > c_7$. The formation of B-

B bond should be a major driving force for relatively high formation energy in configurations c_5 and c_6 shown in Fig 2(a). Configuration c_7 has three B dopants that do not form a B-B bond as shown in Fig 2(f), and thus has a relatively low $E_f(B_{CH})$ of about 1.04 eV. We observe in Fig 2(a) that from configuration c_8 to c_{12}, the formation energy increases drastically, revealing that the substitution of B atoms in a contiguous sequence or in a compact sequence on a single hexagonal ring in a *graphane* monolayer is highly unstable. In summary, configuration c_7 competes very well with configurations c_1, c_3 and c_4 in terms of thermodynamic stability and are the lowest energy competing configurations. Therefore, experimental investigation of these B doped systems is required. In the next section, we investigated the influence of B dopants on the electronic structure of *graphane*.

3.3. Electronic properties

We further investigated how the effect of B dopant alters the electronic structures of *graphane* in all identified configurations using the density of states (DOS) plots shown in Fig. 3. We briefly discuss the DOS for a pristine *graphane* system shown in Fig. 3(a). A *graphane* monolayer is a non-spin polarized system, since its majority DOS (spin-up) are invertedly symmetrical to minority DOS (spin-down) for the entire plot. The valence band maximum (VBM) and conduction band minimum (CBM) are separated by a band gap of 3.35 eV. This band gap value is in good agreement with the values previously reported in Refs [2, 3, 23, 24, 25] revealing that *graphane* is a wide band gap semiconductor.

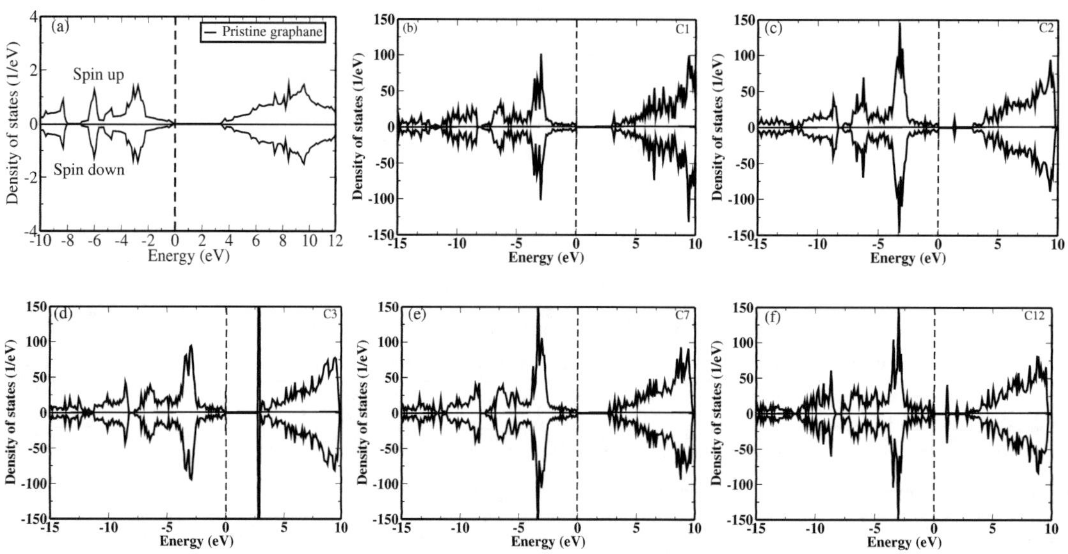

Figure 3. The density of states (DOS) for (a) a pristine *graphane* system, (b) c_1, (b) c_2, (d) c_3, (e) c_7 and (f) c_{12}. The Fermi level is marked by the dashed line.

All the DOS plots show that the B doped *graphane* systems are non-magnetic, even if the spin is imposed before the calculations they turn out to be non-spin polarized after relaxation. The DOS plot for c_1 shows that a B atom induces the non-spin polarized states with the sharp peaks just below the CBM, slightly reducing the band gap of *graphane*. Fig. 4 shows that doping a *graphane* system with a single B atom reduces a band gap to 2.98 eV. The DOS plot for c_2 shows that the two B atoms that are bonded to together induce the non-spin polarized states in the vicinity of the band gap as shown in Fig. 3(c). Revealing that formation of a homo-elemental B-B bond in a *graphane* monolayer reduces a band gap from 3.35 eV to 1.19 eV.

However, for the same concentration, the DOS plot for c_2 shows that the two B atoms that are not bonded together but doped in the same hexagonal ring induce a non-spin polarized pronounced peak states at the CBM (see Fig. 3(d)). This configuration has a band gap of about 2.66 eV. Configuration c_{12} that comprises of six B dopants forming a hexagonal ring in a *graphane* monolayer has a shortest band gap of 1.08 eV. This is because a hexagonal ring of B atoms doped in a *graphane* system induces the multiple non-spin polarized states in the band gap (see Fig. 3(f)).

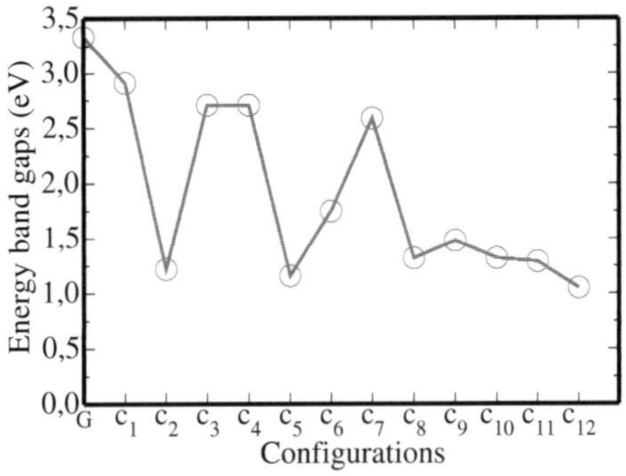

Figure 4. The calculated band gaps for all identified B dopants configurations in a *graphane* monolayer.

4. Conclusions

Using the first-principles density functional theory, we successfully studied energetics, structural and electronic properties of B doped *graphane*. To understand the role of B-B interactions in a *graphane* monolayer, the number of B dopants was increased from 1 to 6, but restricted to a single hexagonal ring in a *graphane* system. At each number of B substitution dopants, different configurations were studied. A total number of 12 distinct B dopants configurations were identified and examined. The lowest energy competing B dopants configurations such as c_1, c_3, c_4 and c_7 were noted, other structural configurations have high formation energies.

Our results suggest that the B substitutions would prefer to be distributed randomly but avoiding the formation of B-B bonds in a *graphane* system at any concentration. Although such kind of systems have positive formation energies, they should be plausible structures remembering that the zero temperature and pressure conditions assumed for our DFT calculations could also have a significant contribution on the obtained positive enthalpy of formation. Besides that, these configurations c_1, c_3, c_4 and c_7 should occur under high energy deposition techniques such as ion implantation.

Our results further show that the effects of B dopants reduce the maximum band gap of *graphane*. The low energy competing configurations such as c_1, c_3, c_4 and c_7 have a relative wide band gap, whereas those that have energetically unfavourable homo-elemental B-B bond possess a small band gap. This suggests that the band gap of *graphane* can successfully be fine tuned through the variation in the B concentration. The adequate control of the electronic structure of *graphane* is essential for technological device applications. Experimental investigation of various dopants in a *graphane* system should be encouraged.

Acknowledgments

The work was financially supported by National Research Foundation (NRF) of South Africa (Grant number 88246). The authors would like to thank the University of Pretoria for resources.

References

[1] D. C. Elias, R. R. Nair, T. M. G. Mohiuddin, S. V. Morozov, P. Blake, M. P. Halsall. A. C. Ferrari, D. W Boukhvalov, M. I. Katsnelson, A. K. Geim, and K. S. Novoselov 2009, *Science* **323**, 610.
[2] J. O. Sofo, A. S. Chaudhari, and G. D. Barber 2007, *Phys. Rev. B* **75**, 153401.
[3] D. W. Boukhvalov, M. I. Katsnelson, and A. I. Lichtenstein 2008, *Phys. Rev. B* **77**, 035427.
[4] D. W. Boukhvalov and M. I. Katsnelson 2009, *J. Phys. Condens. Mat.* **21**, 344205.
[5] H. Sahin, O. Leenaerts, S. K. Singh, and F.M. Peeters 2015, *WIREs: Comp. Mol. Sci.* **5**, 255.
[6] A. Savchenko 2009, *Science* **323**, 589.
[7] P. Chandrachud, B. S. Pujari, S Haldar, B. Sanyal, and D G Kanhere 2010, *J. Phys. Condens. Matter* **22**, 465502.
[8] H. Sahin, C. Ataca, and S. Ciraci 2009, *Appl. Phys. Lett.* **95**, 222510.
[9] B. Wu, and C. Yang 2012, *APL. Adv.* **2**, 012173.
[10] B. S. Pujari and D.G. Kanhere 2009, *J. Phys. Chem. C* **113**, 21063-7.
[11] Y. Wang, Y. Ding, S. Shi, and W. Tang 2011, *Appl. Phys. Lett.* **98**, 163104.
[12] A. Z. AlZahrani 2012, *Phys. B* **407** 992-1002.
[13] U. Bangert, A. Bleloch, M. H. Gass, A. Seepujak, and J. van den Berg 2010, *Phys. Rev. B* **81**, 245423.
[14] G. Kresse and J. Hafner 1993, *Phys. Rev. B* **47**, 558.
[15] G. Kresse and J. Hafner 1994, *Phys. Rev. B* **49**, 14251.
[16] G. Kresse and J. Furthmuller 1996, *Phys. Rev. B* **54**, 11169.
[17] G. Kresse and J. Furthmuller 1996, *Comput. Mater. Sci.* **6**, 11169.
[18] J. P. Perdew, K. Burke, and M. Ernzerhof 1996, *Phys. Rev. Lett.* **77**, 3865.
[19] P. E. Blochl 1994, *Phys. Rev. B* **50**, 17953.
[20] H. J. Monkhorst and J. D. Pack 1976, *Phys. Rev. B* **13**, 5188.
[21] M. Methfessel, and A. T. Paxton 1989, *Phys. Rev. B* **40**, 3616.
[22] R.E. Mapasha, M.P. Molepo, R.C. Andrew, and N. Chetty 2016, *Phys. E.* **79**, 52-58.
[23] S. Lebegue, M. Klintenberg, O. Eriksson, and M. I. Katsnelson 2009, *Phys. Rev. B* **79**, 245117.
[24] O. Leenaerts, H. Peelaers, A. D. Hernandez-Nieves, B. Partoens, and F. M. Peeters 2010, *Phys. Rev. B* **82**, 195436.
[25] F. Karlicky, R. Zboril and M. Otyepka 2012, *J.Chem. Phys.* **137**, 034709.

IOP Conf. Series: Journal of Physics: Conf. Series **905** (2017) 012033 doi:10.1088/1742-6596/905/1/012033

Clustering and spin interactions in Fe-doped diamond

E M Benecha[1] and E B Lombardi[1]

[1]College of Graduate Studies, University of South Africa, P.O. Box 392, UNISA 0003, Pretoria, South Africa

E-mail: ebenecha@gmail.com

Abstract. The origin of high Curie temperature ferromagnetism in dilute magnetic semiconductors and oxides has often been attributed to clustering and crystallographic phase separation of magnetic atoms, which may have a detrimental impact on the properties of the host material for target applications. We present Density Functional Theory calculations on the stability and magnetic interactions of embedded Fe^{+1} ions in diamond by considering various possible cluster configurations. We find that Fe ions have a strong tendency to form embedded clusters in diamond, with larger cluster sizes (n>3) suppressing the induced spin moment. Also, we find that the electronic structure of the stable embedded Fe^{+1} clusters is insulating, in contrast to homogeneous distribution, where a half metallic character with 100% spin polarisation at the Fermi level has previously been predicted. These results present important implications to the understanding of the properties of transition metal dopants in diamond, as well as in other dilute magnetic semiconductors where the effect of aggregation of dopants has generally been neglected.

1. Introduction

Ferromagnetic ordering of dopants in semiconductors continues to attract intense interest [1] for the development of a new generation of multifunctional devices capable of utilizing spin degrees of freedom in semiconductor materials [2, 3]. However, despite recent experimental successes on doping techniques and characterization of the ensuing properties of dopants in semiconductors, the origin and control of ferromagnetism in dilute magnetic semiconductor materials (DMSs) is not well understood [4]. Particularly, segregation and clustering effects of transition metal dopants (as opposed to homogeneous distribution) in semiconductors has been found to play an important role in determining the properties of the host material [4, 5].

Transition metal point defects in diamond have been a subject of significant research interest for many years since the discovery of transition metal-related active centers arising from use of transition metal solvent catalysts during high pressure-high temperature (HPHT) growth of synthetic diamond [6]. More recently, it has been demonstrated that defect centers in diamond, particularly the $(\text{N-V})^{-1}$ center exhibits remarkable spin properties [7] suitable for quantum computing and optoelectronic applications.

Following the prediction by Dietl *et al* [8] that the Curie temperature of a DMS scales as the reciprocal of its lattice constant, there has been renewed interest in diamond aimed at finding suitable transition metal defect centers which may lead to room temperature ferromagnetic ordering due to diamond's small lattice constant (3.569 Å), compared to other semiconductors. In addition, diamond is a wide band gap semiconductor with attractive properties which makes it suitable for a wide range of electrical, electronic [6] and photonic [9] applications. So far, substitutional Fe^{+1} [10] and divacancy Cr^{+2} [11] have been predicted to order ferromagnetically in diamond, while ferromagnetic ordering in Mn-doped diamond has been predicted to be unlikely [12] – in contrast to its behaviour in other dilute magnetic semiconductors such as GaAs:Mn [13] and ZnO:Mn [14]. However, the role of metallic clusters in determining the magnetic and electronic properties of transition metals in diamond has not been considered yet.

While models based on homogeneous distribution of magnetic dopants in a semiconducting host provide insight into DMS systems, it is important to also consider the effects of disorder and other configurations, including clustering of dopant atoms [5, 4]. We have previously [10] shown that substitutional Fe^{+1} in *p*-type diamond exhibits a half-metallic character, with a magnetic moment of 1.0 μ_B per Fe ion and a large ferromagnetic stabilization energy of 33 meV. In this study, therefore, we investigate the properties of Fe^{+1} clusters in diamond using *ab initio* Density Functional Theory electronic structure methods. We consider the stability of different cluster configurations of Fe^{+1} and show that Fe^{+1} ions have a strong tendency to form clusters in diamond, with the electronic structure of the most stable configuration being insulating. This is in contrast to the results of homogeneous distribution of Fe^{+1}, where half metallic magnetic ordering is predicted. Further, we find that larger clusters suppress the induced magnetic moment, while being energetically more favourable compared to smaller clusters. These findings explain various previously perplexing experimental results in other dilute magnetic semiconductor such as the observed strong dependence of the magnetization on growth conditions such as temperature and annealing in Cu:ZnO [15], Cr:GaN [16] and Mn:GaAs [17].

2. Method

The properties of embedded Fe^{+1} clusters in diamond have been studied using *ab initio* Density Functional Theory calculations, as implemented in the CASTEP code [18]. The Perdew-Burke-Ernzerhof generalized gradient approximation [19], incorporating the on-site coulomb interaction parameter (GGA+U) was used to treat exchange-correlation effects, while Vanderbilt ultrasoft pseudopotentials [20] were used for valence-core interactions.

A well converged $4 \times 4 \times 4$ Monkhorst-Pack k-grid sampling [21] for integration over the Brillouin zone (32 k-points in the irreducible wedge of the Brillouin zone) was employed for spin and geometry optimization, with an optimized plane wave cut-off energy of 310 eV. Increasing the plane wave cut-off energy or number of k-points resulted in negligible changes in the energy, structural relaxation and magnetic effects (less than 10^{-4} eV, 10^{-3} Å, and $10^{-1}\mu_B$, respectively).

Geometry optimization, binding energy, and electronic structure calculations of the embedded Fe^{+1} clusters were carried out using a 64-atom diamond supercell constructed from $2 \times 2 \times 2$ conventional fcc diamond unit cells with an optimized lattice constant of 3.569 Å, in close agreement with the experimental value of 3.567 Å [22]. Various clusters sizes (Figure 1) consisting of up to four Fe atoms at substitutional sites in the diamond supercell were considered, and Full geometry optimization was performed without any symmetry or spin restrictions for each cluster in order to determine the most stable cluster configuration. Also, various initial spins from $0\mu_B$ to $6\mu_B$ were considered for each cluster configuration inorder to determine the most stable spin states corresponding to the ground state geometries.

The binding energy E^b of each cluster complex configuration was determined by comparing its complex formation energy with that of its constituent defects relative to the limit in which the defects are well separated as [23, 24]

$$E^b_{A_xB_y} = E_{A_xB_y} + E_N - \{E_{A_x} + E_{B_y}\} \tag{1}$$

where $E_{A_xB_y}$ is the total energy of the cluster complex consisting of x atoms of defect A and y atoms of defect B, while E_N is the total Energy of pristine diamond, calculated in the same defect charge state, i.e. $q = +1$. In this definition, a negative binding energy implies that the

IOP Conf. Series: Journal of Physics: Conf. Series **905** (2017) 012033 doi:10.1088/1742-6596/905/1/012033

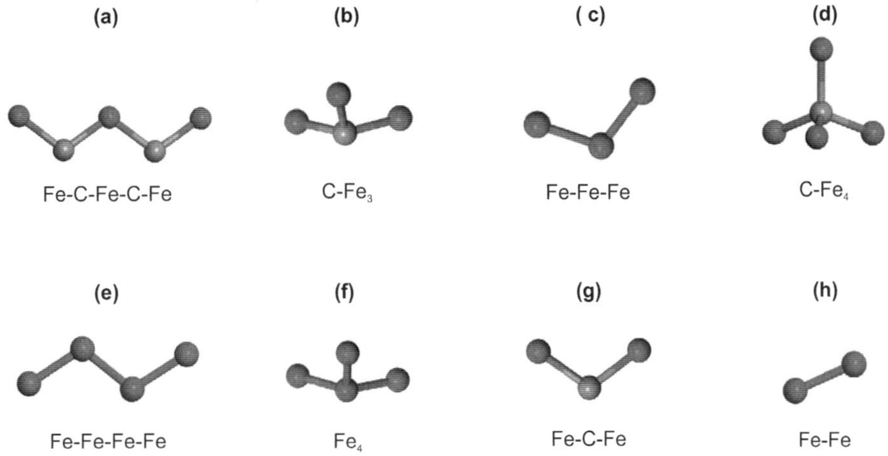

Figure 1. (Color online) Various possible cluster configurations of substitutional Fe^{+1} in diamond (red balls represent Fe atoms, while grey balls represent C atoms).

energy needed to create isolated defects is higher than that for forming the complex, i.e. cluster complex formation is thermodynamically advantageous.

3. Results and discussion

Table 1. Binding energies of various Fe^{+1} cluster configurations in diamond together with point symmetry and net induced spin per Fe^{+1} ion.

Configuration	Structure (Figure 1)	Number of Fe ions	Symmetry	Net spin per Fe^{+1} ion (μ_B)	Binding energy (eV)
Fe-C-Fe-C-Fe (chain)	(a)	3	C_{2v}	0.5	-3.05
C-Fe-Fe-Fe (C-Fe_3	(b)	3	C_{3v}	0.9	-3.05
Fe-Fe-Fe (chain)	(c)	3	C_s	0.3	-2.07
C-Fe-Fe-Fe-Fe (C-Fe_4)	(d)	4	C_{2v}	0.1	-1.84
Fe-Fe-Fe-Fe (chain)	(e)	4	C_s	0.0	-1.75
Fe-Fe-Fe-Fe (Fe_4)	(f)	4	C_s	0.0	-1.75
Fe-C-Fe (chain)	(g)	2	C_{2v}	0.9	-1.84
Fe-Fe (chain)	(h)	2	C_{3v}	0.8	-0.49

Table 1 summarizes the calculated binding energies of the various Fe^{+1} cluster configurations (Figure 1) as well as their point symmetries and net induced magnetic moments in diamond. We see that all the considered cluster configurations have negative binding energies, ranging from -3.05 eV to -0.49 eV, indicating that Fe ions have a tendency to form clusters in diamond, as opposed to a homogeneous spatial distribution. The magnitude of calculated binding energies is dependent on the cluster size and bonding of Fe ions within the cluster, with small clusters

IOP Conf. Series: Journal of Physics: Conf. Series **905** (2017) 012033 doi:10.1088/1742-6596/905/1/012033

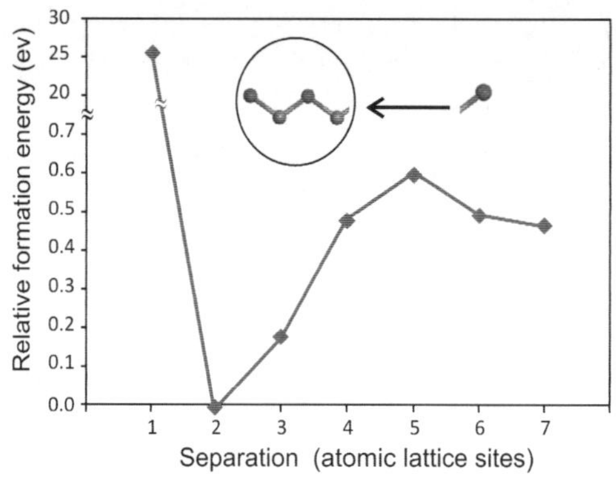

Figure 2. Formation energies of Fe-C-Fe ion cluster as a function of separation distance with a third Fe ion.

of two Fe ions being energetically unfavourable (by up to 2.6 eV) compared to those with three or four Fe ions. We find that clusters containing interspersed or alternating chains of Fe-C atoms are energetically most favorable. Also, the difference in binding energy between some cluster configurations is similar, suggesting that various cluster sizes consisting of different number of Fe atoms and arrangement can co-exist, particularly considering that dilute magnetic semiconductor systems are often grown under non-equilibrium conditions. Therefore, the statistical distribution, and consequently the electronic and magnetic properties of Fe-doped diamond, will depend sensitively on details of the growth conditions and other factors such as the mobility of dopants.

Importantly, the induced magnetic moment per Fe ion also depends sensitively on the cluster size and arrangement of Fe atoms, with larger cluster sizes ($n > 3$) suppressing the induced spin moments. Therefore, Fe is likely to form clusters in diamond during annealing while quenching the induced magnetic moment. This result is consistent with the experimentally observed decay in saturation magnetic moments upon annealing in some dilute magnetic semiconductors, such as Cu:ZnO [15], Cr:GaN [16] and Mn:GaAs [17]. This trend is expected based on increased dopant mobility during annealing, which in turn enhances formation of clusters, leading to suppressed magnetization.

While the binding energy indicates only that a defect complex can in principle be formed or not, the formation energy on the other hand provides information on the energetic stability of a complex cluster, as well as on the relative stabilities between different configurations of the same cluster. In order to determine the nature of the embedded Fe clusters in diamond, we have therefore also calculated the relative formation energy of the most likely cluster configuration, the Fe-C-Fe-C-Fe chain bonded cluster (with the third Fe ion at varying separations) in a large 128 atom cluster, using the method described in reference [10]. We see that the formation energy reaches a minimum when the third Fe atom is at the 2nd nearest neighbour separation as depited in Figure 2, thus forming the stable Fe-C-Fe-C-Fe chain bonded cluster. As expected, moving the third Fe ion further to the nearest neighbour position causes a significant increase in formation energy (by 25 eV), due to electrostatic repulsive interactions. This indicates that alternating Fe-C clusters, with a C atom in between, are more stable compared to simple Fe-Fe bonded pair formation, consistent with the predictions of Mn:GaN [25], where the formation energies of directly bonded Mn-Mn cluster configurations are found to be energetically unfavourable.

3.1. Electronic properties and magnetic interactions

Figure 3 illustrates the density of states of the most stable embedded Fe cluster bonding configuration (Fe-C-Fe-C-Fe), as well as that of single substitutional Fe in diamond consisting of a uniform spatial dopant distribution. We find that clustering of Fe introduces multiple spin polarized impurity states in the diamond band gap, as opposed to single substitutional doping. We also find that hybridization between the carbon sp^3 orbitals and Fe $3d$ orbitals is stronger in the case of embedded Fe clusters as compared to single substitutional Fe. This strong orbital hybridization in the case of embedded cluster leads to band splitting and the formation of a deep localized Fe-$3d$ orbital, which may explain the origin of the observed clustering tendency.

Further, we find that while single substitutional Fe has an electronic structure with half metallic character and 100% spin polarization at the Fermi level, the aggregation of Fe and formation of embedded Fe clusters in diamond renders the system insulating. This observation helps to elucidate hitherto puzzling discrepancies between experimental data and theoretical predictions in other dilute magnetic semiconductors. For example, in Mn: GaN, theoretical investigations predict the system to be half metallic, whereas experimental results show that the system is highly resistive and insulating [26]. This discrepancy is likely due to neglect of random distribution of dopants in the host semiconductor and clustering effects. Thus, these results highlight the importance of considering defect dynamics, distribution and aggregation to the understanding of the properties of transition metal dopants in diamond as well in other dilute magnetic semiconductors.

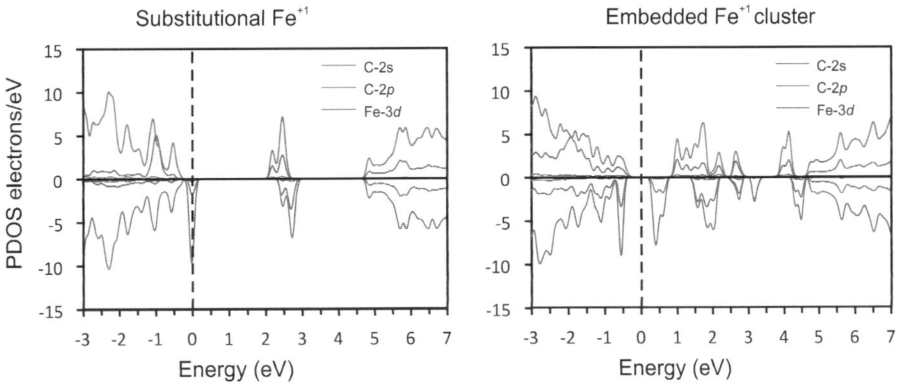

Figure 3. Partial Density of states of single substitutional Fe^{+1} and embedded Fe-C-Fe-C-Fe cluster in diamond.

4. Conclusion

We have systematically investigated the stability and spin interactions of embedded substitutional Fe^{+1} clusters in diamond using DFT+U electronic structure methods. Our results show that Fe ions have a strong tendency to form clusters in diamond. We find that larger clusters (of more than three Fe ions) are more stable compared to smaller clusters, while suppressing the induced magnetic moments. Therefore, Fe is likely to form clusters in diamond during annealing while quenching the induced magnetic moment. We also find that, while the electronic structure of single substitutional Fe has a half metallic character with 100% spin polarization at the Fermi level, the aggregation of Fe and formation of embedded Fe clusters in diamond renders the system insulating.

These results may explain the experimentally observed decay in saturation magnetic moments after annealing in some dilute magnetic semiconductors [15, 16, 17], which can be understood based on increased dopant mobility during annealing, which in turn enhances formation of clusters, leading to suppressed magnetization. Further, our results shed light on previous discrepancies between experimental data and theoretical predictions in other dilute magnetic semiconductors, for example, in Mn: GaN [26] where theoretical investigations predict the system to be half metallic, whereas experimental results show that the system is highly resistive and insulating. These results highlight the importance of considering the effect of distribution and aggregation of dopants to the understanding of the properties of transition metal dopants in dilute magnetic semiconductors.

References

[1] Dietl T and Ohno H 2014 *Rev. Mod. Phys.* **86**187
[2] Ohno H 1998 *Science* **281** 951
[3] Wolf S A , Awschalom D D, R. A. Buhrman R A, Daughton J M, von Molnar S, Roukes M L, Chtchelkanova A Y and Treger D M 2001 *Science* **294** 1488
[4] Dietl T 2010 *Nature Mater.* **9** 965
[5] Chambers S 2010 *Nature Mater.* **9** 956
[6] Sussmann R S 2009 *CVD Diamond for Electronic Devices and Sensors* (John Wiley and Sons Ltd., United Kingdom)
[7] Balasubramanian G, Neumann P, Twitchen D, Markham M, Kolesov R, Mizuochi N, Isoya J, Achard J, Beck J, Tissler J, Jacques V, Hemmer P R, Jelezko F and Wrachtrup J 2009 *Nature Mater.* **8** 383
[8] Dietl T, Ohno H and Matsukura F 2001 *Phys. Rev.* B **63** 195205
[9] Greentree A D, Fairchild B A, Hossain F M and Prawer S 2008 *Mater. Today* **11** 22
[10] Benecha E M and Lombardi E B 2013 *J. Appl. Phys.* **114** 223703
[11] Benecha E M and Lombardi E B 2011 *Phys. Rev.* B **84** 235201
[12] Erwin S C and Hellberg C S 2003 *Phys. Rev.* B **68** 245206
[13] Macdonald A H, Schiffer P and Samartha N 2005 *Nature Mater.* **4**, 195
[14] Sharma P, Gupta A, Rao K V, Owens F J, Sharma R, Ahuja R, Guillen J M O, Johansson B and Gehring G A 2003 *Nature Mater.* **2** 673
[15] Buchholz D B, Changa R P H, Song J H and Ketterson J B 2005 *Appl. Phys. Lett.* **87** 082504
[16] Singh R K, Wu S Y, Liu H X, Gu L, Smith D J and Newman N 2005 *J. Appl. Phys.* **86** 012504
[17] Moreno M, Trampert A, Jenichen B, Dweritz V and Ploog K H 2002 *J. Appl. Phys.* **105** 073911
[18] Clark S J, Segall M D, Pickard C J, Hasnip P J, Probert M J, Refson K and Payne M C 2005 *Z. Kristallogr.* **220** 567
[19] Perdew J P, Burke K and Ernzerhof M 1996 *Phys. Rev. Lett.* **77** 3865
[20] Vanderbilt D 1990 *Phys. Rev.* B **41** 7892
[21] Monkhorst H J and Pack J D 1976 *Phys. Rev.* B **13** 5188
[22] Donnay J D H and Ondik H M 1973 *Crystal Data: Determinative Tables* 3rd ed. (U.S. Department of CommerceNational Bureau of Standards JCPDS) Vol. 2.
[23] Raebiger R, Nakayama H and Fujita T 2014 *J. Appl. Phys.* **115** 012008
[24] Mahadevan P, Osorio-Guillen J M and Zunger A 2005 *Appl. Phys. Lett.* **86** 172504
[25] Cui X Y, Delley B, Freeman A J and Stampfl C 2007 *Phys. Rev.* B **76** 045201
[26] Dhar S, Brandt O, Trampert A, Friedland K J, Sun Y J, and Ploog K H 2003 *Phys. Rev.* B **67** 165205; Dhar S, Brandt O, Trampert A, Daweritz L, Friedland K J, Ploog K H, Keller J, Beschoten B and Guntherodt G 2003 *Appl. Phys. Lett.* **82** 2077

IOP Conf. Series: Journal of Physics: Conf. Series **905** (2017) 012034 doi:10.1088/1742-6596/905/1/012034

Computational study of dye adsorption onto Brookite TiO_2 surfaces for the applications in dye sensitized solar cells

N.E. Maluta [1,2], Dima R.S.[1,2], Nemudzivhadi H.[1], Maphanga R.R.[2,3], Sankaran [1],

[1]University of Venda, Department of physics, P/Bag X 5050, Thohoyandou, 0950
[2]National Institute for Theoretical Physics (NITheP), Gauteng, South Africa
[3]Materials Modelling Centre, University of Limpopo, P/Bag X 5050, Sovenga, 0727

E-mail: eric.maluta@univen.ac.za

Abstract. The theoretical and computational studies of dye sensitized solar cells (DSSCs) can contribute to a deeper understanding of these type of solar cells. In the current study the density functional theory (DFT) is used to understand the electronic properties of low index brookite (1 0 0) surface doped with ruthenium. The structural optimizations, band structure, and electronic density of states of doped and undoped titanium dioxide (TiO_2) brookite surface was performed using the first-principles calculations based on DFT emplotying a plane-wave pseudopotential method. The generalized gradient approximation (GGA) was used in the scheme of Perdew-Burke-Ernzerhof (PBE) to describe the exchange-correlation functional. All calculations were carried out with CASTEP (Cambridge Sequential Total Energy Package) code in Materials Studio of Accelrys Inc. The two different doping methods employed in the current work are, doping by replacement and adsorption. The overlap among the Ruthenium (Ru) 3d, Titanium (Ti) 3d, and Oxygen (O) 2p states enhance photocatalytic activity in the visible light region. The adsorption method shows that an equilibrium position is reached for ruthenium element after optimization. All the methods show that the TiO_2 brookite (1 0 0) surface reduces its band gap after been doped with the ruthenium element. From the two techniques used, the total energy of the doped structures show that they are energetically favorable, with the band gap being reduced to 0.263 eV compared to 2.376 eV of the pure system.

1. Introduction

The escalating costs of crude oil in international markets have created an urgency for utilization of alternative sources of energy. In developing countries like South Africa, efforts are underway to make use of renewable energy as the alternative way of energy production to replace the coal which is deplecting. Several solar plants have been installed in different provinces of South Africa, e.g. Limpopo, Northern Cape, etc. The use of sunlight for energy production is an added advantage to South Africa as a country, as many of its places receive high global solar radiation. Currently, the most useful part of solar spectrum is the visible light which is about 46 % of the sunlight,the other percentage belongs to infrared light 47 % and 5-7 % of the ultra violet light [1,2]. In photovoltaics technologies researchers have put more efforts on improving the efficiency of low cost solar cells which have high photon to energy conversion efficiency, like dye sensitized solar cells, perovskites, polymer, etc [1, 3-5].

In dye sensitized solar cells (DSSCs), the TiO_2 nanostructure used as a semiconductor is a very important component of the solar cell, as it acts as the electron transporter for the efficiency of the solar cell. The dye molecules are adsorbed on its surface, to be used as photon absorber. These molecules adsorbed onto TiO_2 semiconductor are the source of the photocurrent of the solar cell [6-8]. In general the dye sensitizer absorbs the solar radiation, becomes excited and transfers the photoexcited electron to the wide band gap TiO_2 semiconductor layer of nanometer sized particles. The charge hole which is created in the dye after excited state charge injection, is then transferred to a redox couple liquid electrolyte (usually iodine tri iodide) or solid hole conductor.

However, the electron band gap of TiO_2 is relatively large (3.2 eV for anatase, 3.0 eV for rutile and 3.4 eV for brookite) and limit its application in DSSCs [9-12]. To improve the efficiency of the DSSCs, the requirements will be: (i) to reduce the TiO_2 band gap, so that its absorption properties will match the visible solar spectrum, (ii) to modify the sensitizer or dye molecules so that it can absorb on the near infrared spectrum [1,6]. Therefore the modification of different electronic and optical properties of TiO_2 polymorphs (anatase, rutile and brookite) is very important for application in DSSCs and related technologies such as photo-catalyst. Surface doping of TiO_2 polymorphs has been considered as a possible solution to overcome the limitations of using TiO_2 in solar cells [9,10,13]. However, there is still limited on doping of brookite TiO_2 surfaces.

Brookite is one of the structural conformations of titanium dioxide polymorphs. Amongst TiO_2 phases brookite is the least studied TiO_2 photocatalyst due to the difficulties usually encountered in order to obtain it as a pure phase [14]. Experimentally it is difficult to prepare brookite with high purity and large surface area and this is one of the reasons for limited literature on its photocatalytic properties. Despite the difficulties encountered when synthesising pure brookite TiO_2, computer simulation methods have proved to be helpful in studying photocatalytic properties of brookite TiO_2. Thus, density functional theory calculations have shown that the commonly exposed brookite (210) surface is more reactive than anatase (101) surface and might be useful for photocatalytic applications [14]. Li et al. studied the brookite TiO_2 surfaces and showed that the brookite TiO_2 (210) has the same structural building block as anatase TiO_2 (101) [14]. Their findings elaborate that, the differences in surface structure significantly change the reactivity toward adsorption of various molecules, and most importantly, generate highly active sites at the junction between different structural units on brookite TiO_2 (210). The study further suggested that brookite TiO_2 (210) surface exhibits distinct activity, which may be useful in catalytic and photocatalytic applications.

The doping mechanism of different materials can be evaluated easily by electronic structure and related properties calculations. In this paper, we report our investigation of doping brookite TiO_2 (1 0 0) surface with ruthenium as a photocatalyst for application in DSSCs. Density functional theory method is used to calculate the band structures and densities of states (DOS) near the Fermi energy for doped brookite (1 0 0) surface. The discussion is focused on how the electronic properties of this system changes when doped (atom replacement and adsorption) with ruthenium metals.

2. Computational Method

Brookite has an orthorhombic crystalline structure with a unit cell described by the space group (Pbca). The structure is composed of octahedra, each with a titanium atom at its center and

Table 1. Optimized structural parameters for bulk brookite TiO_2 compared with experimental and previous theoretical results.

	Experimental [18]	This work		Literature [9]	
		Results	Deviation (%)	Results	Deviation (%)
a(Å)	9.184	9.166	0.196	9.157	0.294
b(Å)	5.447	5.436	0.202	5.430	0.202
c(Å)	5.145	5.135	0.195	5.122	0.448

oxygen atoms at its corners [15]. In the current study the brookite TiO_2 (1 0 0) surface doped with ruthenium is investigated. The study is geared towards the engineering of the brookite TiO_2 band gap which is higher and need to be tuned to adsorp in the visible spectrum for application in DSSCs. All results computed are based on first principle method, DFT using CASTEP code implemented in Materials Studio package [16, 17]. The structural optimizations, band structure, and electronic density of states for doped and undoped brookite TiO_2 (100) are performed using the generalized gradient approximation (GGA) in the scheme of Perdew-Burke-Ernzerhof (PBE) describing the exchange-correlation functional. The convergence criteria for structural optimization were set to be of medium quality with the tolerance for the self-consistent field (SCF), energy, maximum force, and maximum displacement of $2x10^{-6}$ eV/atom, $2x10^{-5}$ eV/atom, 0.05 eV/Å and $2x10^{-3}$ Å, respectively. The cut-off energy selected was 650 eV and Monkhorst-Pack scheme k-point grid sampling was set as 4x4x1 for the irreducible Brillouin zone. Two approaches were considered for ruthenium doping, i.e., replacement of one Ti atom with one Ru1 atom and adsorption of one Ru atom adjacent to Ti atom. Various distances between ruthenium atom and adjacent titanium atom were considered and they ranged from 1 to 4 Å, with a 0.5 Å interval.

3. Results and discussions

The theoretical computer simulation techniques are able for determination of the stability and structure properties of materials under conditions that are difficult to reproduce in the laboratory. The optimized structural properties of doped brookite (1 0 0) surface were calculated and are listed in Table 1. The listed parameters show that our computational results obtained are in good agreement with the experimental and other theoretical literature. The calculated deviations from the experimental results are in better agreement compared to those reported by Beltran et al. [9]. In particular, our models produced better values for lattice parameters a and c, with b producing same results.

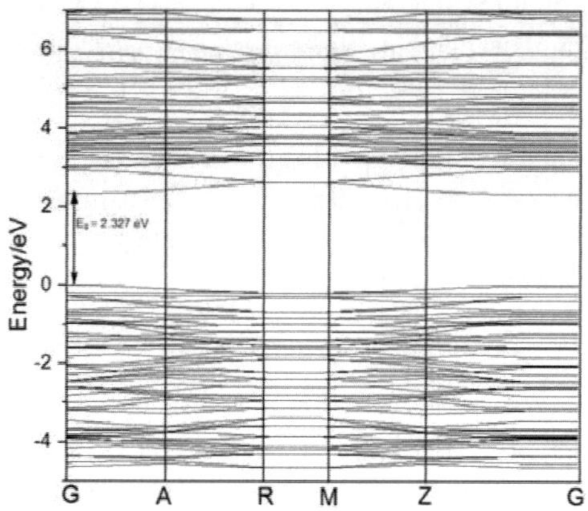

Figure 1. Calculated energy band gap curves for the bulk structure of (1 0 0) TiO_2 surface.

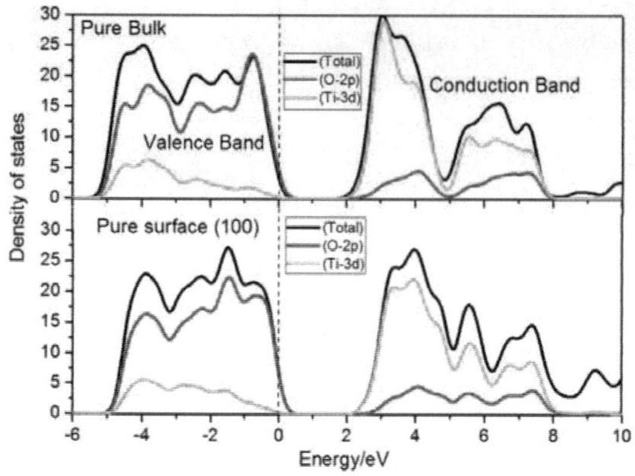

Figure 2. Calculated density of state of bulk and pure TiO_2 (1 0 0) brookite surface.

In order to conveniently investigate the electronic structures of Ru-doped brookite TiO_2, we set the same k-points mesh to sample the first Brillouin zone for pure and Ru-doped models. Figure 1, shows the band gap structure of pure (1 0 0) TiO_2 surface. It can be observed from Figure 1, that the band gap is calculated to be 2.327 eV, which is less than the experimental results of 3.2eV. It is well known that GGA tends to underestimate the band gap due to the choice of the exchange correlation energy. The valence band of pure TiO_2 surface consists of 2p states of oxygen (O) and 3d state of titanium (Ti). The conduction band is dominated by Ti 3d states and is illustrated in Figure 2. The conduction band minimum (CBM) and the valence band maximum (VBM) are located at G, which illustrate that the brookite TiO_2 can be considered as a direct band gap semiconductor.

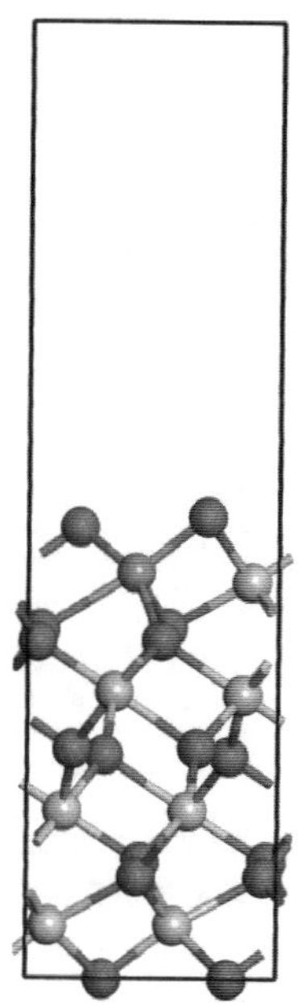

Figure 3. Atomic structure of brookite TiO_2 surface doped with ruthenium element

3.1. Substitutional doping

Figure 3 shows the doped TiO_2 (1 0 0) surface with Ti atom at the top of the surface being substituted by Ru atom. The density of state for the doped system are presented in Figure 4. The doped surface showed that the 3d orbitals for Ru atom generate a coordination fields to interact with the nearest O. The decrease in the band gap as shown in Figure 5, illustrates that there is a shift in the absorbance of the doped TiO_2 to visible spectrum.

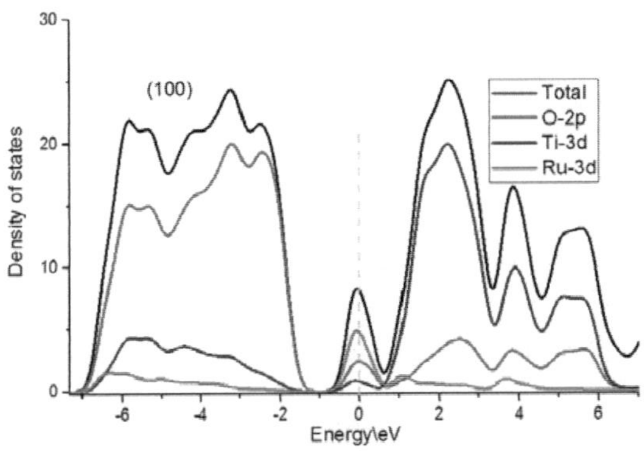

Figure 4. Calculated density of state of doped TiO_2 (1 0 0) brookite surface by substitution of one Ti atom.

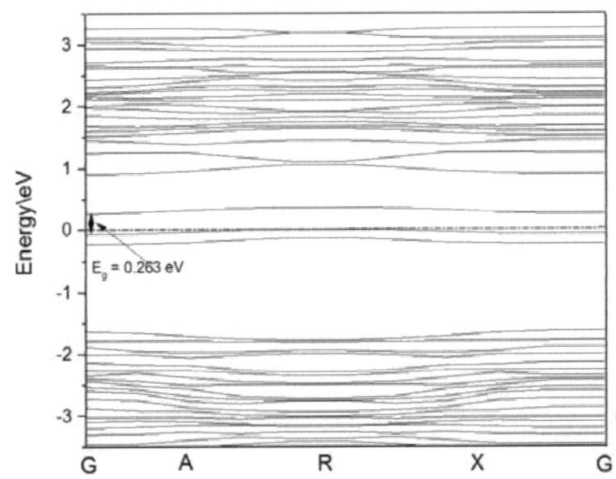

Figure 5. Calculated energy band gap curves of the doped (1 0 0) brookite TiO_2 surface by substitution of one Ti atom

3.2. Adsorption

The procedure followed during the dye uptake in dye sensitized solar cell is of dipping the TiO_2 nano particle film into the solution of the dye molecules. In this paper, we mimic the adsorption of the dye molecules in DSSCs, by adsorbing the ruthenium atom by placing it onto TiO_2 surface. The Ru atom is placed adjacent the Ti atom at the top of the surface at various distances as shown in Figures 6 and 7. The pictures show doped surfaces before and after optimization with the variation of Ru atomic displacement. The two figures depict that after optimization of the doped surface, Ru atom prefers to bond with Ti at an average distance of 2.21 Å. When Ru is placed closer (1 to 2 Å) to the surface it moves away from the surface to a distance of 2.21 Å. And when is placed away (2.5 to 4 Å) from Ti atom it moves closer to the surface to an average bond distance of 2.21 Å. The calculated bond distances are summarised in Table 2.

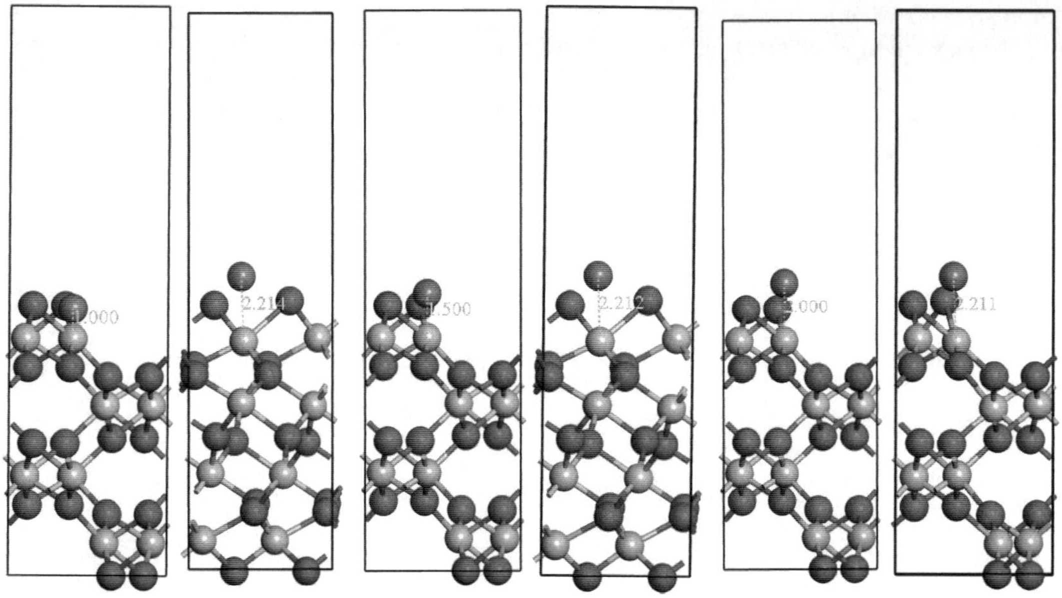

Figure 6. Atomic structure of brookite TiO_2 doped at the distance of 1–2 Å before and after optimisation (the distances are illustrated on the figures).

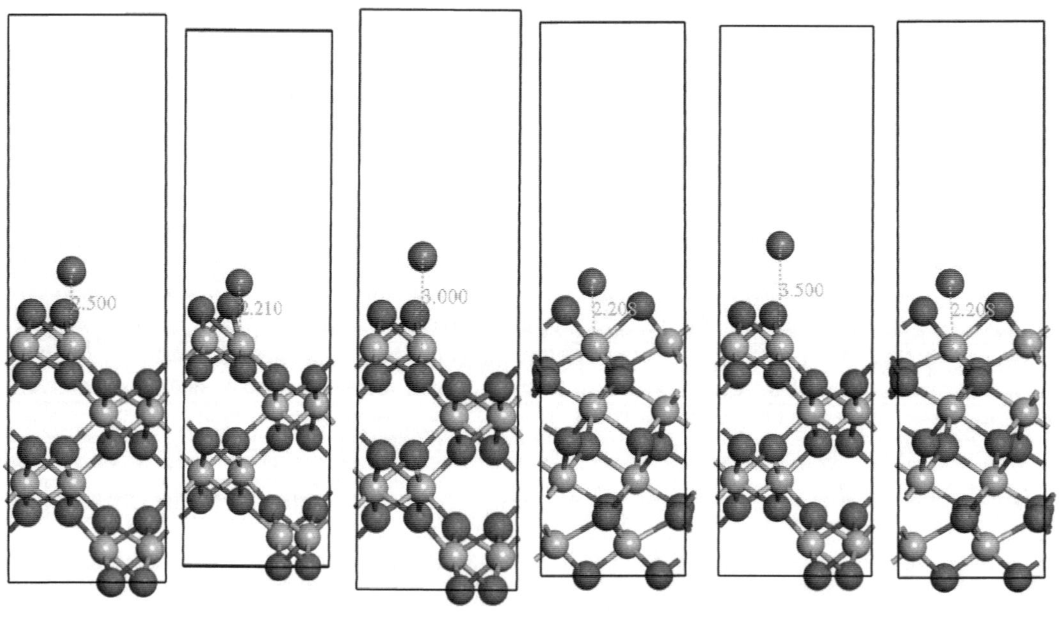

Figure 7. Atomic structure of brookite TiO_2 doped at the distance of 2.5 – 3.5 Å before and after optimisation (the distances are illustrated on the figures).

Figure 8 illustrates calculated energy band gap as a function of Ru position adsorbed on brookite surface. The band gap is defined as the separation between the VBM and CBM. The

Table 2. Final energy of optimized structural parameters for TiO_2 brookite (100) surface as the distance is varied.

Before relaxation (Å)	After relaxation (Å)	Final energy (eV)
1.00	2.214	-22447.819
1.50	2.212	-22447.825
2.00	2.211	-22447.825
2.50	2.210	-22447.823
3.00	2.208	-22447.818
3.50	2.211	-22447.821
4.00	2.209	-22447.821

calculated band gap at various Ru positions is found to be an average of 2.70 eV, which is in agreement with reported results by Mo et al [19]. However, both energy band gaps are underestimated compared to the experimental value of 3.4 eV [20]. It can be observed that the band gap varies slightly as Ru distance is varied. The lowest total energy of the system was obtained to be -22447.824 eV and corresponding to Ru adsorbed at 1.5 Å and 2.5 Å from Ti atom.

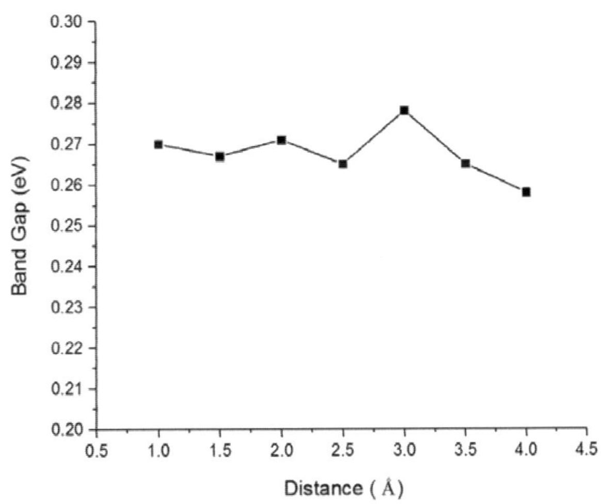

Figure 8. Dependency of the band gap on the doping distance of the ruthenium element

The total density of states (TDOS) and partial density of states (PDOS) of Ru-adsorbed onto brookite TiO_2 are shown in Figure 9. The TDOS shape of Ru-doped TiO_2 becomes broader than that of pure TiO_2, which indicates that the electronic non locality is more obvious, owing to the reduction of crystal symmetry. The Ru 3d states are somewhat delocalized, which contributes to the formation of impurity energy levels (IELs) by hybridizationg with O 2p states or Ti 3d states. Such hybrid effect may form energy levels in the band gap or hybrid with CBM/VBM, providing trapping potential well for electrons and holes. It gives a contribution to the separation of photogenerated electronhole pairs, as well as in favour of the migration of photoexcited carriers and the process of photocatalysis. The Fermi level in the case of the Ru doping shift from the top of the valence band to the bottom of the conduction band, which is normally attributed to the n type doping. Significant reduction in the band gap is noted and this tends to enhance absorption of the brookite TiO_2 to visible light spectrum.

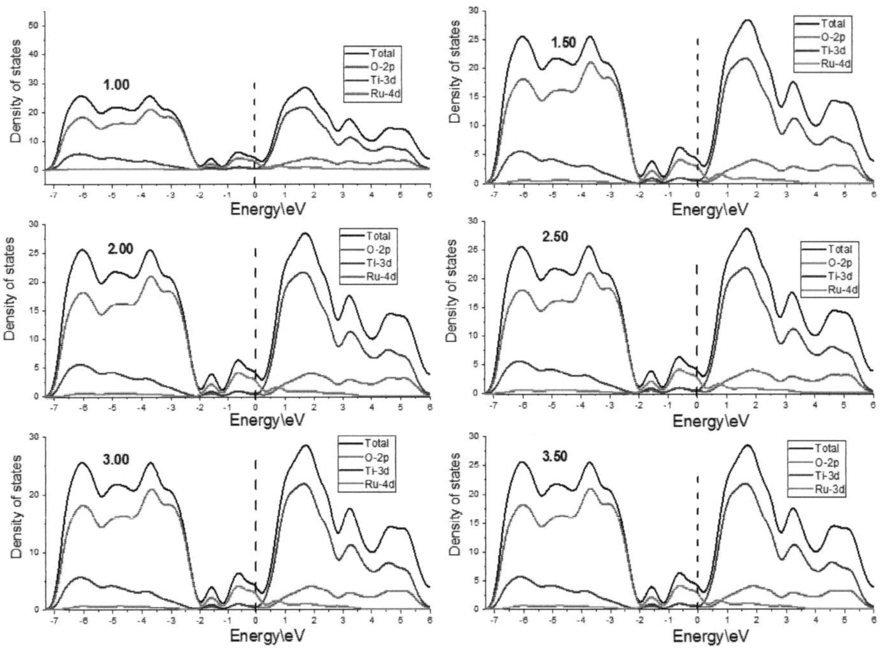

Figure 9. Calculated density of state of doped (1 0 0) brookite TiO_2 surface at different distance as illustrated.

4. Conclusion

First principle calculations were performed successfully in order to investigate the structural and electronic properties of substitutional and adsorbed Ru doped TiO_2 (1 0 0) brookite surface. The calculated lattice parameter are in good agreement with experimental results. The two approaches employed in the study revealed that doping brookite TiO_2 with ruthenium narrow the energy band gap. The decrease in the band gap renders the electronic transfer from valence band to conduction band easier. These decreases in electron transition energy from the valence band to conduction band may be responsible for the visible light absorption. The simulated results showed that doping of TiO_2 semiconductor can improve the efficiency of DSSCs, as the TiO_2 band gap has shifted to absorb on the visible spectrum.

Acknowledgments

This work is financially supported through the program NRF Thuthuka and also by National institute for theoretical Physics (NiTheP). We would like to also thank the University of Venda for the support to carry out this research and Centre for High performance Computing (CHPC) for using their computing resources.

References

[1] Rehman S, Ullah R, Butt AM and Gohar ND 2009 *J. Hazard. Mater.* **170** (2) 560
[2] Wang Y, Zhang R, Li J, Li L and Lin S 2014 *Nanoscale Res. Lett.* **9**(1) 1.
[3] Rawal N, Vaishaly AG, Sharma H and Mathew BB 2015 *EPES.* **2**(2) 46-52
[4] Magne C, Cassaignon S, Lancel G. and Pauport T 2011 *Chem. Phys. Chem.* **12**(13) 2461

IOP Conf. Series: Journal of Physics: Conf. Series **905** (2017) 012034 doi:10.1088/1742-6596/905/1/012034

[5] Sasani A, Baktash A, Mirabbaszadeh K and Khoshnevisan B 2016 *Appl. Surf. Sci.* **384** 298
[6] Arjunan AT and Senthil TS 2013 *Mater. Tech.* **28** (1-2) 9
[7] Suhaimi S, Shahimin MM, Alahmed ZA, Chysk J and Reshak AH 2015 *Int. J. Electrochem. Sci.* **10** 28
[8] Zulkifili ANB, Kento T, Daiki M and Fujiki A 2015 *JOCET* **3**(5) 382
[9] Beltran A, Gracia L and Andres J 2006 *J. Phys. Chem. B* **110**(46) 23417
[10] Zallen R, and Moret MP 2006 *Solid State Commun.* **137** (3) 154
[11] Ma J.G, Zhang CR, Gong JJ, Wu YZ, Kou SZ, Yang H, Chen YH, Liu ZJ and Chen HS 2015 *Mater.* **8**(8) 5508
[12] Li H, Chen L, Liu S, Li C, Meng J and Wang Z 2015 *MATER SCI-POLAND* **33**(3)549
[13] Park, JY, Lee C, Jung KW and Jung D 2009 *Bull. Korean Chem. Soc.* **30**(2) 402
[14] Lin H, Li L, Zhao M, Huang X, Chen X, Li G and Yu R 2012 *J. Amer. Chem. Soc.* **134** (20) 8328
[15] Di Paola A, Bellardita M and Palmisano L 2013 *Catalysts* **3**(1) 36
[16] Clark SJ, Segall MD, Pickard CJ, Hasnip PJ, Probert MIJ, Refson K and Payne MC 2005 Z *Kristallogr.* **220** 567
[17] Segall M, Lindan PJD, Probert M, Pickard C, Hasnip P, Clark S and Payne M 2002 *J. Phys. Condens. Matter* **14** 2717
[18] Meagher EP and Lager GA 1979 *Mineral* **17** 77
[19] Mo SD and Ching WY 1995 *Phys. Rev. B* **55** 13023
[20] Somartin PI and Schwarz K 1992 *Inorg. Chem.* **31** 567

IOP Conf. Series: Journal of Physics: Conf. Series **905** (2017) 012035 doi:10.1088/1742-6596/905/1/012035

Frictional Cooling of Granular Gases: A Molecular Dynamics Study

Prasenjit Das[1], Moshe Schwartz[2,3], Sanjay Puri[1]

[1]School of Physical Sciences, Jawaharlal Nehru University, New Delhi 110067, India.

[2]Beverly and Raymond Sackler School of Physics and Astronomy, Tel Aviv University, Ramat Aviv 69934, Israel.

[3]Faculty of Engineering, Holon Institute of Technology, Golomb 52 Holon 5810201, Holon, Israel.

E-mail: purijnu@gmail.com

Abstract. We study the free evolution of frictional granular gases using large scale molecular dynamics simulation in three dimensions. The system cools due to solid friction among the interacting particles. At early stages of evolution, the density field remains homogeneous and the velocity field follows the Maxwell-Boltzmann (MB) distribution. However, at later times, the density field shows clustering and the velocity field shows local ordering. The ordering in the velocity field is studied by invoking analogy from phase ordering systems. The equal-time correlation function of velocity field follows dynamical scaling. The correlation length of velocity field, $L_v(t)$, exhibits power law growth: $L_v(t) \sim t^{1/3}$.

1. Introduction

Granular materials are one of the most significant and simple forms of matter [1]. A granular system consists of a large number of mesoscopic particles of regular and irregular shapes with smooth or rough surfaces. The typical size of a particle in a granular system ranges from few μm to few cm [2]. These particles loss their kinetic energy due inelasticity and friction, when they collide and slide past each other, and makes a granular system different from ordinary liquids or gases, where it is assumed that energy remains conserved during the collisions between the atoms and molecules. Also, the typical scale of the energy of a granular system is much higher than the thermal energy, i.e., $k_B T$. Thus, granular systems are athermal in nature, and temperature has no role in the dynamics of granular particles. Due to dissipation and athermal nature, a granular system shows distinct properties from usual solids, liquids, and gases [3]. For example, a sand pile at rest with a slope less than the angle of repose produces solid like behavior due to static friction. Also, the flow of dry sand through the narrow neck of a sand clock is an example of liquid-like behavior. Low-density granular gases appear in sand storm, dust, and smoke etc. The flow of granular matters is important from the industrial application point of view. For example, pharmaceutical industries rely on the processing of powder and pills. In agricultural industries and food processing industries, seeds, grains, and food stuff are transported and manipulated. Understanding the dynamics of granular matters has importance in studying the early stage evolution of stars, galaxies, and astrophysical dust [4]. Recently, Brilliantov *et al.* described the size distribution of particles in Saturn's ring using the concept of granular aggregation and fragmentation [5].

Theoretically, the problem of granular gases has been extensively studied in past two decades [6]. In the leading theoretical approach, a granular gas is modeled by a system of inelastic hard spheres, where the collision between a pair of spheres is characterized by a constant coefficient of restitution less than unity [7, 8]. One of the most interesting problem studied within that approach is cooling and pattern formation in density and velocity fields. Particles in the system loss their kinetic energy due to the inelastic binary collisions. In the early stage of evolution, velocity remains random and density remains uniform. However, system cools due to inelastic collisions, is known as homogeneous cooling state (HCS). Haff's law describes the decay of kinetic energy in HCS [9]. However, later in time, the density field becomes unstable to fluctuations and the system enters an inhomogeneous cooling state (ICS), where the particle-rich clusters are formed and grow. Temperature of the system shows algebraic decay. Non-unified results are reported in literature for velocity distribution in the ICS, e.g., power laws, stretched exponential, etc., have been reported in various studies [6]. Puri *et al.* studied the complex pattern formation dynamics of density and velocity fields by invoking analogies from studies of phase ordering dynamics. The growth kinetics of the clusters in density and velocity fields have also been studied [10, 11]. Recently, Schwartz *et al.* presented detailed study of pattern formation in two dimensional granular gases where energy dissipation is modeled by solid friction [12].

In the case of dense flows, many particles interact among each other simultaneously and stay in prolonged contact. Thus, the concept of collision is not meaningful in this case. In case of granular gases cooling by inelastic collisions, it is known that evolution of the density, velocity, and granular temperature fields is well described by the macroscopic hydrodynamic equations [13]. However, finding a macroscopic continuous description of such dense granular systems has been an important and challenging issue in the physics of granular flows. In this paper, we consider the solid friction as the only dissipation mechanism in granular gases. Although we believe that frictional dissipation plays significant role in dense systems, we apply it here to gaseous granular matter, as the only mechanism of mechanical energy dissipation, to study its effect on the cooling properties of such dilute systems. Also, we present results for the phase ordering velocity fields and the corresponding growth dynamics. This work will serve also as a reference point for future work on denser systems.

This paper is organized as follows. In Sec. 2, we describe details of our molecular dynamics simulation. In Sec. 3, we present detailed numerical results. Finally in Sec. 4, we conclude with a brief summary and discussion.

2. Details of Molecular Dynamics Simulation

In our molecular dynamics model simulation, we consider a system of particles confined in a cubic box of size L^3. Particles are spherical in shape, identical in size, and of equal mass m. Any two particles with position vectors $\vec{r}_i$ and $\vec{r}_j$ interact with via a two-body potential with a hardcore of diameter R_1 and a thin shell repulsive potential. To be specific, we consider the following form of interaction potential:

$$V(r) = \begin{cases} \infty & : r < R_1 \\ V_0 \frac{(r-R_2)^2}{(r-R_1)^2} & : R_1 \leq r < R_2 \\ 0 & : r \geq R_2 \end{cases} \tag{1}$$

where $r = \mid \vec{r}_i - \vec{r}_j \mid$ is the separation between the two particles, V_0 is the amplitude of the potential, has the dimension of potential energy and $R_2 - R_1 < R_1$, is the typical thickness of the thin repulsive shell. Here, Eq. (1) is to be taken only as a model of repulsive potential that rises steeply from zero at the outer boundary of the shell to infinity at the hardcore. The normal

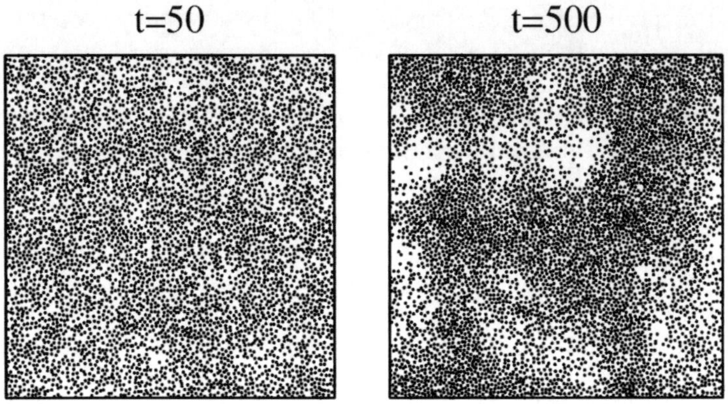

Figure 1. Two-dimensional (x-y plane) cross section of evolution snapshots of density field at different times, as mentioned. Here, we plot the center of the particles whose z coordinate lies in the range $z \in (31.4, 32.6)$. The number of particles in the actual system is $N = 400000$ and friction coefficient $\mu = 0.10$. The size of the system is $V = 92^3$. Void spaces in the density field represents particle free regions.

force applied by particle i to particle j is given by

$$\vec{F}^n_{ij}(r) = -\vec{\nabla}V(r), \tag{2}$$

where the gradient is taken with respect to r_j. The corresponding solid friction force is given by

$$\vec{F}^f_{ij}(r) = \mu \mid F^n_{ij} \mid \frac{\vec{v}_1 - \vec{v}_2}{\mid \vec{v}_1 - \vec{v}_2 \mid}, \tag{3}$$

where $\vec{v}_i$ and $\vec{v}_j$ are the linear velocities of particle i and particle j respectively. Eq. (3) reduces to well known Coulomb's friction force when the thickness of the repulsive shell tends to zero. In this case, the normal component of relative velocity goes to zero and the frictional force becomes purely tangential to the normal force. For simplicity, we did not consider rotational motion of the grains. We use the following units for various relevant quantities: lengths are expressed in units of R_1, temperature in V_0/k_B and time in $\sqrt{mR_1^2/V_0}$. For the sake of convenience and numerical stability, we set $R_1 = 1$, $R_2 = 1.1$, $V_0 = 10$, $k_B = 1$ and $m = 1$. We implemented the standard velocity Verlet algorithm [14] to update positions and velocities of MD simulations. The integration time step is $\Delta t = 0.0005$. The granular gas consisted of N = 400000 particles confined in a 3D box with periodic boundary conditions. We consider two different friction coefficients $\mu = 0.10$ and 0.15. The volume fraction is $\rho \approx 0.269$, which corresponds to number density $\phi \approx 0.514$. This means that the box size is $L = 92$.

We initialize the system by randomly placing particles in the simulation box, such that there is no overlap between the cores of any two particles. We assign equal velocity to all the particles but the velocity vector points in random directions so that $\sum_{i=1}^{N} \vec{v}_i = 0$, i.e., we fix the centre of mass velocity equal to zero. The system is allowed to evolve till $t = 50$ with $\mu = 0$, i.e., the elastic limit. The system is relaxed to a uniform density state and the velocity follows Maxwell-Boltzmann (MB) velocity distribution, which serves as the initial condition (at $t = 0$) for our MD simulation of inelastic spheres with $\mu \neq 0$.

3. Numerical Results

At $t = 0$, we start with the homogeneous initial condition for density and velocity fields. Particles dissipates their kinetic energy due to frictional interaction with $\mu \neq 0$. In Fig. 1, we show the evolution snapshots of the density field for $\mu = 0.10$. Details are given in the figure caption.

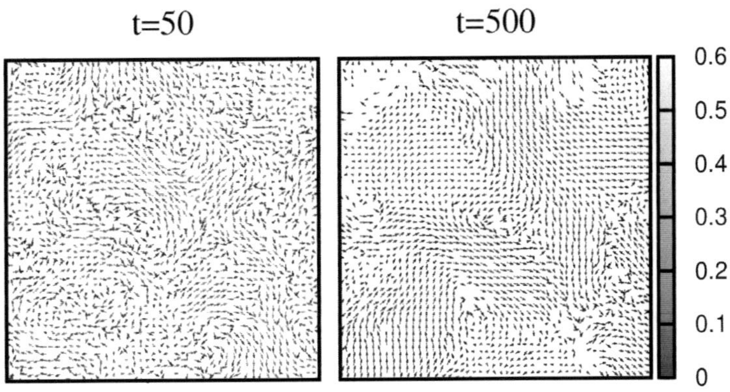

Figure 2. Evolution snapshots of the coarse-grained velocity field in the x-y plane at different times, as mentioned for $z = 32$ layer. The coarse-grained system size is 64^3. For the sake of clarity, we only plot 48^2 of $z = 32$ layer.

At early times, the density remains roughly homogeneous. However, at later times, the formation of clusters is observed. This cluster formation can be explained as follows. Consider some fluctuations in the homogeneous phase of the density field. Particles in the high-density regions lose more energy than those in the low-density regions. Thus the effective temperature and hence pressure are lesser in the clustered phase than those in the homogeneous phase. This gradient of pressure drives particles from low-density to high-density regions, increasing the density fluctuations. At the late stage, the system becomes sluggish due to the dissipation of energy.

At the early stage of evolution, the velocity field remains random. However, at the late stage, local correlation builds up. Due to frictional interaction among the particles, the normal component of relative velocity is dissipated, and the tangential component of relative velocity remains unaltered causing the local parallelization of velocities of two interacting particles. In Fig. 2, we plot the two dimensional cross-section of the coarse-grained velocity field at different times in the x-y plane. We divide the system into non-overlapping cubes of size 1.4375^3 and the average value of velocity is calculated in each cube. The direction of the average velocity is plotted as an arrow starting at the center of each square, using x- and y- components of the velocity. The magnitude of the velocity in each cube is represented by a shade of gray scale. Darker the shade of gray smaller is the velocity. Void spaces in the velocity field are the particle free regions of density field.

We studied ordering in velocity field by invoking analogies from phase ordering kinetics with vector order-parameter field [15, 16, 17]. We introduce an order-parameter $\vec{v}(\vec{r}, t)$, which is the normalized average velocity in each cell, i.e., we only take the direction of average velocity. This is analogous to the vector order-parameter of dynamical X-Y model in three dimensions [18]. The morphology of clustering is studied by using equal-time correlation function $C_{vv}(r, t)$, defined as follows

$$C_{vv}(r,t) = \left[\langle \vec{v}(\vec{R},t) \cdot \vec{v}(\vec{R}+\vec{r},t) \rangle - \langle \vec{v}(\vec{R},t) \rangle \cdot \langle \vec{v}(\vec{R}+\vec{r},t) \rangle \right], \tag{4}$$

where r is the distance between the two points and angular brackets represent the average over different initial conditions [15, 19]. We calculate the characteristic length scale of ordering $L_v(t)$ from $C_{vv}(r, t)$. It is defined as the distance at which correlation decays from 1(at r = 0) to 0.5. The variation of average domain size $L_v(t)$ *vs.* t for different μ's are shown in Fig. 3. Details are given in the figure caption. Line with growth exponent $\alpha = 1/3$ in Fig. 3, corresponds to growth law for plug formation in dense granular flow [12].

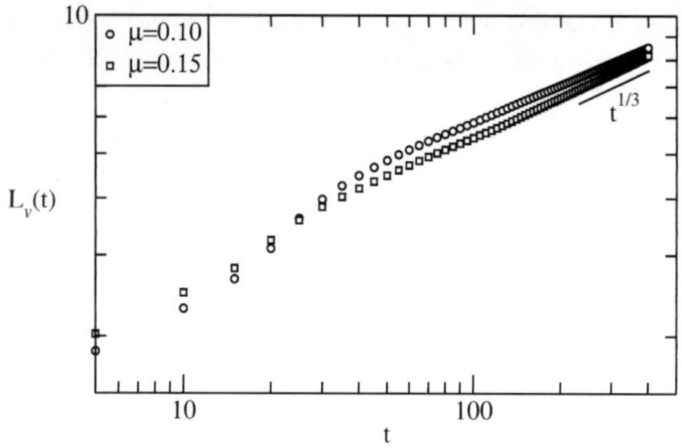

Figure 3. Time-dependence of the characteristic length scales $L_v(t)$ of $\vec{v}(\vec{r},t)$-field for $\rho \approx 0.269$ and for different μ, as mentioned. Solid line with exponent $\frac{1}{3}$ is shown, corresponds to growth law predicted in the context of plug formation in dense granular flow.

If the clustering is characterized by a single length scale $L_v(t)$, $C_{vv}(r,t)$ shows dynamical scaling given by

$$C_{vv}(r,t) = g_v \left[\frac{r}{L_v(t)}\right], \tag{5}$$

where $g_v(x)$ is the master function and x is the time independent scaling variable. Similarly, we calculate *structure factor* $S_{vv}(k,t)$, which is defined as the Fourier transform of the correlation function $C_{vv}(r,t)$ as follows

$$S_{vv}\left(\vec{k},t\right) = \int d\vec{r} e^{i\vec{k}.\vec{r}} C_{vv}(\vec{r},t), \tag{6}$$

at wave vector $\vec{k}$. The corresponding scaling form for $S_{vv}(k,t)$ is given by

$$S_{vv}(k,t) = L_v^d \tilde{S}_{vv}\left[kL_v(t)\right], \tag{7}$$

where $\tilde{S}_{vv}(p)$ is the scaling function, d is the spatial dimensionality and p is the scaling variable [15, 19]. In Fig. 4(a), we plot $C_{vv}(r,t)$ *vs.* $r/L_v(t)$ at different times, as mentioned. We see that numerical data for different times are indistinguishable, which confirms dynamical scaling of ordering. In Fig. 4(b), we plot $S_{vv}(k,t)L_v^{-d}$ *vs.* $kL_v(t)$ at different times, denoted by the symbols indicated. In the limit $k \to \infty$, $S_{vv}(k,t)$ decays as k^{-6}, following the *generalized Porod's law* as $S(k,t) \sim k^{-(d+n)}$ with spatial dimensionality $d = 3$ and the number of order parameter components $n = 3$ [18, 20]. This power law decay of $S_{vv}(k,t)$ implies scattering from vortex like defects in the velocity field.

4. Conclusion

Let us conclude this paper with the summary and discussion of our results. We carried out large-scale molecular dynamics (MD) simulation to study the cooling of granular gases, where friction among the interacting particles is considered as the only source of dissipation. At the early stage of evolution, density field remains uniform and velocity field remains random. However, at later times, as the system cools, clustering in density and local ordering in velocity fields have been observed. The equal-time correlation function for velocity field show dynamical scaling, indicating the morphological similarity of ordering. In the limit $k \to \infty$, structure factor tails for velocity field follows *Generalized Porod's law*: $S_{vv}(k,t) \sim k^{-(d+n)}$, with $d = 3$ and $n = 3$. The average domain size in velocity field follows power law growth: $L_v(t) \sim t^{1/3}$, proposed in the context of plug formation in dense granular flow.

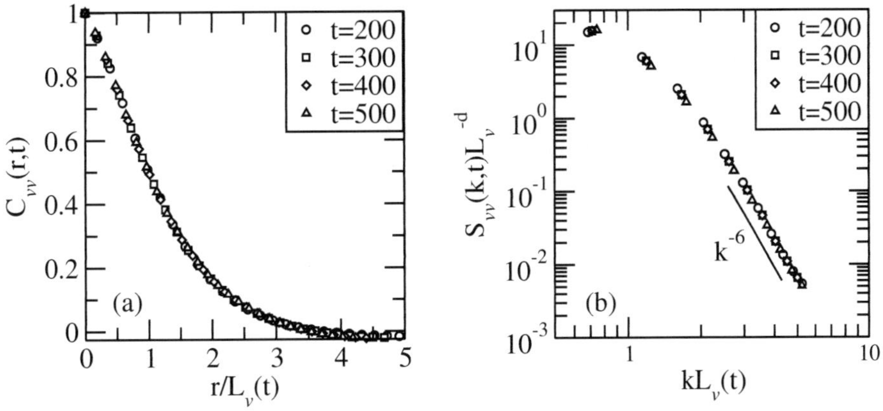

Figure 4. Scaling plot of spherically averaged correlation functions $C_{vv}(r,t)$ and structure factors $S_{vv}(k,t)$ of coarse-grained $\vec{v}(\vec{r},t)$-field at different times, denoted by the indicated symbols. The coarse-grained system size is 64^3 and all results are averaged over ten independent runs for $\mu = 0.10$ and $\rho \approx 0.269$. (a) Plot of $C_{vv}(r,t)$ *vs.* $r/L_v(t)$ at different times, as mentioned. Numerical data at different times are indifferent, which confirms dynamical scaling. (b) Plot of $S_{vv}(k,t)L_v^{-d}$ *vs.* $kL_v(t)$ on log-log scale. Numerical data at different times scaled appropriately, indicate morphological similarity. The line labeled with k^{-6} represents the *generalized Porod's law* with $d = 3$ and $n = 3$. Rest of the details are same as Fig. 1.

5. References

[1] J. Duran, *Sands, Powders and Grains: An Introduction to the Physics of Granular Materials*, Springer-Verlag, New York (1994).
[2] de Gennes P. G., Rev. Mod. Phys. **71**, S374 (1999).
[3] Jaeger H. M. and Nagel S. R., Rev. Mod. Phys. **68**, 1259 (1996).
[4] Aranson I. S. and Tsimring L. S., Rev. Mod. Phys. **78**, 641 (2006).
[5] Brilliantov N. V., Krapivsky P. L., Bodrova A., Spahn F., Hayakawa H., Stadnichuk V., and Schmidt J., Proc. Natl. Acad. Sci. USA **112**, 9536 (2015).
[6] Brilliantov N. V. and Poschel T., *Kinetic Theory of Granular Gases*, Oxford University Press, Oxford (2004).
[7] Goldhirsch I. and Zanetti G., Phys. Rev. Lett. **70**, 1619 (1993); Goldhirsch I., Tan M-L. and Zanetti G., I. Sci. Comp. **8**, 1 (1993).
[8] Sela N., Goldhirsch I. and Noskowicz S. H., Phys. Fluids **8**, 2337 (1996).
[9] Haff P. K., J. Fluid. Mech. **134**, 401 (1983).
[10] Das S. K. and Puri S., Phys. Rev. E **68**, 011302 (2003); Das S. K. and Puri S., Euro. Phys. Lett **61**, 749 (2003).
[11] Ahmad S. R. and Puri S., Euro. Phys. Lett **61**, 749 (2003); Phys. Rev. E **75**, 031302 (2007).
[12] Das P., Puri S. and Schwartz M., unpublished.
[13] van Noije T. P. C., Ernst M. H., Brito R. and Orza J. A. G., Phys. Rev. Lett. **89**, 411 (1997); Phys. Rev. E **61**, 1765 (2000).
[14] Allen M.P. and Tildesley D.J., *Computer Simulation of Liquids* (Oxford University Press), Oxford, (1987).
[15] *Kinetics of Phase Transitions*, edited by Puri S. and Wadhawan V. (CRC Press, Boca Raton, 2009).
[16] Das S. K. and Puri S., Physica A **318**, 55 (2003).
[17] Bray A.J., Adv. Phys. **43**, 357 (1994).
[18] Das P., Roy M. K., Puri S. and Dattagupta S., Euro. Phys. Lett. **104** 66005 (2013).
[19] Dattagupta S. and Puri S., *Dissipative Phenomena in Condensed Matter: Some Applications* (Springer-Verleg Heidelberg 2004).
[20] Oono Y. and Puri S., Mod. Phys. Lett. B **2** 861 (1988); Bray A.J. and Puri S., Phys. Rev. Lett. **67**, 2670 (1991).

IOP Conf. Series: Journal of Physics: Conf. Series **905** (2017) 012036 doi:10.1088/1742-6596/905/1/012036

An effective chaos-geometric computational approach to analysis and prediction of evolutionary dynamics of the environmental systems: Atmospheric pollution dynamics

V V Buyadzhi, A V Glushkov, O Yu Khetselius, Yu Ya Bunyakova, T A Florko E V Agayar and E P Solyanikova

Odessa State Environmental University, L'vovskaya str. 15, Odessa-16, 65016, Ukraine

E-mail: vbuyad@mail.ru

Abstract. The present paper concerns the results of computational studying dynamics of the atmospheric pollutants (dioxide of nitrogen, sulphur etc) concentrations in an atmosphere of the industrial cities (Odessa) by using the dynamical systems and chaos theory methods. A chaotic behaviour in the nitrogen dioxide and sulphurous anhydride concentration time series at several sites of the Odessa city is numerically investigated. As usually, to reconstruct the corresponding attractor, the time delay and embedding dimension are needed. The former is determined by the methods of autocorrelation function and average mutual information, and the latter is calculated by means of a correlation dimension method and algorithm of false nearest neighbours. Further, the Lyapunov's exponents spectrum, Kaplan-Yorke dimension and Kolmogorov entropy are computed. It has been found an existence of a low-D chaos in the time series of the atmospheric pollutants concentrations.

1. Introduction

The last decades have seen a great progress in the understanding, analysis, modelling and evet prediction of the evolutionary dynamics of nonlinear complex systems. Various methods and algorithms of the modern theory of dynamical systems and a chaos theory became a powerful tool in computational studying complex non-linear statistical systems [1-14]. Many studies in different fields of science and technique have appeared, where the chaos theory methods were applied to a great number of dynamical systems. The studies concerning non-linear behaviour in the time series of atmospheric constituent concentrations are sparse, and their outcomes are ambiguous. In ref. [5] there is an analysis of the NO_2, CO, O_3 concentrations time series and is not received an evidence of chaos. Also, it was shown that O_3 concentrations in Cincinnati (Ohio) and Istanbul are evidently chaotic, and non-linear approach provides satisfactory results [6]. In Ref. [14] it has been fulfilled the detailed analysis of the NO_2, CO, CO_2 concentration time series in the Gdansk region (Polland) and it has been definitely obtained the evidence of a chaos. Moreover it has been given a short-range forecast of atmospheric pollutants time evolution using non-linear prediction method. These studies show that chaos theory methodology can be applied and the short-range forecast by the non-linear prediction method can be satisfactory. Time series of concentrations are however not always chaotic, and chaotic behaviour must be examined for each time series.

In this work we study the temporal dynamics of the atmospheric constituents concentration in the Odessa region by using the non-linear prediction and chaos theory methods [3,4,12-27]. A chaotic behaviour in the nitrogen dioxide and sulphurous anhydride concentration time series is numerically investigated. The topological and dynamical invariants, in particular, the Lyapunov's exponents spectrum, Kaplan-Yorke dimension, Kolmogorov entropy etc are computed. It has been found an existence of a low-D chaos in the time series of the atmospheric pollutants concentrations.

2. The data for computational studying and method of testing chaos in time series

In our study, the nitrogen dioxide (NO_2) and sulphurous anhydride (SO_2) concentration data observed in the atmosphere of the Odessa city from 1976 till 2000 years [3]. The multi-year hourly concentrations (one year total of 20x8760 data points, 1990) are analyzed. The temporal series of concentrations (in mg/m^3) of the NO_2 and SO_2 are presented in figure 1 and 2.

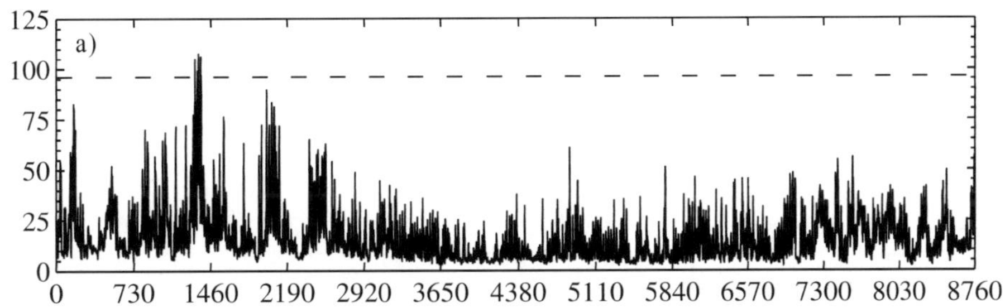

Figure 1. The temporal series of concentrations (in mg/m^3) of the NO_2

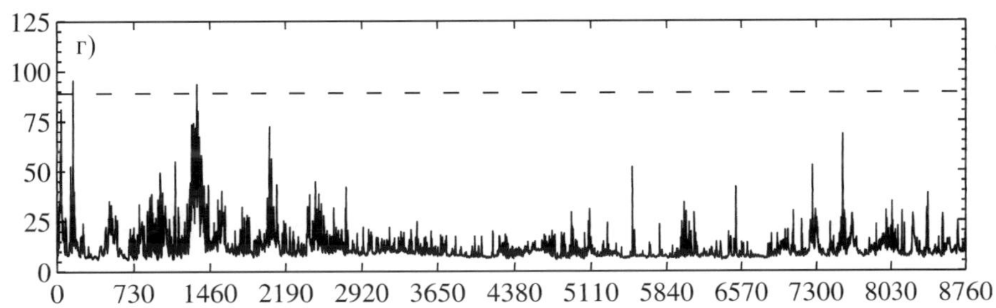

Figure 2. The temporal series of concentrations (in mg/m^3) of the of the SO_2

In Refs. [2-4,11-18] it has been developed computational code for studying chaotic features of the complex non-linear systems and in details described a procedure of testing of the chaos elements in the corresponding time series. Here we are limited only by the key aspects. As usually, we consider scalar measurements $s(n)=s(t_0+ n\Delta t) = s(n)$, where t_0 is a start time, Δt is time step, and n is number of the measurements. In a general case, $s(n)$ is any time series, but here $s(n)$ corresponds to an atmospheric pollutant concentration. The first fundamental step of modelling is in reconstruction of the corresponding phase space using as well as possible information contained in $s(n)$. From the mathematical viewpoint, this procedure results in set of d-dimensional vectors $\mathbf{y}(n)$ replacing scalar measurements. One should further to operate with lagged variables $s(n+\tau)$, where τ is some integer to be defined, results in a coordinate system where a structure of orbits in phase space can be captured. Using a set of the time lags to create a vector in d dimensions, $\mathbf{y}(n)=[s(n),\ s(n + \tau),\ s(n + 2\tau),..,s(n +(d-1)\tau)]$, the required coordinates are provided. The dimension d is defined as an embedding dimension, d_E.

In Refs. [2-4] there are presented a few approaches to the choice of proper time lag. This point is important for the subsequent reconstruction of phase space. First approach [2] is to compute the linear autocorrelation function $C_L(\delta)$ and to look for that time lag where $C_L(\delta)$ first passes through 0.

The alternative approach is based on using method of an average mutual information. Let us remind that the mutual information I of two measurements a_i and b_k is symmetric and non-negative, and equals to 0 if only the systems are independent. The average mutual information between any value a_i from system A and b_k from B is the average over all possible measurements of $I_{AB}(a_i, b_k)$. In Ref. [4] it is suggested, as a prescription, that it is necessary to choose that τ where the first minimum of $I(\tau)$ occurs.

The fundamental goal of the d_E calculation is in the further reconstruction of the Euclidean space R^d large enough so that the set of points d_A can be unfolded without ambiguity. The embedding dimension, d_E, must be greater, or at least equal, than a dimension of the corresponding chaotic attractor, d_A, i.e. $d_E > d_A$. The correlation integral analysis is one of the widely used techniques to investigate the signatures of chaos in a time series. This method is based on using the correlation integral, $C(r)$. As usually, if the corresponding time series is characterized by an attractor, then the correlation integral $C(r)$ is related to the radius r as

$$d = \lim_{\substack{r \to 0 \\ N \to \infty}} \frac{\log C(r)}{\log r},$$

where d is correlation exponent. In a case of the chaotic system the correlation exponent attains saturation with an increase in the embedding dimension. The saturation value of this exponent is defined as the correlation dimension (d_2) of the attractor (see details in Refs. [2-4,9-22]). The technique of application the correlation integral method (say, the Grassberger-Procaccia algorithm [9]) is presented in Refs. [2-4,8,9]. Another method for determining d_E comes from asking the basic question addressed in the embedding theorem: when has one eliminated false crossing of the orbit with itself which arose by virtue of having projected the attractor into a too low dimensional space? This method is called as the method of false nearest neighbours [11]. As a rule, the simultaneous application of two methods provides more exact determination d_E. It is noteworthy that the nearest integer above the saturation value provides the minimum or optimum embedding dimension for reconstructing the phase-space or the number of variables necessary to model the dynamics of the system. This concept can be applied, since the embedding dimension determined by both the correlation dimension method and the algorithm of false nearest neighbours are identical.

The further important step in studying the chaotic time series of the dynamical system is determination of predictability, which can be estimated by the Kolmogorov entropy. The Kolmogorov entropy is proportional to a sum of the positive Lyapunov's exponents. Let us remind that the Lyapunov's exponents spectrum is one of the fundamental dynamical invariants for non-linear system with chaotic behaviour.

According to definition, the Lyapunov's exponents are related to the eigenvalues of the linearized dynamics across the attractor. These parameters indicate the complexity of dynamics of the studied system. As usually, the positive values the Lyapunov's exponents show local unstable behaviour of the system, and respectively, their negative values show stable behaviour. The largest positive value of the Lyapunov's exponents determines some average prediction limit. Since the Lyapunov's exponents are defined as asymptotic average rates, they are independent of the initial conditions, and hence the choice of trajectory, and they do comprise an invariant measure of the attractor. An estimate of this measure is a sum of the positive Lyapunov's exponents. The estimate of the attractor dimension is provided by the conjecture d_L and the Lyapunov's exponents are taken in descending order. The dimension d_L gives values close to the dimension estimates discussed earlier and is preferable when estimating high dimensions. To compute Lyapunov's exponents, we use a method with linear fitted map [2,4,20], although the maps with higher order polynomials can be used too.

3. Results

Table 1 summarizes the results for the time lag, which is computed for first ~10^3 values of time series. The autocorrelation function crosses 0 only for the NO_2 time series, whereas this statistic for other time series remains positive. The values, where the autocorrelation function first crosses 0.1, can be chosen as τ, but in [1] it has been showed that an attractor cannot be adequately reconstructed for very large values of τ. So, before making up final decision we calculate the dimension of attractor for all values in Table 1.

The outcome is explained not only inappropriate values of τ but also shortcomings of correlation dimension method [2]. If algorithm [1] is used, then a percentages of false nearest neighbours are comparatively large in a case of large τ. If time lags determined by average mutual information are used, then algorithm of false nearest neighbours provides $d_E = 6$ for all air pollutants.

Table 1. Time lags (hours) subject to different values of C_L and first minima of average mutual information (I_{min1}) for the time series of NO_2, SO_2 concentrations for the sites of the Odessa city (1990)

	NO_2	SO_2
$C_L = 0$	-	-
$C_L = 0.1$	142	239
$C_L = 0.5$	7	14
I_{min1}	10	20

Table 2 shows the calculated parameters: correlation dimension (d_2), embedding dimension (d_E), two Lyapunov exponents, E(λ_1,λ_2), Kaplan-Yorke dimension (d_L), and average limit of predictability (Pr_{max}, hours) for the NO_2, SO_2 concentration time series in the Odessa region (for two measurement sites) during the period: Jan.-Dec., 1989-1990). From the table 2 it can be noted that the Kaplan-Yorke dimensions, which are also the attractor dimensions, are smaller than the dimensions obtained by the algorithm of false nearest neighbours. It is very important to pay the attention on the presence of the two (from six) positive Lyapunov's exponents λ_i . This fact suggests that the system broadens in the line of two axes and converges along four axes that in the six-dimensional space. The time series of SO_2 at the site 2 have the highest predictability (more than 2 days), and other time series (in particular, the NO_2 concentration) have the predictabilities slightly less than 2 days.

Table 2. The correlation dimension (d_2), embedding dimension (d_E), first two Lyapunov exponents, E(λ_1,λ_2), Kaplan-Yorke dimension (d_L), and the Kolmogorov entropy, average limit of predictability (Pr_{max}, hours) for the time series of the NO_2 and SO_2 concentrations (Odessa city, 1990)

	Site 1 (Odessa) NO_2	Site 1 (Odessa) SO_2	Site 2 (Odessa) NO_2	Site 2 (Odessa) SO_2
λ_1	0.0187	0.0166	0.0191	0.0153
λ_2	0.0059	0.0062	0.0049	0.0048
d_2	5.28	1.62	5.26	3.48
d_E	6	6	6	6
d_L	4.09	5.04	3.92	4.63
K_{entr}	0.025	0.023	0.024	0.020
Pr_{max}	41	46	42	48

4. Conclusions

To conclude, in this work we have studied a dynamics of variations of the atmospheric pollutants (the time series of the dioxide of nitrogen, sulphur etc) concentration in atmosphere of the Odessa city by using the dynamical systems and chaos theory methods. A chaotic behaviour in the nitrogen dioxide and sulphurous anhydride concentration time series at several sites of the Odessa city is numerically investigated for the first time. As usually, to reconstruct the corresponding attractor, the time delay and embedding dimension were determined. The time delay was calculated on the basis of methods of autocorrelation function and average mutual information, and the embedding dimension was calculated by means of the correlation dimension method and algorithm of false nearest neighbours. Further, the Lyapunov's exponents spectrum, Kaplan-Yorke dimension and Kolmogorov entropy are calculated. Our computational study has shown an existence of a low-and high-D chaos in the atmospheric pollutants fluctuations dynamics in the Odessa. This conclusion allows further to develop the corresponding prediction models [13-18] for description of the temporal evolutionary dynamics of the air pollutants concentration in atmosphere of the industrial city.

Acknowledgements

The authors would like to thank Prof. Nithaya Chetty, Prof. Marius Potgieter and colleagues for invitation to make contributions to CCP-2016 (Gauteng, South Africa). The useful comments of the anonymous referees are very much acknowledged too.

References

[1] Kenneth F 2003 *Fractal Geometry: Mathematical Foundations and Applications* (Chichester: John Wiley & Sons)
[2] Abarbanel H, Brown R, Sidorowich J and Tsimring L 1993 *Rev. Mod. Phys.* **65** 1331
[3] Glushkov A V 2008 *Modern theory of a chaos* (Odessa: OSENU)
[4] Glushkov A V and Bunyakova Yu Ya 2010 *Analysis and prediction of antropogenic effect on ait basin of industrial city* (Odessa: Ecology)
[5] Gallager R 1986 *Information theory and reliable communication* (N.-Y.: Wiley)
[6] Lanfredi M and Macchiato M 1997 *Europhys. Lett.* **1997** 589
[7] Koçak K, Şaylan L and Şen O 2000 *Atm. Env.* **34** 1267
[8] Glushkov A V 2006 *Relativistic and correlation effects in spectra of atomic systems* (Odessa: Astroprint)
[9] Grassberger P and Procaccia I 1983 *Physica D* **9** 189
[10] Fraser A M, Swinney H L 1986 *Phys. Rev. A.* **33**(2) 1134
[11] Kennel M B, Brown R and Abarbanel 1992 *Phys. Rev. A* **45** 3403
[12] Glushkov A V, Khokhlov V N, Prepelitsa G P and Tsenenko I A 2004 Opt*ics Atm. and Ocean* **14** 219
[13] Khokhlov V N, Glushkov A V and Tsenenko I A 2004 *Nonlinear Proc. Geophys.* **11**(3) 295
[14] Khokhlov V N, Glushkov A V, Loboda N S and Bunyakova Y Y *2008 Atm. Env.* **42** 7284
[15] Glushkov A V, Khetselius O, Brusentseva S, Zaichko P and Ternovsky V 2014 *Advances in Neural Networks, Fuzzy Systems and Artificial Intelligence* (Recent Advances in Computer Engineering vol 21) ed J Balicki (Gdansk: WSEAS Press) pp 69-75
[16] Glushkov A V, Svinarenko A, Buyadzhi V, Zaichko P and Ternovsky V 2014 *Advances in Neural Networks, Fuzzy Systems and Artificial Intelligence* (Recent Advances in Computer Engineering vol 21) ed J Balicki (Gdansk: WSEAS) pp 143-150
[17] Glushkov A V, Buyadzhi V V and Ponomarenko E L 2014 *Proc. of Int. Geometry Center* **7** 24
[18] Glushkov A V, Bunyakova Yu Ya, Fedchuk A P, Serbov N G, Svinarenko A A and Tsenenko I A 2007 *Sensor Electr. and Microsyst. Techn.* **3** 14
[19] Glushkov A V, Kuzakon V, Ternovsky V B and Buyadzhi V V 2013 *Dynamics of laser systems with absorbing cell and backward-wave tubes with elements of a chaos* (Dynamical Systems Theory vol T1) ed J Awrejcewicz, M Kazmierczak, P Olejnik and J Mrozowski (Lodz: Wyd. Politechniki Lodzkiej) pp 461-466
[20] Glushkov A V, Prepelitsa G P, Svinarenko A A and Zaichko P A 2013 *Studying interaction dynamics of the non-linear vibrational systems within non-linear prediction method:*

Application to quantum autogenerators (Dynamical Systems Theory vol T1) ed J Awrejcewicz, M Kazmierczak, P Olejnik and J Mrozowski (Lodz: Wyd. Politechniki Lodzkiej) chapter 8 pp 467-477

[21] Prepelitsa G P, Glushkov A V, Lepikh Ya I, Buyadzhi V V, Ternovsky V B and Zaichko P A 2014 *Sensor Electr. and Microsyst. Techn.* **11** 43

[22] Packard N, Crutchfield J, Farmer J and Shaw R 1988 *Phys. Rev. Lett.* **45** 712

[23] Rusov V D, Glushkov A V, Vaschenko V N, Myhalus O T, Bondartchuk Yu A, Smolyar V P, Linnik E P, Mavrodiev S C and Vachev B I 2010 *Journ. Atm. and Solar-Terrestrial Phys.* **72** 498

[24] Svinarenko A A 2014 *J. Phys.: Conf. Series* **548** 012039

[25] Glushkov A V, Khetselius O, Svinarenko A and Prepelitsa G P 2011 *Energy Approach to Atoms in a Laser Field and Quantum Dynamics with Laser Pulses of Different Shape* (Coherence and Ultrashort Pulsed Emission vol 1) ed F J Duarte (Rijeka: InTech) chapter 8 pp 159-186

[26] Glushkov A V, Khetselius O Yu, Kruglyak Yu A and Ternovsky V B 2015 *Calculational Methods in Quantum Geometry and Chaos theory P3* (Odessa: TEC)

[27] Glushkov A V, Khetselius O Yu, Svinarenko A A and Buyadzhi V V 2015 *Methods of computational mathematics and mathematicsl physics P1* (Odessa: TEC)

IOP Conf. Series: Journal of Physics: Conf. Series **905** (2017) 012037 doi:10.1088/1742-6596/905/1/012037

The Schrödinger equation on a Lagrange mesh

Gaotsiwe J Rampho

Department of Physics, University of South Africa, P.O. Box 392, Pretoria 0003, South Africa

E-mail: ramphjg@unisa.ac.za

Abstract. In this paper the Lagrange-mesh method is used to solve the two- and three-dimensional Schrödinger equation in spherical coordinates. Matrix elements that involve few free parameters and depend on the grid points for the Hamiltonian, and can be used with any quantum mechanical system, are employed. The accuracy of the matrix elements is tested with the quartic anharmonic oscillator and a double ring-shaped harmonic oscillator potentials. Eigenvalues for the systems are reproduced to within machine accuracy when the free parameters of the matrix elements are chosen appropriately.

1. Introduction

Atomic, molecular and nuclear systems are quantum mechanical systems that are accurately described by the Schrödinger equation. Theoretical studies of these systems are based on the solution to the many-body Schrödinger equation for the systems. However, the Schrödinger equation can be solve exactly only for a few types of potentials describing interactions in quantum systems [1, 2]. As a result, only numerical solutions of the equation with potentials of a general form are possible, even for systems with few constituents. A number of numerical methods is available to generate accurate numerical solutions to the Schrödinger equation [3, 4, 5, 6, 7, 8]. Except for the Lagrange-mesh method [8], many of the methods are computationally demanding and suffer from slow convergence. The Lagrange-mesh method is a variational method that employs specialized basis functions on a specialized grid to construct variational solutions for quantum dynamical equations. The method was shown to converge rapidly and generated simple and accurate numerical solutions for different quantum mechanical systems [9, 10, 11]. In the present work the Lagrange-mesh method is used to construct accurate numerical solutions to the two- and three-dimensional Schrödinger equation in polar coordinates. Not many instructive implementation of the method in polar coordinates, without the use of the spherical harmonics, are available.

The Lagrange-mesh method combines the variational technique through the Lagrange basis expansion with the collocation technique through quadrature rules to construct numerical solutions to quantum mechanical dynamical equations [8]. The method generates matrix elements, for the equations, that depend only on the collocation abscissas, resulting in a set of algebraic equations for the variational parameters that can be solved efficiently and accurately. The use of the Lagrange-mesh method in solving the multi-dimensional Schrödinger equation in polar coordinates has, so far, been generally restricted to the constant-mesh Lagrange functions [12] and the Lagrange-Legendre functions [8]. Matrix elements for the more general Lagrange-Jacobi mesh, from which the matrix elements for the Legendre mesh and other meshes can be obtained, are derived in [13]. This work discusses the specialization of the Lagrange-Jacobi

IOP Conf. Series: Journal of Physics: Conf. Series **905** (2017) 012037 doi:10.1088/1742-6596/905/1/012037

functions [13] to treat the angular components of the solution to the Schrödinger equation in polar coordinates.

This paper is organized as follows. The Schrödinger equation in two- and three-dimensions in spherical polar coordinates is presented in section 2. In section 3 the Lagrange meshes and Lagrange functions applicable to the three-dimensional Schrödinger equation in spherical polar coordinates are discussed. Matrix elements for the Hamiltonian of the equation are also developed. The simplicity and accuracy of the method is demonstrated by solving the Schrödinger equation for the two-dimensional quartic anharmonic oscillator potential as well as for the double ring-shaped harmonic oscillator potential. The results are discussed in section 4 and conclusions are given in section 5.

2. The Schrödinger equation

The configuration-space Schrödinger equation ($\hbar = 2m = 1$) in the two-dimensional cylindrical polar coordinates $\vec{r}(r, \phi)$ can be cast in the form

$$\left[-\frac{\partial^2}{\partial r^2} - \frac{1}{r^2}\left(\frac{\partial^2}{\partial \phi^2} + \frac{1}{4}\right) + V(r,\phi) \right] \Phi(r,\phi) = E\,\Phi(r,\phi) \tag{1}$$

where E denotes the energy, $\Phi(\vec{r})$ the reduced wave function and $V(\vec{r})$ the potential function. In the three-dimensional spherical coordinates $\vec{r}(r, \theta, \phi)$ the equation can be written in the form

$$\left\{ -\frac{\partial^2}{\partial r^2} - \frac{1}{r^2}\left[\frac{\partial^2}{\partial \theta^2} + \frac{1}{\sin^2\theta}\left(\frac{\partial^2}{\partial \phi^2} + \frac{1}{4}\right) + \frac{1}{4}\right] + V(r,\theta,\phi) \right\} \Phi(r,\theta,\phi) = E\,\Phi(r,\theta,\phi) \tag{2}$$

where $r \in [0, \infty]$, $\theta \in [0, \pi]$ and $\phi \in [0, 2\pi]$. The square-integrable solutions $\Phi(\vec{r})$ to the equations are constructed with Dirichlet boundary conditions. A more general discussion of the Schrödinger equation in a D-dimensional spherical coordinate system will be given elsewhere.

In the Lagrange-mesh method the angular domains $(\theta, \phi) \in [0, \pi] \times [0, 2\pi]$ are mapped onto the domains $(u, \eta) \in [-1, +1] \times [-1, +1]$ through suitable coordinate transformations. Common transformations for this purpose include $u = \cos n_u\theta$ ($\eta = \cos n_\eta\phi$) where (n_u, n_η) are real numbers determined by the structure of the potential. These transformations are applicable to many forms of the potential with the popular values of n being either 1 or 2. For $n_\eta = 1$, for example, (1) assumes the form

$$\left[-\hat{T}_r - \frac{1}{r^2}\left(\hat{T}_\eta + \frac{1}{4}\right) + V(r,\eta) \right] \Phi(r,\eta) = E\,\Phi(r,\eta)\,, \tag{3}$$

whereas equation (2) takes the form

$$\left\{ -\hat{T}_r - \frac{1}{r^2}\left[4\,\hat{T}_u + \frac{2}{1-u}\left(4\,\hat{T}_\eta + \frac{1}{4}\right) + \frac{1}{4}\right] + V(r,u,\eta) \right\} \Phi(r,u,\eta) = E\,\Phi(r,u,\eta)\,, \tag{4}$$

when $n_u = 2$ and $n_\eta = 2$. In these equations the notation

$$\hat{T}_r = \frac{\partial^2}{\partial r^2}\,, \qquad \hat{T}_u = \left(1-u^2\right)\frac{\partial^2}{\partial u^2} - u\frac{\partial}{\partial u}\,, \tag{5}$$

is used, with $\hat{T}_\eta$ similar to $\hat{T}_u$. Note that $\hat{T}_u$ has the exact form of the operator related to the Chebyshev polynomials of the first kind, which also are a special case of the classical Jacobi polynomials [14]. The Lagrange basis functions used to evaluate the matrix elements of the Hamiltonian in (2) are discussed in the next section.

3. The Lagrange-mesh matrix elements

Bases expansions in the Lagrange-Laguerre and Lagrange-Jacobi functions, defined respectively on the Laguerre and Jacobi meshes, are used to approximate the solution to (2). Only the properties of the Lagrange-mesh method relevant to this work are indicated in this section. Details of these meshes as well as the principles of the Lagrange-mesh method are discussed by Baye [8].

The Jacobi mesh is defined by the zeros z_j $(j = 1, 2, 3, \ldots, N)$ of the Jacobi polynomials $P_N^{\alpha,\beta}(z)$ of order N. The Lagrange-Jacobi functions defined on this mesh can be regularized by a factor $(1-z^2)^{\mu/2}$ [13], where μ is a free regularization parameter, to suppress singularities in the potentials that diverge at the boundaries. The resulting regularized Lagrange-Jacobi functions have the form

$$U_j(z) = (-1)^{N-j} \left[\frac{(1-z)^{\alpha+\mu}(1+z)^{\beta+\mu}}{(1-z_j^2)^{\mu-1}\,(2N+\alpha+\beta+1)\,h_N^{\alpha,\beta}} \right]^{1/2} \frac{P_N^{\alpha,\beta}(z)}{z-z_j} \tag{6}$$

where $h_N^{\alpha,\beta}$ is the normalizing constant of the Jacobi polynomials. The matrix elements for the angular operator (5), constructed with the regularized Lagrange-Jacobi functions $U_j(z)$, have the form [13]

$$\begin{aligned} T_{ij}^z &= \left\langle U_i(z) \middle| \left[(1-z^2)\frac{\mathrm{d}^2}{\mathrm{d}z^2} - z\frac{\mathrm{d}}{\mathrm{d}z} \right] \middle| U_j(z) \right\rangle \\ &= \begin{cases} (-1)^{i-j} \left(\dfrac{1-z_i^2}{1-z_j^2} \right)^{(\mu-1)/2} \left[\dfrac{(2\,\mu-1)\,z_i}{z_i - z_j} - \dfrac{2\,(1-z_i^2)}{(z_i-z_j)^2} \right] & ; \quad i \neq j \\ \dfrac{1-(2\,N+\alpha+\beta+1)^2}{12} + \dfrac{\omega_0 + \omega_1\,z_i + \omega_2\,z_i^2}{6\,(1-z_i^2)} & ; \quad i = j \end{cases} \end{aligned} \tag{7}$$

where $\omega_0 = \alpha^2 + \beta^2 - 6\,\mu + 4$, $\omega_1 = \alpha^2 - \beta^2$ and $\omega_2 = 6\,(\mu-1)^2$. The parameters (α, β) define the Lagrange-Jacobi functions (6). These matrix elements are a special case of those of the Lagrange-Jacobi given in [13]. An appropriate choice of the (α, β) customizes the matrix elements to the Lagrange-Legendre, Lagrange-Chebyshev as well as the Lagrange-Gegenbauer functions.

There are two main types of Laguerre meshes determined by whether the Laguerre polynomials are functions of either r or r^2. For potentials of harmonic oscillator-type the r^2-dependence is convenient. The modified Laguerre mesh is defined by the zeros r_i^2 $(i = 1, 2, 3, \ldots, K)$ of the Laguerre polynomials $L_K^{\sigma}(r^2)$ of order K. The Lagrange-Laguerre functions defined on this mesh are [8]

$$R_i(r) = (-1)^i \sqrt{\frac{2\,r_i^2}{h_K^{\sigma}}}\,\left(r^2\right)^{(2\,\sigma+1)/4}\,\mathrm{e}^{-r^2/2}\,\frac{L_K^{\sigma}(r^2)}{r^2 - r_i^2} \tag{8}$$

where h_K^{σ} is the normalizing constant of the Laguerre polynomials. Note that this mesh depends on the parameter σ, which is usually chosen so that (8) displays the desired form at small values of r. The matrix elements for the radial kinetic energy operator constructed with the basis (8) are [8]

$$T_{ij}^r = \left\langle R_i(r) \middle| -\frac{\mathrm{d}^2}{\mathrm{d}r^2} \middle| R_j(r) \right\rangle = \begin{cases} (-1)^{i-j} \dfrac{8\,r_i\,r_j}{(r_i^2 - r_j^2)^2} & ; \quad i \neq j \\ \dfrac{1}{3} \left[2\,(2\,K+\sigma+1) - r_i^2 - \dfrac{\sigma^2 + 5/4}{r_i^2} \right] & ; \quad i = j \end{cases} . \tag{9}$$

The parameter σ depends on the parameters (α, β) of the Lagrange-Jacobi functions discussed above. Note that the regularized modified Laguerre meshes [8] can also be used.

The solutions to (1) and (2) are developed through the basis expansion of the wave function. The following discussion focuses on (2) but can be readily reduced to that of (1) by simply omitting the dependency on u. To this end, the reduced wave function $\Phi(r, u, \eta)$ is expanded in the Lagrange functions bases of sizes (K, N, M),

$$\Phi(r,u,\eta) = \sum_{i=1}^{K}\sum_{j=1}^{N}\sum_{k=1}^{M} C_{ijk}\, R_i(r)\, U_j(u)\, U_k(\eta) \tag{10}$$

where C_{ijk} are the variational parameters, $R_i(r)$ the Lagrange-Laguerre functions (8), and $U_j(u)$ (and $U_k(\eta)$) the Lagrange-Jacobi functions (6). Pre-multiplying by Φ^*, applying the expansion (10) to equation (2), and integrating over the problem domain lead to the coupled algebraic equations

$$\sum_{i'=1}^{K}\sum_{j'=1}^{N}\sum_{k'=1}^{M} \left[H^0_{ijk,i'j'k'} + V(r_i, u_j, \eta_k)\, \delta_{ii'}\, \delta_{jj'}\, \delta_{kk'} \right] C_{i'j'k'} = E\, C_{ijk} \tag{11}$$

for the variational parameters. In these equations

$$H^0_{ijk,i'j'k'} = T^r_{ii'}\, \delta_{jj'}\, \delta_{kk'} - \frac{1}{r_i^2}\left[\left(4\, T^u_{jj'} + \frac{1}{4}\, \delta_{jj'} \right) \delta_{kk'} + \frac{2}{1-u_j} \left(4\, T^\eta_{kk'} + \frac{1}{4}\, \delta_{kk'} \right) \delta_{jj'} \right] \delta_{ii'} \tag{12}$$

are the kinetic energy matrix elements where the elements $T^u_{jj'}$ and $T^\eta_{kk'}$ are both given by (7) while $T^r_{ii'}$ are given by (9). Note that the elements $H^0_{ijk,i'j'k'}$ and $V(r_i, u_j, \eta_k)$ are the Gauss quadrature approximations of, respectively, the exact kinetic and potential energy matrix elements. Therefore, both matrices contribute integration errors to the numerical solutions [15]. Illustrative applications of and the accuracy attainable with the matrix elements (11) are discussed in the next section.

4. Applications

Illustrative applications of the presented Lagrange-mesh matrix elements are discussed for the quartic anharmonic oscillator in two dimensions and the double ring-shaped harmonic oscillator in three dimensions. Only the Gauss approximations of the matrix elements for the Hamiltonian are used in all the calculations presented. All the Lagrange meshes used are constructed with Gauss quadrature abscissas determined with accuracy of about 3×10^{-14} and the regularization parameter is set to $\mu = \frac{1}{2}$.

4.1. Quartic anharmonic oscillator

The two-dimensional quartic anharmonic oscillator potential in cylindrical polar coordinates has the form

$$V(r,\phi) = V_1\, r^2 + V_2\, r^4 \left[1 - 2\,(1-\nu)\, \sin^2\phi\, \cos^2\phi \right]; \qquad r \in [0,\infty], \quad \phi \in [0, 2\,\pi], \tag{13}$$

where V_1, V_2 and ν are constants. Using the transformation $\eta = \cos 2\phi$ the potential is cast in the form

$$V(r,\eta) = V_1\, r^2 + V_2\, r^4 \left[1 - \tfrac{1}{2}\,(1-\nu)(1-\eta^2) \right]; \qquad r \in [0,\infty], \quad \eta \in [-1, +1]. \tag{14}$$

The system is solved using the Lagrange-Laguerre and Lagrange-Chebyshev functions where σ is determined by α_θ and β_θ. Note that the Chebyshev functions are the eigenfunctions of the

Table 1. Convergence of the energies for the first two states of the quartic anharmonic oscillator (13) with bases sizes (K, N). $V_1 = 1$ and $V_2 = 0.001$.

ν	(K,N)	E_{00}	E_{01}
-1	(3,6)	2.000 998 505 3	6.006 971 8
	(6,6)	2.000 998 505 469 824	6.006 970 242 132 808
	(7,3)[a]	2.000 998 505 469 810	6.006 970 242 132 821
0	(3,6)	2.001 497 385	6.010 463 0
	(6,6)	2.001 497 385 346 379	6.010 460 565 461 274
	(7,3)[a]	2.001 497 385 346 371	6.010 460 565 461 293
1	(3,6)	2.001 995 521 5	6.013 940 2
	(6,6)	2.001 995 522 094 719	6.013 936 098 189 642
	(7,3)[a]	2.001 995 522 094 708	6.013 936 098 189 653

[a] Results of [16] where N represents the size of the sub-problem.

kinetic energy operator of the system. To facilitate the comparison with the results of [16], the Schrödinger equation for this potential is solved with the parameters $V_1 = 1$, $V_2 = 0.001$ and $\nu = \{-1, 0, 1\}$, although the parameters (K, N) have a slightly different meaning [a]. Convergence tests with the bases sizes were conducted and the results are shown in table 1.

As can be seen in the table, the calculated energies converge rapidly with the increasing bases size. Results comparable with those of [16] are obtained with a basis of comparable size, $(K, N) = (6, 6)$. Note that in the treatment presented in [16] preconditioning specific to this potential is used before constructing the eigenvalue problem for the system. However, in the present work the system is readily transformed to a standard eigenvalue problem. In addition, the kinetic energy matrix elements presented in this work are applicable not only to the Chebyshev functions of the first kind. The application of the matrix elements with the Lagrange-Jacobi and the Lagrange-Gegenbauer functions is illustrated in the next subsection.

4.2. Double ring-shaped harmonic oscillator

The harmonic oscillator coupled with the Pöschl-Teller double ring-shaped potential is written in the form

$$V(r,\theta,\phi) = r^2 + \frac{1}{r^2}\left[\frac{1}{\sin^2\theta}\left(\frac{a^2-\frac{1}{4}}{\sin^2\phi} + \frac{b^2-\frac{1}{4}}{\cos^2\phi}\right) + \frac{c^2-\frac{1}{4}}{\cos^2\theta}\right]; \quad r \in (0,\infty], \quad \theta \in [0,\tfrac{1}{2}\pi] \quad \text{and} \quad \phi \in [0,\tfrac{1}{2}\pi], \tag{15}$$

where a, b and c are constants. The Schrödinger equation for this potential admits exact solutions [2] with eigenvalues

$$E_{ijk} = 2\,[\,2\,(\,i+j+k\,)+a+b+c+3\,] \tag{16}$$

for $i = 0, 1, 2, \ldots$, $j = 0, 1, 2, \ldots$, and $k = 0, 1, 2, \ldots$. Through the transformations $u = \cos 2\theta$ and $\eta = \cos 2\phi$ the potential is cast in the form

$$V(r,u,\eta) = r^2 + \frac{2}{r^2}\left[\frac{2}{1-u}\left(\frac{1}{1-\eta} + \frac{1}{1+\eta}\right) + \frac{1}{1+u}\right]; \quad r \in (0,\infty], \quad u \in [-1,+1] \quad \text{and} \quad \eta \in [-1,+1] \tag{17}$$

where, for this work, the potential parameters are set to $a = b = c = 3/2$. The system is solved using the Lagrange-Laguerre, Lagrange-Jacobi and Lagrange-Gegenbauer functions ($\alpha_\phi = \beta_\phi$). Note that α_θ is determined by α_ϕ and β_ϕ while σ is determined by α_θ and β_θ. Convergence tests with bases sizes were conducted and the results are shown in table 2 for $\alpha_\phi = \beta_\phi = 3/2$ and $\beta_\theta = 1$ and in table 3 for $\alpha_\phi = \beta_\phi = \beta_\theta = 3/2$. The errors

$$\epsilon_n = \left| \frac{E_n(\text{num}) - E_n(\text{anal})}{E_n(\text{anal})} \right| ,$$

of the numerical energies $E_n(\text{num})$ relative to the analytical energies $E_n(\text{anal})$, are also shown in the tables.

Table 2. Convergence of the energies for the first three states of the double ring-shaped harmonic oscillator with bases sizes $(K, N, M) = (15, 15, 15)$. $\alpha_\phi = \beta_\phi = 3/2$ and $\beta_\theta = 1$.

n	E_n(num)	E_n(anal)	ϵ_n
0	14.998	15.0	1.10(-04)
1	18.995	19.0	2.85(-04)
1	18.997	19.0	1.41(-04)
1	18.998	19.0	8.77(-05)
2	22.988	23.0	5.27(-04)
2	22.992	23.0	3.62(-04)
2	22.995	23.0	2.36(-04)
2	22.996	23.0	1.69(-04)
2	22.997	23.0	1.17(-04)
2	22.998	23.0	7.62(-05)

Practically no computational time is spend on constructing the matrices and it is immediately noticed, in the tables, that the degeneracies of the states are correctly reproduced. In general

Table 3. Same as in table 2, with bases sizes $(K, N, M) = (3, 3, 3)$ and $\alpha_\phi = \beta_\phi = \beta_\theta = 3/2$.

n	E_n(num)	E_n(anal)	ϵ_n
0	14.999 999 999 999 978	15.0	1.42(-15)
1	18.999 999 999 999 978	19.0	1.12(-15)
1	19.000 000 000 000 000	19.0	0.00(+0)
1	19.000 000 000 000 017	19.0	9.35(-16)
2	22.999 999 999 999 986	23.0	6.18(-16)
2	22.999 999 999 999 996	23.0	1.54(-16)
2	23.000 000 000 000 000	23.0	0.00(+0)
2	23.000 000 000 000 004	23.0	1.54(-16)
2	23.000 000 000 000 004	23.0	1.54(-16)
2	23.000 000 000 000 004	23.0	1.54(-16)

the numerical energies converge, though quite slowly, with the increasing basis size when the parameters of the Lagrange-Jacobi functions are chosen arbitrarily. For example, with the parameter set $\alpha_\phi = \beta_\phi = 3/2$ and $\beta_\theta = 1$, all the calculated energies, given in table 2, are reproduced to about $\epsilon_n \approx 10^{-4}$ with the bases sizes $(K, N, M) = (15, 15, 15)$. However, when the correct values of the parameters of the Lagrange functions $(\alpha_\phi = \beta_\phi = \beta_\theta = 3/2)$ are used, the calculated energies, shown in table 3, are reproduced to machine accuracy with very low bases sizes $(K, N, M) = (3, 3, 3)$.

5. Conclusions

The Lagrange-mesh method was used to solve the Schrödinger equation in polar coordinates. The method readily transforms the Schrödinger equation to an eigenvalue problem and is applicable to any well behaved potential functions. The recently developed Lagrange-Jacobi matrix elements were specialized to the Lagrange-Chebyshev elements to represent the angular kinetic energy matrix elements. The method was then applied to the quartic anharmonic oscillator in two-dimensions as well as the double ring-shaped harmonic oscillator in three-dimensions. The results show that convergence to the correct solutions with increasing basis size is generally slow but reasonable when the basis parameters (α, β) are chosen consistent with the eigenfunctions of the kinetic energy operator. However, eigenvalues for the systems are reproduced to within machine accuracy when the parameters are chosen appropriately.

References

[1] Flügge S 1999 *Practical quantum mechanics* (Berlin: Springer-Verlag)
[2] Bachkhaznadji A, Lassaut M and Lombard R J 2009 *J. Phys. A: Math. Theor.* **42** 065301
[3] Colbert D T and Miller W H 1992 *J. Chem. Phys.* **96** 1982–1991
[4] Chen Z and Goldman S P 1992 *Phys. Rev. A* **45**(3) 1722–1731
[5] Ram-Mohan L 2002 *Finite Element and Boundary Element Applications in Quantum Mechanics* (New York: Oxford University Press)
[6] Amore P and Fernández F M 2010 *Phys. Scr* **81** 045011
[7] Al-Jaber S M 2010 *Nuovo Cimento* **125** 1099–1108
[8] Baye D 2015 *Phys. Rep.* **565** 1–107
[9] Descouvemont P, Daniel C and Baye D 2003 *Phys. Rev. C* **67**(4) 044309
[10] Lacroix G, Semay C and Buisseret F 2012 *Phys. Rev. E* **86**(2) 026705
[11] Baye D, Filippin L and Godefroid M 2014 *Phys. Rev. E* **89**(4) 043305
[12] Vincke M, Malegat L and Baye D 1993 *J. Phys. B: At. Mol. Opt. Phys.* **26** 811
[13] Rampho G J 2016 *J. Phys. A: Math. Theor.* **49** 27501
[14] Abramowitz M and Stegun I 1965 *Handbook of mathematical functions* (New York: Dover)
[15] Szalay V 1993 *J. Chem. Phys.* **99** 1978–1984
[16] Taşeli H, Erhan İ M and Uğur Ö 2002 *J. Math. Chem.* **32** 323–338

IOP Conf. Series: Journal of Physics: Conf. Series **905** (2017) 012038 doi:10.1088/1742-6596/905/1/012038

Superfluid-Mott glass quantum multicritical point on a percolating lattice

Martin Puschmann[1,2] **and Thomas Vojta**[1]

[1] Department of Physics, Missouri University of Science & Technology, Rolla, MO 65409, USA
[2] Institute of Physics, Technische Universität Chemnitz, 09107 Chemnitz, Germany

E-mail: vojtat@mst.edu

Abstract. We employ large-scale Monte Carlo simulations to study a particle-hole symmetric site-diluted quantum rotor model in two dimensions. The ground state phase diagram of this system features two distinct quantum phase transitions between the superfluid and the insulating (Mott glass) phases. They are separated by a multicritical point. The generic transition for dilutions below the lattice percolation threshold is driven by quantum fluctuations while the transition across the percolation threshold is due to the geometric fluctuations of the lattice. We determine the location of the multicritical point between these two transitions and find its critical behavior. The multicritical exponents read $z = 1.72(2)$, $\beta/\nu = 0.41(2)$, and $\gamma/\nu = 2.90(5)$. We compare our results to other quantum phase transitions in disordered systems, and we discuss experiments.

1. Introduction

Randomly diluted quantum many-particle systems feature two kinds of fluctuations at zero temperature. First, geometric fluctuations due to bond and/or site dilution of the lattice lead to a percolation problem with a geometric phase transition at the corresponding percolation threshold [1]. Second, quantum fluctuations are caused by non-commuting operators in the many-particle Hamiltonian. Combining both often leads to rich ground-state phase diagrams with several different quantum phase transitions.

Prototypical examples of such behavior can be found in magnetic systems (for a review see, e.g., Ref. [2]). The diluted transverse-field Ising model displays long-range order for dilutions p up to the percolation threshold p_c provided that the transverse-field h_x is below the critical field $h_c(p)$. For dilutions above the percolation threshold, long-range order is not possible since the system consists of decoupled finite-size clusters. As a result, the p-h_x ground state phase diagram displays two quantum phase transitions, separated by a multicritical point at (p_c, h^*) [3, 4, 5, 6, 7]. The generic transition across $h_c(p)$ for $0 < p < p_c$ is driven by quantum fluctuations while the percolation transition across p_c for $h_x < h^*$ is driven by the geometric criticality of the lattice. Other quantum magnets behave similarly. The diluted square lattice Heisenberg quantum antiferromagnet is long-range ordered at the percolation threshold [8, 9]. If the quantum fluctuations are strengthened, for example via an interlayer coupling in a bilayer quantum antiferromagnet, a phase diagram with a multicritical point appears, as in the transverse-field Ising case [10, 11, 12, 13].

In the present paper, we investigate the interplay of geometric and quantum fluctuations in a system of interacting bosons on a diluted lattice that undergoes a quantum phase transition between a superfluid and an insulating ground state. Experimental applications include Josephson junction arrays [14, 15], granular superconducting films [16, 17], helium absorbed in vycor [18, 19], as well as doped quantum magnets in high fields [20, 21, 22]. In the presence of disorder, the superfluid phase and the Mott insulator are separated by a "glassy" quantum Griffiths phase [23, 24, 25] in which rare regions of superfluid order exist in an insulating bulk system. For generic disorder (without special symmetries), this intermediate phase is the Bose glass, a compressible gapless insulator [26, 27, 28]. If the Hamiltonian is particle-hole symmetric even in the presence of disorder, the intermediate phase between the superfluid and the Mott insulator is instead the so-called Mott glass (or random-rod glass) [29, 30] which is an incompressible gapless insulator.

While the quantum phase transition between superfluid and Bose glass has been studied in great detail, the quantum phase transition between superfluid and Mott glass has attracted comparatively less attention. Early numerical work in two dimensions using quantum Monte Carlo simulations [31, 32] and a strong-disorder renormalization group [33] produced inconclusive and partially contradicting results for the critical behavior of this transition. To resolve this issue, we recently performed large-scale Monte Carlo simulations of a site-diluted two-dimensional quantum rotor model with particle-hole symmetry [34]. Similar to the quantum magnets discussed above, this system features two distinct quantum phase transitions. For dilutions p below the percolation threshold p_c of the lattice, we find a superfluid-Mott glass transition characterized by a universal (dilution-independent) critical behavior. The transition across the lattice percolation threshold p_c falls into a different universality class.

In the present paper, we briefly summarize these results and then focus on the multicritical point separating the generic quantum phase transition ($p < p_c$) from the percolation quantum phase transition across the lattice percolation threshold. The paper is organized as follows. We introduce in Sec. 2 the quantum rotor Hamiltonian and its mapping onto a classical XY model with columnar disorder. Section 3 is devoted to the Monte Carlo method and the data analysis techniques. Results for the generic transition, the percolation transition, and the multicritical point are discussed in Sec. 4. We conclude in Sec. 5.

2. Diluted quantum rotor model

The quantum rotor Hamiltonian is a prototypical model of interacting bosons. We consider a diluted square-lattice array of superfluid grains, coupled by Josephson junctions. The Hamiltonian reads

$$H = \frac{U}{2}\sum_i \epsilon_i(\hat{n}_i - \bar{n}_i)^2 - J\sum_{\langle ij\rangle} \epsilon_i\epsilon_j \cos(\hat{\phi}_i - \hat{\phi}_j) \tag{1}$$

where $\hat{\phi}_i$ is the phase operator of site (grain) i, and $\hat{n}_i$ is the number operator canonically conjugate to the phase. The sum over $\langle ij\rangle$ comprises pairs of nearest neighbors on the square lattice. The parameters U and J are the charging energy and the Josephson coupling, respectively. $\bar{n}_i$ is the offset charge at site i. Site dilution is introduced via the quenched random variables ϵ_i which are independent of each other. They can take the values 0 (vacancy) with probability p and 1 (occupied site) with probability $1 - p$.

For generic values of the offset charges $\bar{n}_i$ or generic non-integer filling, the Hamiltonian (1) is not particle-hole symmetric, and the intermediate phase between superfluid and Mott insulator is a Bose glass. To preserve the particle-hole symmetry, we set all offset charges $\bar{n}_i = 0$ and fix the particle number $\langle \hat{n}\rangle$ at a large integer value. In this case, the disorder introduced by the site dilution is off-diagonal, and the intermediate phase is a Mott glass.

The qualitative behavior of the particle-hole symmetric site-diluted quantum rotor model can be easily understood [27, 30, 34]. If the Josephson energy J dominates over the charging energy U and if the site dilution p is below the lattice percolation threshold p_c, the ground state is a long-range ordered superfluid. Long-range order can be destroyed either by increasing the charging energy (which enhances the quantum fluctuations) or by raising the dilution beyond the percolation threshold (where the lattice decomposes into disconnected finite-size clusters).

Because we plan to perform Monte Carlo simulations using a highly efficient classical cluster algorithm, we now map the quantum rotor Hamiltonian (1) onto a classical model. In the presence of particle hole symmetry, the mapping leads to a classical XY model on a cubic lattice [35]. The classical Hamiltonian reads

$$H_{\text{cl}} = -J_s \sum_{\langle i,j\rangle,t} \epsilon_i \epsilon_j \mathbf{S}_{i,t} \cdot \mathbf{S}_{j,t} - J_\tau \sum_{i,t} \epsilon_i \mathbf{S}_{i,t} \cdot \mathbf{S}_{i,t+1} \ . \tag{2}$$

Here i denotes a position in two-dimensional real space and t is the "imaginary time" coordinate. The dynamical variable $\mathbf{S}_{i,t}$ at each lattice site (i,t) is an O(2) unit vector. Note that the vacancy positions do not depend on the imaginary time coordinate t. The classical XY model (2) therefore features columnar defects, i.e., the disorder is perfectly correlated in the imaginary time direction. The values J_s and J_τ of the interaction energies as well as the "classical" temperature T are determined by the mapping from the original quantum rotor Hamiltonian (the classical temperature differs from the real physical temperature which is zero). As we are interested in the critical behavior which is expected to be universal, the precise values of J_s and J_τ are not important. Consequently, we fix $J_s = J_\tau = 1$ and tune the classical XY model (2) through its phase transitions by varying either the classical temperature T or the dilution p.

In the absence of dilution, the Hamiltonian (2) represents the usual three-dimensional XY model. The clean two-dimensional superfluid-Mott insulator quantum critical point is therefore in the three-dimensional classical XY universality class which features a correlation length exponent $\nu \approx 0.6717$ [36]. The stability of the clean critical behavior against disorder is governed by the Harris criterion [37] $d\nu > 2$. Here, d is the number of dimensions in which there is randomness; for columnar defects in a cubic lattice this implies $d = 2$. The Harris criterion is therefore violated, and the three-dimensional clean XY critical point is unstable against columnar defects.

3. Monte Carlo method

To determine the critical behavior of the classical XY Hamiltonian (2), we perform Monte Carlo simulations using the highly efficient Wolff cluster algorithm [38] which greatly reduces the critical slowing down close to the phase transition. In addition to the Wolff updates, we also perform conventional Metropolis updates [39] to equilibrate small disconnected clusters of lattice site that can occur at higher dilutions p and may be missed by the Wolff algorithm.

Our simulations cover several dilutions $p = 0$, 1/8, 1/5, 2/7, 1/3, 9/25 and the lattice percolation threshold $p_c = 0.407253$. Linear sizes range from $L = 10$ to 150 in the two space directions and $L_\tau = 6$ to 1792 in the imaginary time direction. All observables are averaged over 10,000 to 50,000 disorder configurations, depending on system size. Each sample is equilibrated for 100 full Monte Carlo sweeps, followed by a measurement period of 500 sweeps with a measurement taken after every sweep. (A full Monte Carlo sweep consists of a Metropolis sweep and a Wolff sweep, the latter is defined as a number of cluster flips such that the total number of flipped spins equals the number of lattice sites.) The choice of rather short runs for a large number of disorder configurations helps reducing the overall statistical error of the data, as discussed in Refs. [13, 40, 41, 42, 43]. To eliminate systematic biases in observables stemming from the short simulation runs, we use improved estimators (see, e.g., the appendix of Ref. [43]).

We analyze the Monte Carlo data by finite-size scaling. As the columnar disorder in the classical XY model breaks the symmetry between the space and imaginary time directions, observables generally do not scale in the same way with the system sizes L and L_τ. We consider, for example, the average Binder cumulant

$$g_{\mathrm{av}} = \left[1 - \frac{\langle|\mathbf{m}|^4\rangle}{3\langle|\mathbf{m}|^2\rangle^2}\right]_{\mathrm{dis}} . \tag{3}$$

Here, $[\ldots]_{\mathrm{dis}}$ refers to the disorder average, $\langle\ldots\rangle$ denotes the Monte Carlo average for each sample, and $\mathbf{m} = (1/N)\sum_{i,\tau}\mathbf{S}_{i,\tau}$ is the order parameter (with N being the number of lattice sites). The finite-size scaling form of the Binder cumulant,

$$g_{\mathrm{av}}(r, L, L_\tau) = X_{g_{\mathrm{av}}}(rL^{1/\nu}, L_\tau/L^z) , \tag{4}$$

depends on two independent arguments. Here, $r = (T - T_c)/T_c$ is the distance from criticality, z is the dynamical critical exponent, and $X_{g_{\mathrm{av}}}$ is the dimensionless scaling function.[1]

In the absence of a value for the dynamical exponent z, the usual analysis of the Binder cumulant (which identifies the critical point with the crossing of the curves for different L) breaks down because the correct sample shapes are not known. We therefore follow the method devised in Refs. [13, 42, 45, 46]. The Binder cumulant has a maximum as function of L_τ at fixed L and T. At the critical temperature, the peak position $L_\tau^{\max}$ scales as L^z. This allows us to use an iterative approach that finds the value of z together with the critical point. The method begins with a guess of the dynamical exponent and the corresponding sample shapes. The approximate crossing of the g_{av} vs. T curves for these shapes results in an estimate for the critical temperature T_c. We then consider g_{av} as a function of L_τ for fixed L at this temperature. The peak positions $L_\tau^{\max}$ provide improved estimates for the optimal sample shapes. We iterate these steps three or four times to converge the values of T_c and z with reasonable accuracy.

Once the optimal sample shapes are known (which fixes the scaling combination L_τ/L^z), the finite-size scaling analysis of other thermodynamic observables such as the order parameter m and its susceptibility χ proceeds normally. Their finite-size scaling forms read

$$m = L^{-\beta/\nu} X_m(rL^{1/\nu}, L_\tau/L^z) , \tag{5}$$
$$\chi = L^{\gamma/\nu} X_\chi(rL^{1/\nu}, L_\tau/L^z) \tag{6}$$

where X_m and X_χ are dimensionless scaling functions, and β and γ are the order parameter and susceptibility critical exponents, respectively. Correlation lengths in space and imaginary time directions are calculated, as usual, from the second moment of the spin correlation function [47, 48, 49]. The finite-size scaling forms of the reduced correlation lengths are given by

$$\xi_s/L = X_{\xi_s}(rL^{1/\nu}, L_\tau/L^z) , \tag{7}$$
$$\xi_\tau/L_\tau = X_{\xi_\tau}(rL^{1/\nu}, L_\tau/L^z) . \tag{8}$$

4. Results

4.1. Phase diagram

The phase diagram of the classical Hamiltonian (2) in the dilution-(classical) temperature plane that arises from our Monte Carlo simulations is presented in Fig. 1. The generic phase boundary (for $p < p_c$) was determined in Ref. [34] by performing simulations at dilutions $p = 0$, 1/8, 1/5, 2/7, 1/3, and 9/25. In that work, we also confirmed the location of the percolation phase transition by carrying out simulations at $p = p_c$ and $T = 1.0$. The position of the multicritical

[1] We have assumed conventional power-law dynamical scaling rather than activated scaling for which the scaling combination would be $\ln L_\tau/L^\psi$ (with ψ the tunneling exponent), in agreement with the general classification of disordered critical points put forward in Refs. [25, 44].

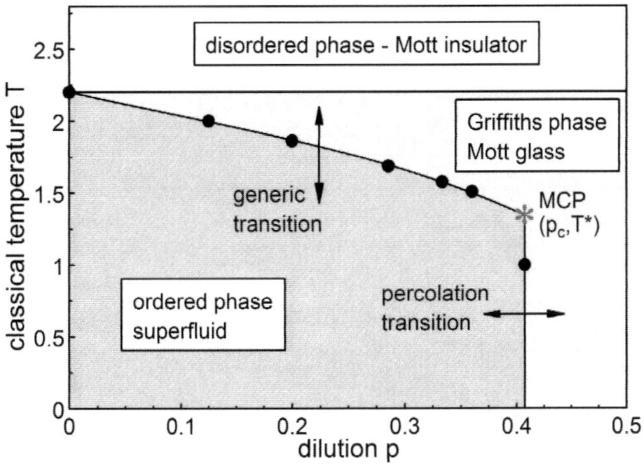

Figure 1. (Color online) Phase diagram of the (2+1)-dimensional XY model (2) as a function of dilution p and classical temperature T. The black dots mark the numerically determined critical points on the generic and percolation phase boundaries. The red star denotes the multicritical point (MCP) at $p = p_c = 0.407253$ and $T = T^* = 1.337$ that separates the generic and percolation transitions. The lines are guides for the eye only.

point follows from our present simulations (see below); it is located at $p = p_c = 0.407253$ and $T = T^* = 1.337$.

Qualitatively, this phase diagram is very similar to the phase diagrams of the diluted quantum magnets discussed in the introductory section. Long-range (superfluid) order is restricted to dilutions $p \le p_c$ and sufficiently weak quantum fluctuations (which are represented by the classical temperature T after the mapping of the rotor Hamiltonian onto the classical XY model). Importantly, as for the quantum magnets, long-range order survives on the critical percolation cluster (at $p = p_c$) up to a nonzero classical temperature T^*. This causes the existence of two distinct phase transitions (generic and percolation) separated by the multicritical point that is the focus of the present paper. Note that the disordered (insulating) phase can be decomposed into two regions. If the classical temperature is below the transition temperature $T_c(0)$ of the undiluted system, large spatial regions without vacancies can show local superfluid order even if the bulk is insulating. This is the Griffiths phase of the transition, i.e., the Mott glass. For classical temperatures above $T_c(0)$, local order is impossible; the system is thus a conventional Mott insulator.

4.2. Generic and percolation transitions

In this subsection, we briefly summarize the critical behaviors of the generic and percolation transitions. To find the critical exponents of the generic transition, we analyzed [34] Monte Carlo data for dilutions $p = 1/8$, $1/5$, $2/7$, $1/3$, and $9/25$. By including correction-to-scaling terms that account for deviations from pure power-law behavior, we could fit all these data using universal, dilution-independent exponents. The values of the dynamical exponent z, the scale dimension β/ν of the order parameter, the corresponding susceptibility exponent γ/ν, and the correlation length exponent ν are listed in Table 1. Other exponents can be found using various scaling relations.

The critical behavior of the transition across the lattice percolation threshold was analyzed earlier by means of a scaling theory [12] that relates its exponents to the classical percolation exponents (which are known exactly in two space dimensions). The resulting exponent values

IOP Conf. Series: Journal of Physics: Conf. Series **905** (2017) 012038 doi:10.1088/1742-6596/905/1/012038

Table 1. Critical behavior of the superfluid-Mott glass quantum phase transitions. The exponents of the generic ($p < p_c$) transition were found numerically in Ref. [34]. The exponents of the percolation transition were determined from a scaling theory [12] and numerically confirmed in Ref. [34]. Their values are exact. The multicritical exponents are the result of the present paper.

Exponent	generic	percolation	multicritical
z	1.52(3)	91/48	1.72(2)
β/ν	0.48(2)	5/48	0.41(2)
γ/ν	2.52(4)	59/16	2.90(5)
ν	1.16(5)	4/3	

are listed in Table 1 as well. Our Monte Carlo data for the percolation transition (taken at $p = p_c$ and $T = 1.0$) agree with these predictions [34].

4.3. Multicritical point

We now turn to the main topic of this paper, viz., the multicritical point that separates the generic and percolation transitions. To find the multicritical temperature T^* and the dynamical exponent z, we employ the iterative procedure described in Sec. 3. It consists of two kinds of Monte Carlo runs: (i) runs right at T_c for systems of different L_τ for each L. The position $L_\tau^{\max}$ of the maximum of the Binder cumulant g_{av} yields the optimal sample shapes and the dynamical exponent. (ii) In the second type of runs, we vary the temperature but only use the optimal sample shapes. The crossing of the g_{av} vs. T curves gives an (improved) estimate of T_c.

Results of the first type of runs are presented in Fig. 2. It shows the Binder cumulant g_{av} as function of L_τ for several $L = 10$ to 86 after the iteration has converged to a multicritical temperature of $T^* = 1.337$. (The error $\Delta T^* \approx 0.003$ to 0.005 is estimated heuristically from

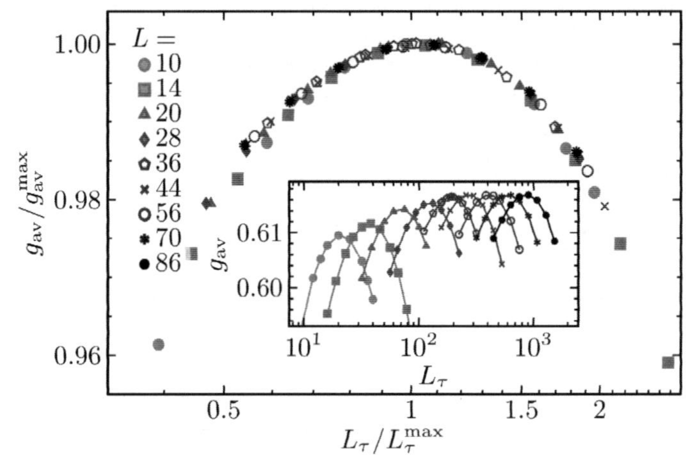

Figure 2. (Color online) Binder cumulant g_{av} as function of L_τ for several L at the multicritical temperature $T^* = 1.337$. The relative statistical error of g_{av} is between 0.05% and 0.08%. Inset: Raw data g_{av} vs. L_τ. Main panel: Scaling plot $g_{\mathrm{av}}/g_{\mathrm{av}}^{\max}$ vs. $L_\tau/L_\tau^{\max}$.

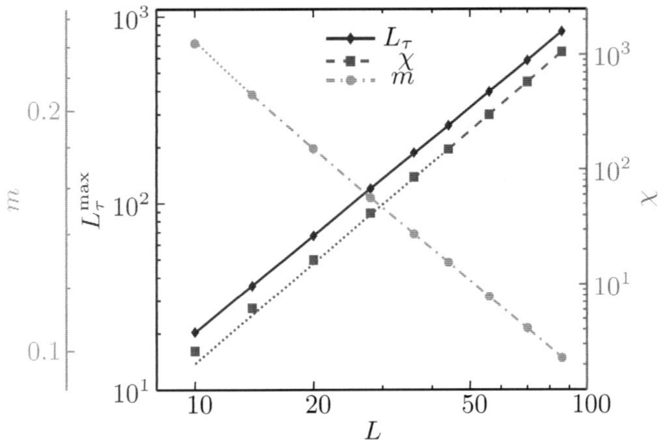

Figure 3. (Color online) Double-logarithmic plot of $L_\tau^{\max}$, m, and χ vs. L for samples of the optimal shape at the multicritical temperature $T^* = 1.337$. The statistical errors are well below the symbol size. The lines represent power-law fits (see text). They are dotted in the regions not included in the fit.

the quality of the crossings of the g_{av} vs. T curves for optimally shaped samples.) The figure demonstrates that the maximum Binder cumulant $g_{\text{av}}^{\max}$ for each of the curves does not depend on L (at least for the larger L), it takes a universal value of about 0.6171(3). This is the behavior expected at a (multi)critical point. The deviations for smaller L can be attributed to finite-size effects, i.e., to corrections to scaling. The scaling plot presented in the main panel of Fig. 2 confirms that g_{av} fulfills the scaling form (4) with high precision (slight deviations for small L and L_τ stem, again, from corrections to scaling).

Once the optimal shapes are known, the dynamical critical exponent z can be found by analyzing the dependence of $L_\tau^{\max}$ on L, as shown in Fig. 3. The data can be fitted well using a pure power-law dependence $L_\tau^{\max} \sim L^z$ giving a value $z = 1.725(3)$. The statistical errors of $L_\tau^{\max}$ are determined from 1000 synthetic data sets. Each set is obtained from the original data by adding Gaussian random noise corresponding the uncertainties of the original data. The power-law fit is of good quality, it yields $\tilde{\chi}^2 \approx 0.74$. [The reduced sum of squared errors of the fit (per degree of freedom) is denoted by $\tilde{\chi}^2$ to distinguish it from the order parameter susceptibility χ.] In addition to the statistical error, z also has an error stemming from the uncertainty in T^*. To estimate it, we repeat the $L_\tau^{\max}$ vs. L analysis at the higher temperature $T^* + 0.01$, significantly further away from T^* than our estimated T^* error. This leads to a shift of z of about 0.025. Our final estimate for the dynamical critical exponent thus reads $z = 1.72(2)$.

The critical exponents β/ν and γ/ν can be determined from the L-dependence of the order parameter and the susceptibility, respectively, for samples of optimal shape at the multicritical temperature. The corresponding plots are also shown in Fig. 3. The order parameter data for $L > 20$ can be fitted well by a pure power law $m \sim L^{-\beta/\nu}$ giving $\beta/\nu = 0.411(1)$ with $\tilde{\chi}^2 \approx 0.99$. If the leading correction-to-scaling term is included via $m = aL^{-\beta/\nu}(1 + bL^{-\omega})$, the fit range can be extended to all $L > 10$, yielding an almost unchanged exponent value $\beta/\nu = 0.406(4)$ with $\tilde{\chi}^2 \approx 1.26$. The largest source of error is again the uncertainty in T^*. To estimate it, we repeat the analysis at $T^* \pm \Delta T^*$ with $\Delta T^* = 0.003$. This leads to shifts of β/ν of about 0.02. Our final estimate therefore is $\beta/\nu = 0.41(2)$.

The susceptibility data show stronger deviations from pure power-law behavior. A fit to $\chi \sim L^{\gamma/\nu}$ works for sizes $L > 36$ and yields $\gamma/\nu = 2.905(4)$ with $\tilde{\chi}^2 \approx 0.66$. Extending the fit range to sizes $L > 28$ gives $\gamma/\nu = 2.880(3)$ but leads to unacceptably large $\tilde{\chi}^2$ values. The fit becomes unstable when a correction-to-scaling term is included via $\chi = aL^{\gamma/\nu}(1 + bL^{-\omega})$. Due to these instabilities γ/ν has a larger error. Our final value is $\gamma/\nu = 2.90(5)$ where the error bar is estimated heuristically from the robustness of the fit; it includes the error due to the uncertainty in T^* obtained from repeating the analysis at $T^* \pm 0.003$.

Note that the multicritical exponents β/ν, γ/ν, and z must fulfill the hyperscaling relation $2\beta/\nu + \gamma/\nu = d + z$ where $d = 2$ is the number of real space dimensions. The Monte Carlo estimates $\beta/\nu = 0.41(2)$, $\gamma/\nu = 2.90(5)$, and $z = 1.72(2)$ fulfill this relation within their error bars. This gives us confidence that they represent true asymptotic rather than effective critical exponents.

5. Conclusions

In conclusion, we have reported the results of large-scale Monte Carlo simulations of a site-diluted particle-hole symmetric quantum rotor model in two dimensions. They were aimed at investigating the quantum phase transitions between the superfluid and insulating (Mott glass) ground states of disordered interacting bosons. The ground state phase diagram in the dilution-quantum fluctuation plane features two distinct quantum phase transitions, (i) the generic transition that occurs for dilutions below the lattice percolation threshold and is driven by quantum fluctuations, and (ii) the transition across the lattice percolation threshold which is driven by geometric fluctuations. The two phase transition lines meet at a multicritical point at which both quantum and geometric fluctuations become critical.

In the absence of disorder, the superfluid-Mott insulator transition belongs to the three-dimensional classical XY universality class. Its correlation length exponent $\nu \approx 0.6717$ violates the Harris criterion $d\nu > 2$. The critical exponents of the diluted system are therefore expected to differ from the clean ones. This expectation is fulfilled by our Monte Carlo results, summarized in Table 1, for the generic transition, the percolation transition, and the multicritical point. In all cases, the transition is of conventional "finite-disorder" type (featuring power-law dynamical scaling) rather than an infinite randomness critical point [50, 51, 52, 53] or a smeared transition [54, 55].

We have considered here the two-component quantum rotor Hamiltonian that is a model of disordered interacting bosons. The analogous three-component quantum rotor Hamiltonian which arises as an effective low-energy theory of a bilayer Heisenberg antiferromagnet was studied in Refs. [13, 42]. Both systems have very similar phase diagrams, but their exponent values differ. For example, the dynamical exponent z of the two-component model considered here takes the values 1.52 and 1.72 for the generic transition and the multicritical point, respectively. The corresponding values for the three-component case are 1.31 and 1.54. Other exponents show similar differences. Note, however, that the exponents of the percolation quantum phase transition do not depend on the number of rotor components, as they are determined by the geometric criticality of the lattice only.

It is also interesting to compare the influence of the space dimensionality. The one-dimensional equivalent of the rotor Hamiltonian (1) undergoes a Kosterlitz-Thouless quantum phase transition [56] in the two-dimensional classical XY universality class in the absence of disorder. This transition fulfills the Harris criterion (as $\nu = \infty$), weak disorder therefore does not change the critical behavior. At larger disorder, the character of the transition does change even though the details of its critical behavior are still controversially discussed (see, e.g., Refs. [57, 58, 59, 60] and references therein). In contrast, the clean rotor model in three space dimensions violates the Harris criterion because its correlation length exponent takes the mean-field value 1/2. We therefore expect a scenario similar to the two-dimensional case considered in the present paper.

Experimentally, there are several ways to realize diluted systems of interacting bosons. These include granular superconductors (whose superconductor-insulator quantum phase transition is believed to be bosonic in nature, at least in some systems), ultracold atoms (where the dilution could perhaps be realized via mixtures of heavy and light atoms), or certain magnetic systems such as diluted anisotropic spin-1 antiferromagnets [61]. The last example is particularly promising because the required particle-hole symmetry appears naturally as a consequence of the up-down symmetry of the spin Hamiltonian in the absence of an external magnetic field.

Acknowledgments
This work was supported in part by the NSF under Grant Nos. DMR-1205803 and DMR-1506152. M.P. acknowledges support by an InProTUC scholarship of the German Academic Exchange Service during the early parts of this work. We thank Daniel Arovas, Jack Crewse, Yury Kiselev, Gil Refael, and Nandini Trivedi for helpful discussions.

References

[1] Stauffer D and Aharony A 1991 *Introduction to Percolation Theory* (Boca Raton: CRC Press)
[2] Vojta T and Hoyos J A 2008 *Recent Progress in Many-Body Theories* ed Boronat J, Astrakharchik G and Mazzanti F (Singapore: World Scientific) p 235
[3] Harris A B 1974 *J. Phys. C* **7** 3082
[4] Stinchcombe R 1981 *J. Phys. C* **14** L263
[5] dos Santos R R 1982 *J. Phys. C* **15** 3141
[6] Senthil T and Sachdev S 1996 *Phys. Rev. Lett.* **77** 5292
[7] Hoyos J A and Vojta T 2006 *Phys. Rev. B* **74** 140401(R)
[8] Sandvik A W 2001 *Phys. Rev. Lett.* **86** 3209
[9] Sandvik A W 2002 *Phys. Rev. B* **66** 024418
[10] Sandvik A W 2002 *Phys. Rev. Lett.* **89** 177201
[11] Vajk O P and Greven M 2002 *Phys. Rev. Lett.* **89** 177202
[12] Vojta T and Schmalian J 2005 *Phys. Rev. Lett.* **95** 237206
[13] Vojta T and Sknepnek R 2006 *Phys. Rev. B.* **74** 094415
[14] van der Zant H S J, Fritschy F C, Elion W J, Geerligs L J and Mooij J E 1992 *Phys. Rev. Lett.* **69** 2971–2974
[15] van der Zant H S J, Elion W J, Geerligs L J and Mooij J E 1996 *Phys. Rev. B* **54** 10081–10093
[16] Haviland D B, Liu Y and Goldman A M 1989 *Phys. Rev. Lett.* **62** 2180–2183
[17] Hebard A F and Paalanen M A 1990 *Phys. Rev. Lett.* **65** 927–930
[18] Crooker B C, Hebral B, Smith E N, Takano Y and Reppy J D 1983 *Phys. Rev. Lett.* **51** 666–669
[19] Reppy J D 1984 *Physica B+C* **126** 335
[20] Oosawa A and Tanaka H 2002 *Phys. Rev. B* **65** 184437
[21] Hong T, Zheludev A, Manaka H and Regnault L P 2010 *Phys. Rev. B* **81** 060410
[22] Yu R, Yin L, Sullivan N S, Xia J S, Huan C, Paduan-Filho A, Jr N F O, Haas S, Steppke A, Miclea C F, Weickert F, Movshovich R, Mun E D, Scott B L, Zapf V S and Roscilde T 2012 *Nature* **489** 379
[23] Griffiths R B 1969 *Phys. Rev. Lett.* **23** 17
[24] Thill M and Huse D A 1995 *Physica A* **214** 321
[25] Vojta T 2006 *J. Phys. A* **39** R143
[26] Fisher D S and Fisher M P A 1988 *Phys. Rev. Lett.* **61** 1847–1850
[27] Fisher M P A, Weichman P B, Grinstein G and Fisher D S 1989 *Phys. Rev. B* **40** 546–570
[28] Pollet L, Prokof'ev N V, Svistunov B V and Troyer M 2009 *Phys. Rev. Lett.* **103** 140402
[29] Giamarchi T, Le Doussal P and Orignac E 2001 *Phys. Rev. B* **64** 245119
[30] Weichman P B and Mukhopadhyay R 2008 *Phys. Rev. B* **77** 214516
[31] Prokof'ev N and Svistunov B 2004 *Phys. Rev. Lett.* **92** 015703
[32] Swanson M, Loh Y L, Randeria M and Trivedi N 2014 *Phys. Rev. X* **4** 021007
[33] Iyer S, Pekker D and Refael G 2012 *Phys. Rev. B* **85** 094202
[34] Vojta T, Crewse J, Puschmann M, Arovas D and Kiselev Y 2016 (*Preprint* arXiv:1607.01860)
[35] Wallin M, Sorensen E S, Girvin S M and Young A P 1994 *Phys. Rev. B* **49** 12115
[36] Campostrini M, Hasenbusch M, Pelissetto A and Vicari E 2006 *Phys. Rev. B* **74** 144506
[37] Harris A B 1974 *J. Phys. C* **7** 1671
[38] Wolff U 1989 *Phys. Rev. Lett.* **62** 361
[39] Metropolis N, Rosenbluth A, Rosenbluth M and Teller A 1953 *J. Chem. Phys.* **21** 1087
[40] Ballesteros H G, Fernández L A, Martín-Mayor V, Muñoz Sudupe A, Parisi G and Ruiz-Lorenzo J J 1998 *Phys. Rev. B* **58** 2740–2747

[41] Ballesteros H G, Fernandez L A, Martin-Mayor V, Munoz Sudupe A, Parisi G and Ruiz-Lorenzo J J 1998 *Nucl. Phys. B* **512** 681
[42] Sknepnek R, Vojta T and Vojta M 2004 *Phys. Rev. Lett.* **93** 097201
[43] Zhu Q, Wan X, Narayanan R, Hoyos J A and Vojta T 2015 *Phys. Rev. B* **91** 224201
[44] Vojta T and Schmalian J 2005 *Phys. Rev. B* **72** 045438
[45] Guo M, Bhatt R N and Huse D A 1994 *Phys. Rev. Lett.* **72** 4137
[46] Rieger H and Young A P 1994 *Phys. Rev. Lett.* **72** 4141
[47] Cooper F, Freedman B and Preston D 1982 *Nucl. Phys. B* **210** 210
[48] Kim J K 1993 *Phys. Rev. Lett.* **70** 1735
[49] Caracciolo S, Gambassi A, Gubinelli M and Pelissetto A 2001 *Eur. Phys. J. B* **20** 255
[50] Fisher D S 1992 *Phys. Rev. Lett.* **69** 534
[51] Motrunich O, Mau S C, Huse D A and Fisher D S 2000 *Phys. Rev. B* **61** 1160
[52] Hoyos J A, Kotabage C and Vojta T 2007 *Phys. Rev. Lett.* **99** 230601
[53] Vojta T, Kotabage C and Hoyos J A 2009 *Phys. Rev. B* **79** 024401
[54] Vojta T 2003 *Phys. Rev. Lett.* **90** 107202
[55] Hoyos J A and Vojta T 2008 *Phys. Rev. Lett.* **100** 240601
[56] Kosterlitz J M and Thouless D J 1973 *J. Phys. C* **6** 1181
[57] Altman E, Kafri Y, Polkovnikov A and Refael G 2004 *Phys. Rev. Lett.* **93** 150402
[58] Altman E, Kafri Y, Polkovnikov A and Refael G 2010 *Phys. Rev. B* **81** 174528
[59] Hrahsheh F and Vojta T 2012 *Phys. Rev. Lett.* **109** 265303
[60] Pollet L, Prokof'ev N V and Svistunov B V 2014 *Phys. Rev. B* **89** 054204
[61] Roscilde T and Haas S 2007 *Phys. Rev. Lett.* **99** 047205